FinFET/GAA Modeling for IC Simulation and Design

FinFET/GAA Modeling for IC Simulation and Design
Using the BSIM-CMG Standard
SECOND EDITION

Yogesh Singh Chauhan
Department of Electrical Engineering, Indian Institute of Technology Kanpur, Kanpur, India

Girish Pahwa
International College of Semiconductor Technology, National Yang-Ming Chiao-Tung University, Hsinchu, Taiwan

Avirup Dasgupta
Department of Electronics and Communication Engineering, Indian Institute of Technology Roorkee, Roorkee, India

Darsen Lu
Department of Electrical Engineering, National Cheng Kung University, Tainan City, Taiwan

Sriramkumar Venugopalan
Skyworks Solutions, Inc., Berkeley, CA, United States

Sourabh Khandelwal
Macquarie University, Sydney, Australia

Juan Pablo Duarte
University of California, Berkeley, CA, United States

Navid Paydavosi
Intel Corp., DuPont, WA, United States

Ali Niknejad
University of California, Berkeley, CA, United States

Chenming Hu
University of California, Berkeley, CA, United States

Sayeef Salahuddin
University of California, Berkeley, CA, United States

ELSEVIER

ACADEMIC PRESS
An imprint of Elsevier

Notices

Knowledge and best practice in this field are constantly changing. As new research and experience broaden our understanding, changes in research methods, professional practices, or medical treatment may become necessary.

Practitioners and researchers must always rely on their own experience and knowledge in evaluating and using any information, methods, compounds, or experiments described herein. In using such information or methods they should be mindful of their own safety and the safety of others, including parties for whom they have a professional responsibility.

To the fullest extent of the law, neither the Publisher nor the authors, contributors, or editors, assume any liability for any injury and/or damage to persons or property as a matter of products liability, negligence or otherwise, or from any use or operation of any methods, products, instructions, or ideas contained in the material herein.

ISBN 978-0-323-95729-8

For information on all Academic Press publications
visit our website at https://www.elsevier.com/books-and-journals

Publisher: Mara Conner
Acquisitions Editor: Tim Pitts
Editorial Project Manager: John Leonard
Production Project Manager: Kamesh R
Cover Designer: Vicky Pearson

Typeset by STRAIVE, India

Working together
to grow libraries in
developing countries

www.elsevier.com • www.bookaid.org

Contents

Author biographies

Yogesh Singh Chauhan is a chair professor in the Department of Electrical Engineering at the Indian Institute of Technology Kanpur, India. He has developed several industry-standard models, including ASM-HEMT, BSIM-BULK (formerly BSIM6), BSIM-CMG, BSIM-IMG, BSIM4, and BSIM-SOI models. His research group is involved in developing compact models for GaN transistors, FinFET, Nanosheet/Gate-All-Around FETs, FDSOI transistors, Negative Capacitance FETs, and 2D FETs. His research interests include RF characterization, modeling, and simulation of semiconductor devices.

He is a fellow of IEEE and Indian National Academy of Engineering. He is the editor of IEEE Transactions on Electron Devices and a distinguished lecturer of the IEEE Electron Devices Society. He is the chairperson of the IEEE UP section and IEEE-EDS Compact Modeling Committee. He has published more than 400 papers in international journals and conferences.

He received Ramanujan fellowship in 2012, IBM faculty award in 2013, PK Kelkar fellowship in 2015, CNR Rao faculty award, Humboldt fellowship, and Swarnajayanti fellowship in 2018. He has served in the technical program committees of the IEEE International Electron Devices Meeting (IEDM), IEEE International Conference on Simulation of Semiconductor Processes and Devices (SISPAD), IEEE European Solid-State Device Research Conference (ESSDERC), IEEE Electron Devices Technology and Manufacturing (EDTM), and IEEE International Conference on VLSI Design and International Conference on Embedded Systems.

Girish Pahwa is an assistant professor at the International College of Semiconductor Technology at National Yang-Ming Chiao Tung University. Girish's research primarily focuses on modeling, simulation, and device-circuit codesign of upcoming and emerging nanoscale technologies.

Before his current role, Girish held multiple positions in the Department of Electrical Engineering and Computer Sciences at the University of California, Berkeley, including assistant professional researcher, postdoctoral scholar, and research and development engineer. He also served as the inaugural executive director of the Berkeley Device Modeling Center (BDMC) and made significant contributions to the BSIM suite of compact models. His notable works include the development of industry-focused compact models for cryogenic CMOS operation, ferroelectric devices, high-voltage MOSFETs, and thin-film transistors.

Girish has received several prestigious awards, including the IEEE EDS Early Career Award (2022), the Outstanding PhD Thesis Award from IIT Kanpur (2020), and the Best Paper Award at the IEEE International Conference on Emerging Electronics (2016). He is an active member of the IEEE EDS Compact Modeling Committee and the IEEE EDS Electronic Materials Committee. He has served as a reviewer for several journals and conferences and has been part of the technical program committee for the IEEE Electron Devices Technology and Manufacturing (EDTM) conference.

Girish earned his PhD and master's degrees in electrical engineering from the Indian Institute of Technology (IIT) Kanpur and his bachelor's degree in electronics and communication engineering from Delhi Technological University.

Avirup Dasgupta is a faculty member in the Department of Electronics and Communication Engineering at the Indian Institute of Technology Roorkee (IITR), where he leads the DiRac Lab. His work primarily involves semiconductor physics, modeling and design of semiconductor devices and EDA. Prof. Dasgupta is a codeveloper of multiple industry standard models including BSIM-BULK, BSIM-IMG, BSIM-CMG, BSIM-SOI, and ASM-HEMT. He also develops various compact models for different industry clients. Prof. Dasgupta is an IEEE Senior Member and the recipient of multiple awards including the IEEE EDS Early Career Award (2021). He has served as a reviewer for multiple journals and conferences and serves as a member of the IEEE EDS Compact Modeling Technical Committee, IEEE EDS Publications and Products Standing Committee, and IEEE UP Section Executive Committee.

Prof. Dasgupta completed his undergraduate, graduate, and doctoral work at the Indian Institute of Technology Kanpur (IITK). He worked as the manager of the Berkeley Device Modeling Center (BDMC) and as a postdoctoral scholar in the BSIM group at the Department of Electrical Engineering and Computer Science, University of California, Berkeley, prior to joining IITR.

Darsen D. Lu was one of the key contributors to the industry standard FinFET compact model, BSIM-CMG, and thin-body SOI compact model, BSIM-IMG. He received his BSc in electrical engineering in 2005 from National Tsing Hua University, Hsinchu, Taiwan, and his MSc and PhD in electrical engineering from the University of California, Berkeley, in 2007 and 2011, respectively. From 2011 to 2015, he was a research scientist at the IBM Thomas J. Watson Research Center, Yorktown Heights, New York. He is currently a Macronix endowed chair (associate) professor at National Cheng Kung University, Tainan, Taiwan. His current research focuses on the fabrication and modeling of ferroelectric memory (ferroelectric Fin-FET) devices, cryogenic CMOS modeling for high-performance and quantum computation, and the design of AI/neuromorphic circuits and systems.

Sriramkumar Venugopalan received his MSc and PhD in electrical engineering from the University of California, Berkeley, and his BSc from the Indian Institute of Technology (IIT), Kanpur. While pursuing his doctoral degree, he contributed to the research and development of multi-gate transistor compact SPICE models. He led the industry standardization effort for the BSIM-CMG model, representing the BSIM Group at the Compact Model Council. He was the recipient of the Outstanding Researcher Award from TSMC for his contributions to multi-gate SPICE models. He has authored and coauthored more than 30 research papers in the areas of semiconductor device SPICE models and RF integrated circuit design. Dr. Venugopalan is currently leading the wireless system design group at Skyworks Solutions, Inc. Prior to that, he cofounded and served as the CEO of RF Pixels, a 5 G mmWave radio startup that was later acquired by Skyworks. Dr. Venugopalan was also with Samsung Electronics, pursuing RF integrated circuit design in advanced semiconductor technology nodes.

Sourabh Khandelwal is an associate professor at Macquarie University. He is the lead author of two industry-standard compact models: ASM-HEMT for GaN RF and power technology, and ASM-ESD for silicon ESD applications. He has also coauthored BSIM-CMG, BSIM-IMG, and BSIM6 compact models during his tenure at the BSIM group at the University of California, Berkeley. Dr. Khandelwal has published three books and over 150 research papers. He regularly serves as a consultant to multinational semiconductor companies.

Juan Pablo Duarte Sepúlveda obtained his PhD from the University of California, Berkeley, in 2018. He received his BSc in 2010 and MSc in 2012, both in electrical engineering from the Korea Advanced Institute of Science and Technology (KAIST). He held a position as a lecturer at the Universidad Tecnica Federico Santa Maria, Valparaiso, Chile, in 2012. He has authored many papers on nanoscale semiconductor device modeling and characterization. He received the Best Student Paper Award at the 2013 International Conference on Simulation of Semiconductor Processes and Devices (SISPAD) for the paper titled Unified FinFET Compact Model: Modelling Trapezoidal Triple-Gate FinFETs.

Navid Paydavosi is a seasoned hardware engineer with a decade of experience in advanced Si process technology and GPU and memory subsystem optimization. He excels in optimizing PPAC for complex SoC systems. He holds a PhD in electrical engineering from the University of Alberta and completed a postdoctoral scholarship at UC Berkeley under the supervision of Prof. Chenming Hu, contributing to the development of FinFET and SOI SPICE compact models. Navid began his Intel career in 2014 as a Logic Technology Development Device Engineer, where he contributed to key advancements in Intel 4 and Intel 3 technology nodes. He then served as a GPU Micro-Arch Power Optimization Engineer, leading innovations such as a novel Glitch minimization algorithm. As a NAND Flash Power and Performance Optimization Engineer, Navid significantly improved the power and performance of Intel's 3D NAND Flash products. Currently, he is a senior staff Intel Foundry Device Engineer, customizing the Intel 3 technology node for customers. Navid's expertise spans device physics, semiconductor manufacturing, and power optimization. His goal is to deliver world-class AI hardware solutions for data center and edge computing environments.

Ali M Niknejad received his BSc in electrical engineering from the University of California, Los Angeles, in 1994, and his MSc and PhD, also in electrical engineering, from the University of California, Berkeley, in 1997 and 2000, respectively. He is currently a professor in the EECS department at UC Berkeley and faculty director of the Berkeley Wireless Research Center (BWRC) Group. He is also the associate director of the Center for Ubiquitous Connectivity (CUbiC) and served as the associate director for the Center for Converged TeraHertz Communications and Sensing (ComSenTer). Prof. Niknejad received the 2020 SIA/SRC University Research Award, recognized "for noteworthy achievements that have advanced analog, RF, and mm-wave circuit design and modeling, which serve as the foundation of 5G+ technologies." Prof, Niknejad was the recipient of the 2012 ASEE Frederick Emmons Terman Award for his work and textbook on electromagnetics and RF integrated circuits. He has coauthored over 400 conference and journal publications in the field of integrated circuits and device compact modeling. His focus areas of research include analog, RF, mixed-signal, mm-wave circuits, device physics and compact modeling, and numerical techniques in electromagnetics.

Chenming Hu is the TSMC distinguished chair professor emeritus at the University of California, Berkeley. He was the chief technology officer of TSMC. He received the US Presidential Medal of Technology and Innovation from President Barack Obama for developing the first 3D thin-body transistor FinFET, MOSFET reliability models, and leading the development of the BSIM industry-standard transistor model used in designing most of the integrated circuits in the world. He is a member of the US Academy of Engineering, the Chinese Academy of Science, and Academia Sinica. He received the highest honor of IEEE, the IEEE Medal of Honor, and its Andrew Grove Award, Solid Circuits Award, and the Nishizawa Medal. He also received the Taiwan Presidential Science Prize and UC Berkeley's highest honor for teaching—the Berkeley Distinguished Teaching Award.

Sayeef Salahuddin is the TSMC distinguished chair professor at the University of California, Berkeley. He received the Presidential Early Career Award for Scientists and Engineers from President Obama, among several other early career awards from multiple federal agencies. He also received the IEEE Andrew S Grove Award for his pioneering work in ferroelectric negative capacitance and its incorporation into the most advanced DRAM memory and back-end capacitors. He is a Fellow of the IEEE, APS, and AAAS and is the editor-in-chief of the IEEE Electron Devices Letters.

Preface

Since you have opened this book, you probably know the 3D thin-body transistors, FinFET, and their successor, GAA transistors. The semiconductor industry's adoption of them, for their power, speed, and density advantages over planar transistors, has been the biggest semiconductor technology news in over 50 years.

Because a book reflects the background of its authors, we would like to introduce ourselves. We are current or former PhD students, postdoctoral researchers, or professors who have participated in the development of BSIM-CMG at the University of California, Berkeley.

BSIM-CMG is the industry-standard model of FinFET and GAA transistors for simulating and designing ICs fabricated with these technologies. The BSIM FinFET model has been used to design very high-speed and low-power ICs that have given the world video smartphones, streaming entertainment, autonomous driving, and generative AI. One may expect the BSIM GAA model to enable even more powerful new technologies, trillion-dollar industries, and new solutions to the world's myriad problems.

We wrote this book for IC designers, EDA engineers, semiconductor technologists, researchers, and students. It presents the what, why, and how of the compact modeling of FinFET and GAA for digital, analog, and RF applications. It is also a thorough handbook of the industry stand model of FinFET and GAA (BSIM-CMG). The University of California, Berkeley, provides royalty-free licenses for all users of BSIM models.

Even if you are familiar with FinFETs, you may be surprised to learn that BSIM-CMG can model FinFETs with arbitrary fin shapes such as trapezoidal, round-cornered, cylindrical, and general gate-around shapes as described in Chapter 3. This Unified Model is the common core model that intimately links the FinFET model with the GAA or Nanosheet FET model, as presented in Chapter 4. Already, other 3D thin-body shapes beyond fin and nanosheet have been proposed. BSIM-CMG will be applicable to model them too. "CMG" in BSIM-CMG stands for Common-Multiple-Gate. BSIM-CMG was designed to model MOSFETs having general 3D thin-body and gate shapes, the restrictions being that the body must be thin and there is only one common gate voltage. BSIM-CMG also models 3D thin-body MOSFETs employing nonsilicon channel materials such as SiGe, III–V, and even noncrystalline materials. You are holding the best handbook on transistor modeling for designing future ICs using advanced MOSFETs.

We did not create all the equations and models presented in this book. Many other members of the BSIM group contributed to BSIM-CMG. Most notably, Chung-Hsun Lin and Mohan Dunga were the early developers of BSIM-CMG, starting in 2004. Other direct or indirect contributors to the book include Walter Li, Wei-Man Lin, Shijin Yao, Muhammed Karim, Chandan Yadav, Harshit Agarwal, Pragya Kushwaha, Dinesh Rajasekharan, and Chien-Ting Tung.

We thank, first of all, the industrial members of Berkeley Device Modeling Center for funding the continuing research on BSIM models (Synopsys, Cadence, TSMC, Samsung, Intel, Globalfoundries, and Qualcomm). We also thank the industry members of the BSIM-CMG Workgroup of Compact Model Coalition who test and use the model and point out its weaknesses in accuracy or robustness. An incomplete list includes A.S. Roy, S. Mudanai (Intel); K.-W. Su, W.-K. Lee, C-K Lin (Taiwan Semiconductor Manufacturing Company); H. Lee, D. Kim (Samsung); J.-S. Goo (Globalfoundries); J. Wang, W. Liu (Synopsys); J. Xie, M. Tian (Cadence); Y. Yue, F. B. Yang (Qualcomm); A. Ramadan (Mentor Graphics); and G. Coram (Analog Devices).

We wish to express our deep gratitude to our families, who tolerated and made tolerable our long hours at the office and in front of computers. We also thank the University of California, Berkeley, for providing royalty-free licenses to all users.

Lastly, we thank you, dear readers, for giving our book meaning by using it.

The authors

3D thin-body FinFET and GAA: From device to compact model

A part of the semiconductor industry's formula for success is to make incremental changes, not drastic changes. Planar MOSFET has served the electronics industry well since the beginning of MOSFET. Aggressive engineering managed to reduce its size again and again without changing its basic two-dimensional (2D) planar structure. Yet, the IC design windows for performance, power consumption, and sensitivity to device variation gradually shrunk to the point that a major switch to a better transistor structure is unavoidable. FinFET is that new three-dimensional (3D) thin-body transistor. Adopting FinFET was hailed by Intel as the most drastic shift in semiconductor technology in 50 years. FinFET and its successor GAA provide relief from the performance, power, and device variation predicaments that the IC industry has struggled with in the decade before and the decade after 2000. More importantly, it redirects device scaling from a path heading for the cliff to a new one that allows scaling to continue onward as long as lithography allows and upward into the new vertical dimension.

After making the FinFET's fin taller and thinner for many technology generations, the semiconductor industry will move on to FinFET's successor, gate-all-around (GAA) transistor. These new transistors require new design infrastructures to enable the design of FinFET- and GAA-based circuits and products. The foundations of the infrastructures are computationally efficient mathematical models that accurately represent FinFET and GAA transistors manufactured by any company. They are called compact models or SPICE models.

This chapter presents FinFET and GAA—what they are, what they do, and what new scaling concept inspired their invention. This chapter also introduces the role of compact models in the semiconductor industry and the industry's first and dominant standard compact model, BSIM. BSIM FinFET and GAA models enable the design of new generations of ICs with higher performance, lower-power consumption, and higher layout density. These high-performance and low-power ICs, in turn, enable the new generations of cell phones, data centers, and artificial intelligence chips.

1.1 The root cause of short-channel effects in the 21st century MOSFETs

What is the basic concept behind the remarkable FinFET and GAA? To understand the pain reliever, we must first understand the pain itself. As gate length shrinks, MOSFET's I_g-V_g characteristics degrade in two major ways as illustrated in

FinFET/GAA Modeling for IC Simulation and Design. https://doi.org/10.1016/B978-0-323-95729-8.00009-X

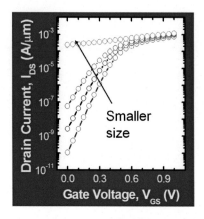

FIG. 1.1

Shrinking gate length makes the sub-V_t swing larger, and V_t and I_{off} more sensitive to gate critical dimension variation and random dopant fluctuation.

Fig. 1.1. First, the subthreshold swing (S) degrades and V_g decreases, therefore, the device cannot be turned off easily by lowering V_g. Second, S and V_g become increasingly sensitive to L_g variations, that is, device variations become more problematic for designing and manufacturing ICs. These problems are known as the short-channel effects.

Fig. 1.2 shows the root cause of short-channel effects in the 20th-century MOSFETs [1]. A transistor is turned on and off when V_g lowers and raises the potential of the channel (and thus the potential barrier between the channel and the source) through the gate to channel capacitance, C_g. In an ideal transistor, the channel potential is only controlled by V_g and C_g. In a real transistor, the channel potential is also subject to the influence of V_g through C_g. When L_g is large, C_g is much smaller than C_g and the drain voltage does not interfere with V_g's role as the sole controlling voltage. As L_g decreases, C_g increases [1, 2] and V_g loses its dominant control. In extreme cases, V_g has less control than V_g and the transistor can be turned on by V_g alone

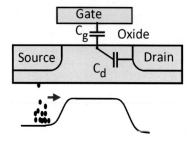

FIG. 1.2

With decreasing L, rising C_d allows V_d to control the channel potential (*lower figure*) just as V_g. Scaling gate oxide thickness was a good solution for this problem in the 20th century.

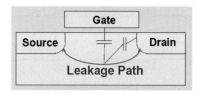

FIG. 1.3

Beyond 20 nm even a 0-nm thin gate dielectric cannot stop V_d from lowering the potential barrier along leakage paths a few nanometers below the interface.

without V_g as shown in the top curve of Fig. 1.1. Before reaching that extreme, we get the other undesirable to unacceptable curves in Fig. 1.1. The solution that has worked well in the 20th century is to increase C_g by reducing the gate oxide thickness in proportion to L_g. The semiconductor industry grew used to depending on this one cure for all ills.

However, almost all researchers missed another root cause that can limit L_g scaling even if an ideal "zero thickness" dielectric was available. Fig. 1.3 shows that the leakage current does not have to flow along the silicon-dielectric interface. Leakage paths that are many nanometers from the gate as illustrated in Fig. 1.3 are more problematic than the surface leakage path because the submerged leakage paths are only weakly controlled by V_g, that is, "C_g" can be small even with zero oxide thickness. As a result, the potential barriers along these weakly controlled paths can be easily lowered by V_g through C_g in a small L_g device.

Having identified this new root cause, researchers at UC Berkeley proposed the thin-body FET concept and the FinFET structure to the US Government's Defense Advanced Research Project Agency (DARPA) in response to a request for the proposal of sub-25 nm switching devices in 1996.

1.2 The thin-body FET concept

Fig. 1.4 shows a MOSFET whose body is a thin piece of semiconductor with gates above and below it. If the body is thin, any lines drawn between the source and drain (any current leakage paths) would not be far from a gate or both gates. The center of the thin body is still the path more leakage current compared to the surface paths. However, a thin-body FET's total leakage current is many orders of magnitude smaller compared to the top curve in Fig. 1.1. In fact, the adoption of FinFET by the semiconductor industry in 2011 basically eliminated the short-channel leakage current problem overnight. The thin-body FET also eliminated the need for heavy channel doping to suppress the short-channel effects [3]. Channel doping's other function of adjusting the threshold voltage is now performed by gate metal work function engineering, such as using SiGe gate rather than Si gate as taught in Huang et al. [3]. In this way, random dopant fluctuation, a major and fundamental contributor to device variation before the invention of thin-body FET, was eliminated. Channel carrier

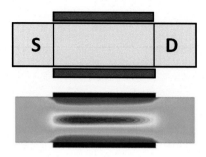

FIG. 1.4

Top: A thin semiconductor (*yellow*) body eliminates the leakage paths in Fig. 1.3. *Bottom*: Leakage current density is low near the gate and significant only in the center of the body.

mobilities were improved. Also, an undoped body reduces the electric field at the semiconductor and oxide interface and normal to the interface. This improved the temperature bias instability (NBTI and PBTI), the gate dielectric tunneling leakage, and oxide reliability.

1.3 3D thin-body FinFET: A new scaling path for MOSFET

FinFET in Fig. 1.5 is a manufacturable version of the thin-body transistor in Fig. 1.4 turned 90 degrees to stand on its edge. The thin body is shaped like the dorsal fin of a shark and created with lithography patterning and etching technologies. The fin is then processed into FinFET in basally the same way as processing planar MOSFET. FinFET occupies less Si area than planar MOSFET because the channel width (W) of FinFET [4] is the peripheral length all sides of the fine covered by the gate. W can be significantly larger than the fin footprint on the Si wafer and can be increased by making the fin taller.

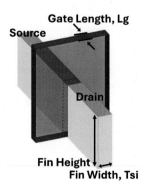

FIG. 1.5

FinFET is the thin-body FET in Fig. 1.4 turned 90 degrees with the gate covering multiple sides of the thin body. A thin-body ($T_{si} \sim L_g$) dramatically reduces short-channel effects including device variation and I_{off}.

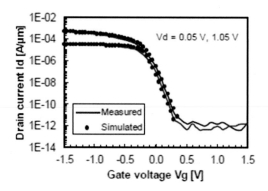

FIG. 1.6

1999 FinFET with excellent subthreshold swing (64 mV/decade) and I_{off} and undoped body eliminating random dopant effect. $L_g = 45$ nm and $T_{si} = 30$ nm [3].

More importantly, a new scaling path was introduced by the thin-body FET invention: L_g can be scaled by scaling the body thickness. As long as the fin thickness (body thickness), T_{si}, is sufficiently smaller than L_g, the short-channel effects are well suppressed and the subthreshold swing is basically the theoretical best case, 62 mV/decade at room temperature (Fig. 1.6) [3]; 18 and 45 nm working FinFETs and 10 nm FinFET simulation results were reported in 1999 [3]. Soon after, 10 nm and then 5 nm FinFETs [4] were reported by IC manufacturers.

Intel was the first company to introduce FinFET into mass production, in 2011, and reported that the manufacturing cost is comparable to the conventional planar MOSFET technology. It also reported that no special manufacturing equipment is required. Soon after Intel adopted the FinFET at 22 nm node [5], TSMC introduced 16 nm FinFET for the foundry market [6]; 14 nm FinFET was introduced at various times by Intel, Samsung, TSMC, Globalfoundries, and UMC. After that, a handful of companies were able to continue to develop more advanced nodes with FinFET—12, 10, 7, 5, 3, and 2 nm.

FinFET brought the semiconductor industry and information and communication technologies into a decade of rapid advances. Chip performance and power consumption improved so much that movie streaming became a routine activity for millions around the world and cell phone-satellite direct communication became reality. Most importantly, artificial intelligence became a reality. The high density of FinFETs enabled massive parallel processing, which is particularly beneficial for graphic processing and matrix operations, which are key to deep learning. Nvidia is one of the companies that benefited from the drastically improved density, speed, and energy of FinFETs. Its 2017 website says under the heading of Nvidia AI Solutions, "16 NANOMETER FINFET WITH UNPRECEDENTED ENERGY EFFICIENCY." With 150 billion transistors built on bleeding-edge 16 nm FinFET fabrication technology. The Pascal CPU is the world's largest FinFET chip ever built. It is engineered to deliver the fastest performance and best energy efficiency for workloads with near-infinite computing needs.

In 2018, *Time Magazine* published a special issue named The Optimists, which contained opinions of experts from various fields expressing their optimism about the future. Investor Warren Buffet expressed the general mood this way, "Today, an average person can enjoy options in travel, medicine, entertainment, and communication that were simply not imaginable to John D. Rockefeller."

Amazingly, the cost per million transistor or logic gates ended its falling trend since the beginning of the industry and became stable or slowly rising at around the 14 nm node. Still, the demand for FinFETs and gates using the more advanced technology nodes continued to grow. Semiconductor growth is no longer driven by falling cost, but by better speed/energy performance. This change has raised the values of many semiconductor companies.

1.4 GAA: The next 3D thin-body FET after FinFET

With each new generation of FinFET technology, the fin was made thinner and taller. Doing so reduced the semiconductor footprint and allowed the gate lengths to shrink. Thinner and taller fins are more difficult to fabricate. Eventually, an alternative way for fabricating the thinner-body FET was needed.

Such an alternative has been known as early as 2004 [7] when Samsung engineers fabricated the multibridge-channel FET by epitaxial growth. They fabricated alternating layers of epitaxial SiGe and Si films on top of the Si wafer. After etching the stack into stripes and protecting the ends of the stripes, they etched away the "sacrificial" SiGe leaving thin Si films hanging between the protected anchors at the ends. Gate dielectric and gate metal were deposited around the Si films. This FET structure is called GAA for gate-all-around. The fabrication method and structure were refined in 2017 and Loubet et al. [8].

One may think of GAA as FinFET turned 90 degrees as shown in Fig. 1.7. GAA resembles the basic thin-body FET in Fig. 1.4 even more than FinFET itself. GAA retains all the benefits of the thin-body FET. Compared to FinFET, GAA in Fig. 1.7 can further reduce the footprint area for the same W. In addition, IC designers can finely vary GAA's W by changing the width of the GAA while FinFET's W can only be changed by changing the fin number in a discrete manner. This flexibility gives GAA a further advantage in reducing the circuit size.

Samsung introduced GAA for mass production at the 3 nm node, calling it also the multibridge-channel FET (MBCFET). TSMC announced its plan to offer GAA at 2 nm node. Intel plans to offer it as "RibbonFET" at its 20 A node with A standing for Angstrom.

The GAA film material can be Si or SiGe for the best mobilities. It is possible to reduce the thickness of the epitaxial film toward 5 and 4 nm or even less with good surface smoothness and thickness uniformity. Therefore, the film-body thickness scaling can proceed for many technology nodes.

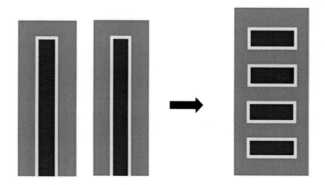

FIG. 1.7

GAA (*right*) resembles the basic thin-body FET in Fig. 1.4 even more than the FinFET (*left*).

GAA FET may start with three or four layers of semiconductor films in each fin. The number of films may increase with time. It is tempting to stack an n-type GAA FET and a p-type GAA FET using different films in the same fin to further reduce footprint. The stacked n- and p-FET is called CFET for complimentary FET.

The 3D trend that was started with FinFET will continue to grow and expand for decades to come. The world will benefit from the further improvement of IC density and performance that result from the 3D trend.

1.5 FinFET/GAA compact models: The bridge between technology and IC design

Designing FinFET-based integrated circuits requires a FinFET model for circuit simulation. In order to design ICs, design teams need two things from their foundry partners or the wafer manufacturing divisions of their companies—design rules and SPICE model. SPICE model is also known as compact model. Design rules are created by each manufacturing company. The SPICE compact model is a set of long equations that are capable of calculating the complex transistor I(V) and Q(V) accurately and fast for SPICE simulation. The compact model that a designer uses is almost certainly a royalty-free industry standard model provided by a university.

The compact model is effectively a "contract" between the wafer fab and its customers. "First wafer success" has become the norm in the past 20 years in spite of the rising technology complexity and circuit size and variety. That is partly due to the excellent accuracy of the compact models. If the model does not describe the transistor characteristics accurately, no amount of design effort can guarantee design success. The model equations contain adjustable parameters. Engineers, with the help of automated parameter extraction tools, choose the parameter values so that the model accurately reproduces the current, capacitance, and noise over nine orders of magnitude of currents for the entire operating range of voltages, gate lengths and

Compact Model is the

Major Bridge for
Information Exchange

Device/Fab Technology   Circuit/Product Design

Example: BSIM
- **First industry standard compact model**
- **~25,000 lines of C code ~8,000 lines of Verilog-A**

- **Accurate: Fit MOSFET data with ~ 1% RMS error**
- **Fast: ~ 10μs per bias point**
- **Smooth**

FIG. 1.8

Complex device and manufacturing technologies determine the electrical behaviors of a transistor. Compact model captures this information and makes it available for computer-based IC circuit and product design.

widths, temperatures, and frequencies. The accuracy and robustness of circuit simulation greatly depend on the model equations. The compact model may be used with SPICE simulators to simulate and design circuits directly. Or it may be used to design and characterize standard cells, which are used to design circuits.

A good compact model must be very accurate to avoid expensive design respins. It also must be fast to support the simulation of large circuits, and robust for fast convergence while simulating a wide range of complex circuits as shown in Fig. 1.8.

1.6 A brief history of the first standard compact model, BSIM

BSIM originally stands for Berkeley short-channel IGFET model but it is often thought of as Berkeley simulation. BSIM is the first industry standard model and continues to be the most popular compact model today. BSIM's genesis may be traced to BSIM1 published in 1984 [8], which was followed by BSIM2 in 1988 [9].

At the same time, the Berkeley research team had a very productive parallel research effort in MOSFET physics and technology. Our research into the many device-physics effects and behaviors of aggressively scaled MOSFETs has gradually built a collection of models for V_t dependence on bias and gate length, mobility degradation, velocity saturation, output conductance, unified flicker noise, etc. Eventually, these models became the building blocks of new versions of BSIM models.

BSIM3 incorporated many of these new original device-physics models. That approach was a marked departure from all previous compact models including BSIM1 and BSIM2. Those models used simplistic device physics and relied heavily on "curve fitting." That approach may be acceptable for modeling I_d versus V_d, for example, but inadequate for modeling the transconductance and its higher-order derivatives.

BSIM3 was such an improvement over the previous models that the Compact Model Council, an industry standard organization formed in 1995 selected BSIM3v3 as the world's first industry standard model. Soon after, BSIM3 replaced the many dozens of SPICE models in use in the semiconductor industry. From the beginning till today, BSIM has been provided by UC Berkeley to all BSIM users royalty-free following the tradition of the Berkeley SPICE. A handful of leading semiconductor companies provide funding for continued research of BSIM models to keep up with changing technology at the Berkeley Device Modeling Center.

1.7 Core and real device models

All compact MOSFET models start with a "simple model" or "core model" that models a prototype long-channel transistor. For the real transistors used in ICs, the model accuracy is achieved with numerous "real device models" added onto the core model as shown in Fig. 1.9. With the CMOS technology aggressively scaled, the real device effects have become the dominant effects that determine the accuracy of circuit simulation. BSIM excels with its accurate real-device models. For example, BSIM3 [10] introduced three separate physical mechanisms to model the output

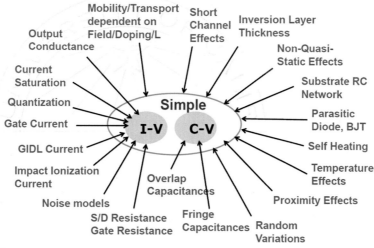

FIG. 1.9

A compact model is a simple (long-channel) model and numerous real device models. The latter is 90% of the model and responsible for the global accuracy.

conductance: channel length modulation, drain-induced barrier lowering, and hot carrier-induced body bias effect. Each of these three mechanisms is modeled with a nonlinear multivariable function of channel length, oxide thickness, V_t, V_{ds}, V_{gs}, and V_{bs}. The BSIM output conductance model was an instant success and continues to be used today. An accurate output conductance model is very important for analog and RF IC circuit design.

Another example is the gate-induced-drain leakage or GIDL. It was introduced into BSIM3 after we discovered this new leakage current and explained it as the band-to-band tunneling current induced by the gate-to-drain voltage [11]. Once the mechanism was clearly understood, a simple analytical model could be developed and it proved to be accurate for all subsequent generations of MOSFET technologies including FinFET and GAA.

Yet another example is the flicker noise or 1/f noise. It unified the noises due to fluctuation in the number of the channel inversion charge carriers and the fluctuation in the Coulombic scattering mobility. They can be unified because both result from the capture and emission of charge by the charge traps in the gate dielectric [12]. We validated the model in detail using random telegraphic noise measurements that can only be observed in transistors with such small length and width that one transistor contains only one or two observable oxide traps. These physics studies led to an accurate BSIM unified flicker noise model based on the assumption of a uniform distribution of electron traps in the gate dielectric. Due to the use of complex thin high-k gate dielectric in modern FinFETs, we have found it necessary to assume spatially nonuniform distribution of oxide electron traps. One consequence of this is that the flicker noise power spectrum over a wide range of frequency may have a nonconstant slope.

Even more complex real device models include the self-heating model and non-quasistatic model. The real device models account for 80% of the model code, simulation time, and model development effort. They are responsible for the accuracy of the compact model and the IC simulation.

1.8 Industry standard FinFET and GAA compact model BSIM-CMG

Anticipating the industry adoption of the FinFET and GAA technologies, the BSIM team started to develop a thin-body compact model in the mid-2000s calling it BSIM-CMG. CMG stands for common-multigate. Multigate is a description for FinFET and GAA in all their flavors: double gate, triple gate, and quadruple gate, with various channel cross-sectional shapes. "Common" means that all the multiple gates are electrically connected and share a common gate voltage.

After the industry embraced the FinFET technology for production, BSIM-CMG was selected as the industry standard FinFET model, without competition, by the Compact Model Council in 2012. The council members are major integrated IC companies, fabless companies, design automation companies, and IC foundries. BSIM-CMG provided a timely foundation for the design infrastructure that enables FinFET-/GAA-based circuit design and product development. The following chapters discuss all the important aspects of the standard FinFET and GAA models.

Initially, the core model of the BSIM-CMG modeled two simple gate shapes—parallel-plate double-gate and cylindrical gate for circular channel cross section. Those were the only two geometries having simple analytical solutions to the Poisson's equation (3.1) involving the gate, oxide, and semiconductor (in accumulation, depletion, or inversion).

It was a major innovation, as explained in Section 3.2, that we found a way to generalize the solutions of the two geometries into a unified core model equation, of which the double-gate and cylindrical-gate solutions are only two special cases. The unified model equation does not assume a specific geometry and therefore does not contain the distance between the double gates nor the diameter of a cylindrical gate. Instead, it contains two other geometric parameters: the cross-sectional area of the channel enclosed by the gate (area of the circle in the cylindrical gate case and area of a rectangle in the double gate case), and the channel length ($2\pi R$ for cylindrical gate and twice the gate width for double gate). We were able to explain the physical meanings of each term in the unified model equation for the two special cases (e.g., the strong inversion charge is proportional to the channel length and the subthreshold current of the fully depleted thin-body FET should be proportion to the channel cross-sectional area). We hypothesized that the unified model may be a good approximate solution for other channel cross sections too.

The next step is to test that hypothesis against TCAD for various channel cross sections. At first, we concentrated on the fin cross sections of the early production FinFET—trapezoid with rounded edges, for which there is no analytical solution of the Poisson's equation [13]. We obtained excellent agreements between the unified model and TCAD as well as the measured data provided by FinFET manufacturer. On the strength of that verification, the Compact Model Council accepted a new version of BSIM-CMG containing the new unified core model.

Fig. 1.10 shows one verification test showing that the unified model is capable of accurately predicting the current of GAA FETs with circular, square, and triangular channels. We tested many increasing odd channel shapes including the shape shown in Fig. 1.11. By comparison, the GAA channel shape, stacked rectangles with round corner, is relatively simple.

Samsung announced its plan to introduce GAA to mass production in 2018. The Compact Model Council adopted a new version of BSIM-CMG as the standard GAA compact model in 2019.

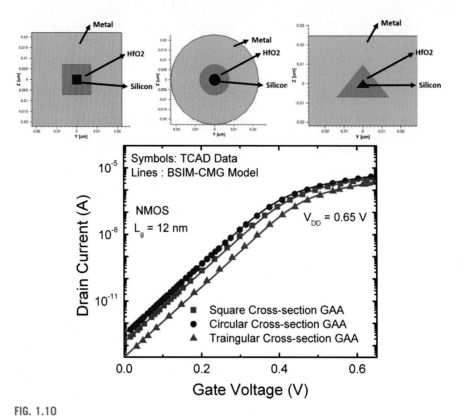

FIG. 1.10

Verification of the range of applicability of the BSIM-CMG unified core model to GAA of various cross sections.

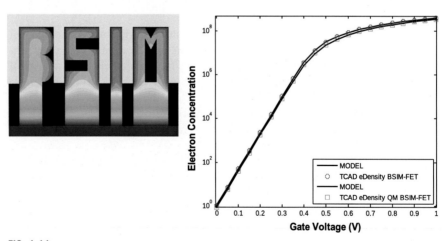

FIG. 1.11

An extreme verification of the range of applicability of the BSIM-CMG unified core model to GAA of various cross sections. *Left figure* shows the cross section of the GAA; *yellow* is metal gate, *brown* is dielectric, the rest is silicon. *Color* is the silicon region that represents TCAD electron density. *Right figure* is a comparison of TCAD electron density with BSIM-CMG unified core model. Two cases are compared: one with the quantum effect turned on and the other with it turned off.

References

[1] C. Hu, Modern Semiconductor Devices for Integrated Circuits, Pearson/Prentice Hall, 2010. Chapter 7.

[2] Z. Liu, et al., Threshold voltage model for deep-submicrometer MOSFET's, IEEE Trans. Electron Devices 40 (1993) 86–95.

[3] X. Huang, et al., Sub 50-nm FinFET: PMOS, in: IEDM Technical Digest, 1999, p. 67.

[4] F.-L. Yang, et al., 5 nm-gate nanowire FinFET, in: VLSI Technology Symposium, 2004, pp. 196–197.

[5] C.-H. Jan, et al., Design-technology co-optimization of anti-fuse memory on Intel 22 nm FinFET technology, in: IEDM Technical Digest, 2012, https://doi.org/10.1109/IEDM.2012.6478969.

[6] S.Y. Wu, et al., A 16 nm FinFET CMOS technology for mobile SoC and computing applications, in: IEDM Technical Digest, 2013, https://doi.org/10.1109/IEDM.2013.6724591.

[7] E.-J. Yoon, et al., Sub 30 nm multi-bridge-channel MOSFET (MBCFET) with metal gate electrode for ultra high performance application, in: IEDM Technical Digest, 2004, https://doi.org/10.1109/IEDM.2004.1419244.

[8] N. Loubet, et al., Stacked nanosheet gate-all-around transistor to enable scaling beyond FinFET, in: IEDM Symposium on VLSI Technology, 2017, https://doi.org/10.23919/VLSIT.2017.7998183.

[9] M.C. Jeng, P.K. Ko, C. Hu, A deep submicron MOSFET model for analog/digital circuit simulations, in: Technical Digest of International Electron Devices Meeting (IEDM), 1988, pp. 114–117. San Francisco, CA.

[10] J.H. Huang, et al., BSIM3 Manual, University of California, Berkeley, 1993.

[11] T.Y. Chan, et al., The impact of gate-induced drain leakage current on MOSFET scaling, in: Technical Digest of International Electron Devices Meeting (IEDM), 1987, pp. 718–721. Washington, DC.

[12] K.K. Hung, et al., A unified model for the flicker noise in metal-oxide-semiconductor field-effect transistors, IEEE Trans. Electron Devices 37 (3) (1990) 654–665.

[13] J.P. Duarte, et al., Unified FinFET compact model: modelling trapezoidal triple-gate Fin-FETs, in: SISPAD, 2013.

Compact models for analog and RF applications

2

2.1 Introduction

A compact model is the interface between the process technology and the circuit designer. The compact model, usually in the form of a design kit, encapsulates all the relevant details of the process in an easily accessible form that allows the designer to run simulations in a familiar environment, assessing the impact of the device on important circuit and system parameters, without having to understand the details of the physics or implementation details of the model. In fact, most designers have no idea what is inside the compact model, but they expect the model to capture the device behavior as accurately as possible. An analog designer will expect accurate prediction of bias points, small-signal parameters such as transconductance, output resistance, capacitance, thermal and flicker noise, and device-matching properties. Since designers have no control over the design of a device, they can only vary the operating point and the device geometry, so they expect good physical scalability in the model. RF design requires accurate prediction of power gain, noise (thermal and flicker), linearity, and power.

FinFET devices come in an age where CMOS technology is playing a vital role in both analog and RF/microwave circuits. While technology scaling has improved the device speed by orders of magnitude, allowing operation in the millimeter-wave and subterahertz regime, voltage scaling and lower intrinsic gain have been a constant challenge to analog design. In such a world, most designers are averse to moving to advanced technology nodes, since often digital blocks are the only ones that gain in area and power. In this regard, a FinFET device can offer a change, improved output resistance (and hence intrinsic gain) and low capacitance. Given the high cost of doing fabrication runs in advanced technology nodes, an analog/RF compatible FinFET model will offer a unique opportunity to explore designs without incurring any costs of fabrication.

In this chapter, we will highlight important performance metrics for analog and RF circuits and relate these to the various compact model modules that are found in the FinFET compact model. This chapter therefore serves as a transition to the rest of the book, highlighting the importance of the various modules and the properties of a compact model.

FinFET/GAA Modeling for IC Simulation and Design. https://doi.org/10.1016/B978-0-323-95729-8.00001-5

2.2 Important compact model metrics

We begin by identifying important metrics that can be used to judge a compact model. Without these metrics, it would be difficult to gauge the accuracy of a compact model with regard to important quantifiable aspects that closely relate to circuit design. Metrics capture the essence of the model and are important for the designer; in other words, they should be related to the performance of important circuit building blocks. The metrics are divided into analog and RF categories. It goes without saying that RF metrics are a superset and depend on analog metrics. In other words, an RF compact model should first and foremost satisfy the analog criteria for evaluation as they will form the core foundation for the model.

2.3 Analog metrics

Classic analog design is concerned primarily with the gain-bandwidth product of amplifiers. This is due to the wide application of amplifiers (operational and transconductance amplifiers) as core building blocks and the use of feedback to linearize these blocks, making their performance predictable over the process and with temperature variation. Device currents are selected to minimize power while meeting speed requirements. Moreover, noise is of paramount importance in analog circuits as it determines the resolution, or lower end of the dynamic range, of the signal processing chain. The upper end of the dynamic range is determined by a voltage swing, which is related to the supply voltage and the linearity of the amplifier (and hence transistors). In many cases, linearity is a secondary concern because of the application of feedback, which inherently linearizes stages by trading gain for linearity. Owing to the lower supply voltage and lower intrinsic gain, the loop gain of a closed-loop system is dropping with technology scaling, making nonlinearity a more prominent concern even in analog applications. In the other extreme, people are exploring open-loop design by relying on digital signal processing techniques to measure and correct analog errors.

In fact, analog design is increasingly mixed-signal design. Pure analog design is very rare, or mostly confined to high-voltage applications. A FinFET device will live in a very mixed-signal environment, where sample time and charge processing will take place side by side with traditional linear voltage/current analog techniques. Moreover, FinFET devices will form the core of switches for the purpose of tuning and calibration, and often devices in these applications experience voltage excursions that invert the device polarity (source-drain switch), requiring model symmetry. In sampled time systems, signal-dependent charge injection is important for quantifying the nonlinearity and systematic errors in analog signal processing.

2.3.1 Quiescent operating point

In many respects, analog design comes down to biasing transistors correctly.[a] Circuit topologies are often selected based on performance requirements, and the job of the circuit designer is to ensure that the transistors remain at the proper quiescent

---

[a]A saying attributed to Barrie Gilbert.

operation point with process variation and temperature variation. This is often a challenging task, as supply voltages scale to 1 V, and there is often very little margin, forcing designers to use biasing schemes that can compensate for variations in the device threshold voltage V_T with counter variations of another device. This means that they will trust the model's ability to track temperature or process variations (say, in the transistor geometric dimensions and doping). This places a high burden on a compact model since its current-voltage (I-V) behavior must be built on a physical foundation that inherently can predict these changes. Often students of compact modeling ask why one cannot simply reduce compact modeling to a mathematical exercise in curve fitting. The answer is partly that a mathematical model that fits the I-V curves from measured data without knowledge of any device physics would fail to predict the I-V curve for a small change in a physical parameter, such as doping or temperature. In theory, all these variations could be captured by a compact model, but technology foundries cannot afford to run thousands of combinatorial experiments to capture these variations, and instead rely on the compact model to reproduce these effects.

This reliance on physical trends is in contrast to the requirement of absolute accuracy. Absolute accuracy is fictional since the actual device of a sample will differ from the test structures used to measure the I-V curves. This process variation is an inevitable fact of fabricating integrated circuits with microscopic dimensions. In fact, it is remarkable that we are able to reproduce device behavior to the level of accuracy measured in the field, given that modern minimum-sized devices are limited by doping variation due to only a small number of dopant atoms. In fact, performance is more fundamentally becoming limited by quantum mechanical fluctuations due to the small dimensions and small number of charge carriers and dopants.

Most compact models are built from a physical core which is derived using a long- channel pseudo-two-dimensional transistor. The transistor charge Q-V characteristics are derived using only vertical fields, and currents are in turn derived by introducing a tangential field along the channel. The advantage of this approach is that it allows a closed-form equation to be derived without resorting to numerical integration. The most accurate solutions are the so-called surface potential solutions, which relate device charge to the surface potential along the channel. The downside is that the relationship between surface potential and device terminal voltages is not explicit, requiring Newton-Raphson style iteration to determine the solution. Moreover, the sensitivity to the surface potential is quite high, requiring very accurate convergence. But careful study of the convergence properties of the core equations allows these iterations to terminate in a fixed number of iterations, in effect unrolling the loops. The advantage of these "toy" models is that they are physically derived, and hence can capture the physical behavior of the device over the voltage excursions, ranging from accumulation, depletion, and weak inversion to moderate and strong inversion. Most importantly, the model can predict I-V and Q-V behavior in various stages of inversion, ranging from weak to moderate inversion, and to strong inversion. The moderate region is increasingly important for analog circuit operation, since it allows the designer to trade off speed for gain and high transconductance efficiency (see below). Another important aspect of the compact model is its ability to self-saturate.

In other words, as the drain voltage is varied from tens of millivolts to the full supply voltage across the device, saturation in the current should happen in a smooth and natural manner. In a classical long-channel transistor, saturation occurs owing to the so-called pinch-off effect. In a modern device, saturation occurs earlier due to velocity saturation of carriers, arising from the high-field effects. Once a device is saturated, variation in I-V curve is attributed to the output conductance of the device, which is an accumulated sum of various complex mechanisms such as channel length modulation, drain-induced barrier lowering (DIBL), and impact ionization. Correct prediction of the curvature in this regime of operation is critical for analog circuits, since the intrinsic gain will be determined by the slope of the I_{ds}-V_{ds} curve.

Finally, leakage currents play a critical role in discrete-time and low-power applications. Similar to digital circuits, leakage currents are a concern when the device is "off," which happens when it is biased well below the threshold voltage. In this regime, a real device will conduct very little current, but such a small current can quickly discharge a small capacitor, which introduces an error when sampling and storing a voltage on the capacitor.

How does one quantify these observations? The I-V curve of a given transistor with fixed dimension (W and L) reveals the most salient features. For example, a plot of I_{ds} versus V_{gs} for a family of V_{ds} (Fig. 2.1A) quickly reveals how well a model can predict the device current with absolute bias voltage. If a device is biased in weak or moderate inversion, then the logarithmic plot is more useful as it expands this regime of operation (Fig. 2.1B). A plot of the current versus drain-source voltage or a plot of I_{ds} versus V_{ds} for a family of V_{gs} (Fig. 2.2) shows the current saturation behavior of a compact model and the accuracy of the model in saturation, which is critical for predicting the output conductance. The output conductance is more easily observed using small-signal parameters, as will be discussed shortly, but certain trends, such as predictions of higher or lower current, can be consistently observed even from the I-V curves. Even though absolute accuracy is not important, such a consistent discrepancy reveals that certain device physics are missing in the compact model (or the device extraction was done incompletely).

If the device has a bulk terminal, then the body bias effect should also be included in these plots by varying the body bias and observing the predictability of the model. In this plot, we observe a shift in the threshold voltage of the device. If two independent gates are available, these curves should be plotted for both gates. In practice, many "back gates" are used for leakage control (threshold shift) and may not be able to invert the channel, so the plots are very similar to those for bulk devices. For "common-gate" devices without a body contact, the plot of V_{gs} is the only relevant curve. In these cases, it may be important to measure the I-V curves using pulsed mode measurements to avoid self-heating. Even though most devices have thin fins that are fully depleted, under the right conditions, partial depletion may introduce kinks in the I-V curve that should be correctly modeled. DC and pulsed measurements with various back-gate/body biasing should reveal these device intricacies and the model should be verified to reproduce these effects.

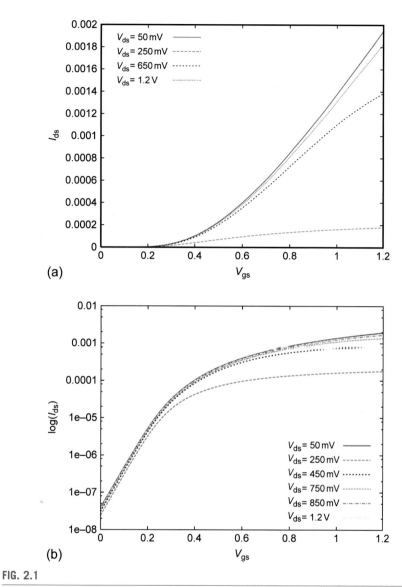

FIG. 2.1

(A) I_{ds} versus V_{gs} for a family of V_{ds} and (B) log I_{ds} versus V_{gs} for a family of V_{ds}.

2.3.2 Geometric scalability

Aside from the bias point, the only other variable under the control of a designer is the transistor geometry. Analog circuits make extensive use of L and W to trade off gain for speed, and unlike digital circuits, nonminimal values of L are used in many circuits. Moreover, it is not uncommon for the circuit designer to sweep W and L of

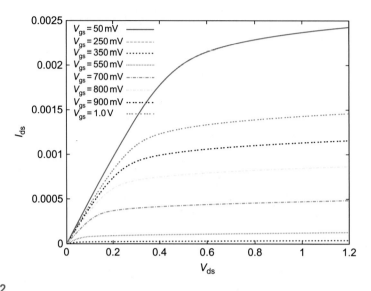

FIG. 2.2

I_{ds} versus V_{ds} for a family of V_{gs}.

key transistors to optimize performance. This means that a compact model must be extremely scalable with both W and L. In fact, a good designer would use a fixed W and sweep the number of transistor fingers N_f, which for FinFETs is equivalent to scaling the number of fins. In this regard, a FinFET scaling is more constrained but naturally follows the preferred technique to scale a device (when matching is a concern).

In the extremes of scaling, short-channel effects are a well-known physical phenomenon that must be very well addressed by any nanoscale model. The complex two-dimensional (or three-dimensional) electrical field pattern along the channel and source/drain region impacts the device's effective threshold voltage V_T, particularly because of nonuniform doping profiles. Scaling the length also gives rise to more pronounced DIBL, which is improved in the double gate structure of a FinFET. A plot of V_T versus the channel length (Fig. 2.3A) is an important metric for a compact model, because the threshold voltage plays a paramount role in circuit design. The importance of V_T is easily appreciated if one observes that the behavior of a FET is directly related to the amount of channel inversion (channel charge), which is directly impacted by V_T. This is why analog designers are concerned with the "overdrive" voltage (V_{gs}-V_T for a long-channel device), since it normalizes the gate bias relative to the threshold, which varies from device to device and as a function of temperature. Interestingly, in most FinFET processes, only a minimum-channel length is allowed, implying that longer-channel-length devices can be realized only by stacking devices. For all short-channel devices, the impact of DIBL is very important and can be understood from Fig. 2.3B, which shows the variation

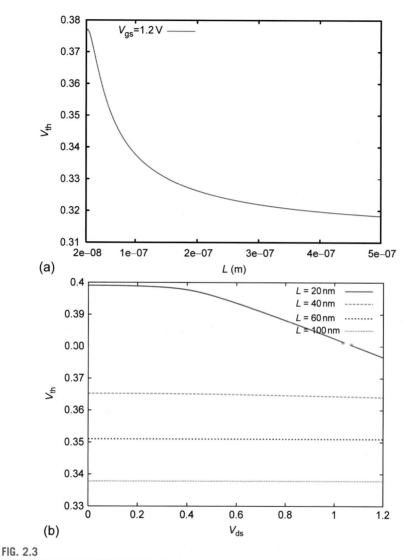

FIG. 2.3

(A) Device threshold voltage V_T versus channel length L. (B) V_T as a function of the drain voltage V_{ds} for a family of channel lengths L.

(reduction) in threshold voltage with increasing drain bias. One of the key advantages of the FinFET device is the reduction in DIBL owing to the strong-channel control afforded by two gates.

Since transistors are three-dimensional structures, the results of geometric scaling are not trivial, especially when one considers the transistor extrinsic parasitic elements such as device resistance and capacitance. In fact, a FinFET device has more capacitance than a comparable bulk device, and it is important to give the designer the maximum flexibility to scale the device in order to optimize the performance. Finally, the impact of well proximity is a well-known phenomenon in bulk devices, which forces designers to use many dummy fingers in their layouts to reduce the impact on the variation in threshold voltage and mobility. These effects, which are mostly extrinsic to the device, must be well modeled and captured. Given the smaller dimensions of a FinFET and its three-dimensional structure, the impact of stress and strain is even more pronounced, and can have both a positive and a detrimental impact on the device.

2.3.3 Variability model

It is extremely important to capture device variation. Variation has many sources, including systematic and random fluctuations. Systematic variations include layout-dependent effects, such as the variation in the threshold voltage V_T as a function of the distance to the well, whereas random variations arise from natural variation in doping profiles and lithographic variations. These variations are usually categorized into device-to-device mismatch for closely spaced transistors, die-to-die variations, and wafer-to-wafer variations, often just called process variation. In other words, we are interested in both the variation for a single transistor from die to die and that between a pair of "matched" transistors within the same die. Since circuit yield is a strong function of the variation in transistor performance, designers go to great lengths to ensure that their circuits operate over all possible process, voltage, and temperature variations. A common practice is to deliver transistor parameters for particular corners, such as "fast," "nominal," and "slow" corners for transistors (and other devices such as resistors and capacitors). If we just take transistor variations into account, since two flavors of transistors are used (n-type MOS and p-type MOS), this results in five separate corners that must be checked, often called SS, FF, SF, FS, and TT ("S" for "slow," "F" for "fast," and "T" for "typical").

While corner models are prevalent in industry, they are actually overpessimistic in most circumstances. The corner cases often represent worst-case conditions that occur very rarely in practice (Six Sigma), and designing circuits to meet corner cases requires overdesign, resulting in larger area and higher current consumption. A more attractive approach is statistical variation, where model cards are derived by varying the actual physical parameters that vary, such as doping levels and lithographic dimensions. This underscores the need for compact models that are physically derived so that statistical variation in basic physical parameters leads to correct variation in the model card parameters. When the model is nonphysical, it is very difficult to come up with a set of basic parameters to vary in the model card to match observed measurements of transistor variations. Even using principal component analysis on compact models with nonphysical origin in equations has not yielded satisfactory results.

An accurate mismatch model is another key model requirement. In fact, one may argue that mismatch is ultimately the biggest enemy of an analog circuit designer.

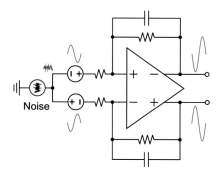

FIG. 2.4

A fully differential circuit can reject supply noise and other common-mode noise if the circuit is perfectly balanced.

The reason for this is the need for high levels of common-mode rejection and supply insensitivity in analog circuits. For example, the circuit in Fig. 2.4 is a typical building block in amplifiers and filters in analog circuits. It is realized as a differential circuit in order to reject variations of the common mode of the signal and the power supply variations in the circuit. Power supply noise is inevitable, especially when circuits live in a mixed-signal environment where digital circuits create supply and substrate noise. If the circuit is perfectly balanced, this noise appears equally on the inverting and noninverting terminals of the amplifier, and it can be rejected when the output signal is measured differentially, requiring good common-mode rejection. Any mismatch between the positive and negative sides of the amplifier creates imbalance, limiting the amount of rejection. Moreover, mismatch in these differential circuits creates a DC offset that must be canceled, which introduces additional complexity in the circuit.

To fight mismatch, analog circuit designers scale transistor dimensions and bias points, or completely reject one technology node in favor of another. A good mismatch model again relies on a physically derived compact model that can accurately predict variations in key parameters such as V_T with variations in doping and lithography.

2.3.4 Intrinsic voltage gain

One of the most important metrics in analog design is the DC gain of an amplifier, which is limited by the device output resistance in modern devices. As shown in Fig. 2.5, a key building block such as a differential stage has a gain that is related to the product of the transconductance g_m and the output resistance r_o of the transistor. In reality, both the output resistance of the n-type MOS and the p-type MOS must be taken into account, so this metric represents a maximum upper bound on the gain of a single-stage amplifier:

$$A_0 = g_m \left(r_{on} \big\| r_{op} \right) < g_m r_{on} = \frac{g_m}{g_{ds}} \tag{2.1}$$

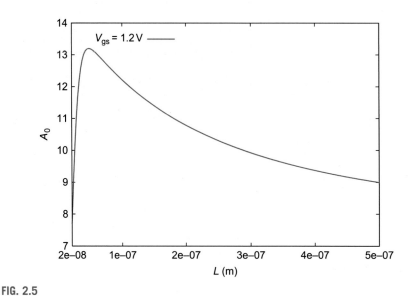

FIG. 2.5

Device intrinsic gain A_0 plotted as a function of channel length L.

DC gain is important because it determines the accuracy of a closed-loop amplifier. One desires a high DC gain so that the loop gain of a multistage feedback amplifier is maximized. A high value of the loop gain translates into a large rejection of unwanted distortion generated by the amplifier. It also ensures high DC accuracy that is largely independent of temperature and the process.

DC gain is a good metric for compact model evaluation since it is the ratio of two small-signal parameters, which depend on the first derivatives of the current/voltage relations in the device. It is not a function of the device width (to first order), since as we scale W, both g_m and g_o increase in the same proportion. It is a strong function of L, the channel length, owing to its impact on both g_m and g_o.

FinFETs offer a clear advantage over bulk devices because they offer higher intrinsic gain for the same channel length, mainly due to the reduction in DIBL as noted above, since a dual gate structure can maintain better channel control over the drain. This is why Eq. (2.1) is recast as the ratio of g_m (gate control) and g_{ds} (drain control), which amplifies this point.

It is interesting to observe that the device intrinsic gain depends on the device bias current. Taking simple square law equations for simplicity, we have

$$A_0 = \frac{g_m}{g_{ds}} = \frac{2I_{ds}/V^*}{I_{ds}/V_A} = \frac{2V_A}{V^*} \tag{2.2}$$

In the above equation, we have related the transconductance to the device overdrive voltage V^* and the "early voltage" V_A. For a long-channel device, V^* is equal to $V_{ds,sat}$, or the drain-source saturation voltage. This would be the value of the drain

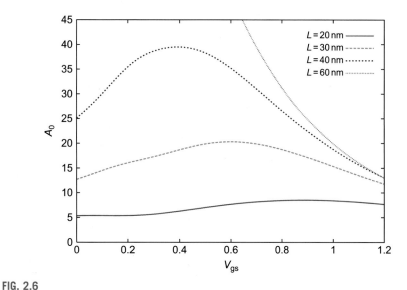

FIG. 2.6

Device intrinsic gain A_0 versus V_{gs}.

voltage which causes pinch off in the channel owing to a zero bias at the drain side given by $V_{ds,sat} = V_{gs} - V_T$. In a short-channel device, owing to high-field effects, $V_{ds,sat} \neq V^*$, so we can define V^* as the ratio of I_{ds} and g_m:

$$V^* = \frac{2I_{ds}}{g_m} \tag{2.3}$$

The important point is that the intrinsic gain is maximized by using the optimum overdrive voltage, as shown in Fig. 2.6. As the current in the device is reduced, it eventually leaves the strong-inversion region, and as we approach weak inversion, the device drain current becomes an exponential function of the gate-source bias, similar to a bipolar device, which implies that V^* approaches a constant related to the thermal voltage:

$$V^* = \frac{2I_{ds}}{(1/(nV_t))I_{ds}} = 2nV_t = 2n\frac{kT}{q} \tag{2.4}$$

where $n < 1$ relates to how the gate voltage controls the surface potential through the voltage divider,

$$n = \frac{C_{ox}}{C_{ox} + C_{dep}} \tag{2.5}$$

In the limit $n \to 1$, V^* approaches about 50 mV at room temperature. A plot of g_m/I_{ds}, sometimes known as transconductance efficiency, versus V_{gs} is shown in Fig. 2.7. As expected, the device bias current normalized transconductance peaks for low values of overdrive, or operation in weak inversion. Plotting the inverse relation (scaled by 2) gives the effective overdrive voltage V^* (Fig. 2.7B). For comparison, the square

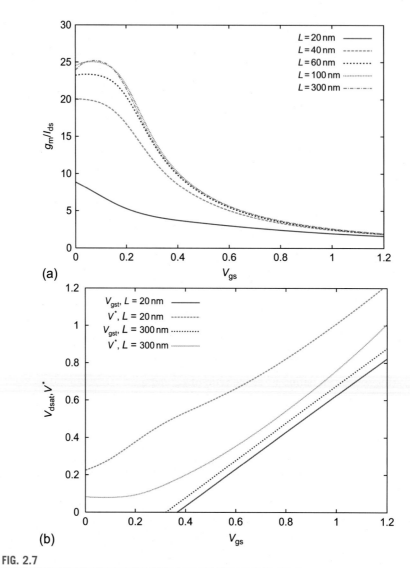

FIG. 2.7

(A) Device transconductance efficiency g_m/I_{ds} versus V_{gs}. (B) The effective overdrive voltage V^* compared with $V_{gs}-V_T$. For the long-channel device, the plots match fairly well, but for a short-channel device, they are very different.

law relationship is also plotted, clearly showing a deviation as V_{gs} drops below the threshold voltage.

The intrinsic gain is often plotted as a function of V_{ds} for a family of channel lengths L (Fig. 2.8). For low values of V_{ds}, the device is in the triode region and the output resistance is low, resulting in low gain. Once V_{ds} causes saturation to

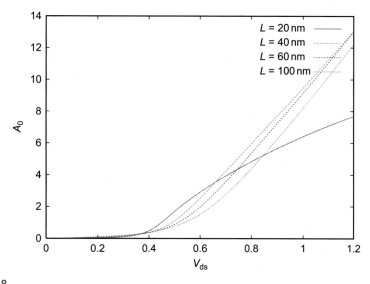

FIG. 2.8

Device intrinsic gain A_0 versus V_{ds} for a family of channel lengths L.

occur, the output resistance increases and the gain maximizes. For large values of V_{ds}, detrimental effects of high drain fields such as DIBL and impact ionization currents once again reduce the gain.

To see these relations more explicitly, we can plot g_m as a function of V_{gs} and V_{ds} as shown in Fig. 2.9. We observe that as the gate overdrive increases, the absolute value of g_m increases. For the diffusion component of current, at low overdrive, the increase in g_m is exponential owing to the exponential dependence of current on gate voltage. Operating with a higher bias increases the slope of the I_{ds}/V_{gs} curve in proportion to the increase in current. On the other hand, if we consider the drift component of the current, we know the current is proportional to the amount of charge in the channel and the carrier velocity. The channel charge scales in proportion to the gate bias, so for a fixed mobility, we expect g_m to increase linearly. In fact, for low fields, owing to Coulomb scattering, the mobility will experience an enhancement owing to screening provided by the inversion layer, and the increase in g_m is faster than linear. On the other hand, owing to high-field effects, we know the mobility will eventually degrade as carriers are pushed closer to the silicon-insulator interface, where surface scattering causes the mobility to drop. The trend for g_m as a function of V_{ds} can be divided into two regions: (1) in the triode region of operation, increasing V_{ds} increases the current, so the overall g_m increases and (2) as the device nears saturation, one would expect g_m to saturate.

In short-channel FETs, the output resistance $r_o = 1/g_{ds}$ has a strong bias dependence, especially on the drain voltage (Fig. 2.10). The device output conductance in the triode region is very large since the drain voltage has a direct impact on the

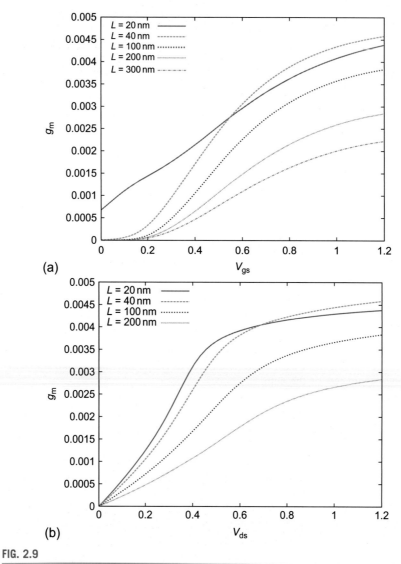

FIG. 2.9

Device transconductance g_m versus (A) V_{gs} and (B) V_{ds}.

channel charge. In fact, to first order, the device g_{ds} is equal to the device g_m in this region, since varying the drain has the same impact as varying the gate owing to the presence of the inversion layer, making the drain terminal just as effective as the gate terminal. In a long-channel device, in the so-called pinch-off region, the device terminal no longer controls the channel charge directly, and the output conductance drops dramatically. In practice, the drain still modulates the depletion

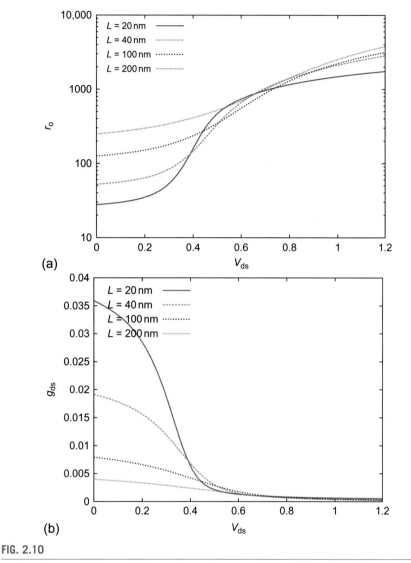

FIG. 2.10

(A) Device output resistance r_o and (B) $g_{ds} = 1/r_o$ versus V_{ds}.

region width, which indirectly impacts the device through channel length modulation, which causes the current to increase with increasing drain bias, thereby increasing g_{ds}. The drain terminal has an even greater impact in short-channel devices, both because of the increase in the relative magnitude of the change in channel length δL relative to L, and because of DIBL. Owing to the complex two-dimensional field pattern in a short-channel device, a high voltage on the drain creates a high-field region that stores charge in the depletion region. This charge sharing in turn means that more electrons can flow into the channel, thereby lowering the threshold of the device. Fortunately, this effect is reduced for FinFETs, resulting in an improvement in intrinsic gain. Finally, as the voltage is further increased, impact ionization in the high-field region causes a further increase in current and thereby a drop in the output resistance.

2.3.5 Speed: Unity gain frequency

In analog circuits, DC gain is only half the story. While operation at subthreshold is beneficial for gain and especially transconductance efficiency, in practice this is rarely acceptable because of the reduction in the device speed. Only extremely slow circuits can tolerate operating in this region. For most other circuits, operation in moderate or strong inversion is desired.

The speed of a transistor is often measured using the device unity gain frequency f_T, or the frequency when the current gain of a transistor is unity. Since $i_d = g_m v_{gs}$, and $v_{gs} j\omega(C_{gs} + C_{gd}) = i_s$, taking the ratio, we have

$$A_i = \frac{i_d}{i_s} = \frac{g_m}{j\omega(C_{gs} + C_{gd})} \tag{2.6}$$

Solving for $|A_i| = 1$, we arrive at the unity gain frequency

$$\omega_T = 2\pi f_T = \frac{g_m}{C_{gs} + C_{gd}} \tag{2.7}$$

It is not at first obvious why f_T plays such an important role in analog (and digital) circuits. To see this, consider the simple cascade amplifier shown in Fig. 2.11A, where a single-stage amplifier drives an identical copy. The gain of this first stage is given by the intrinsic gain of the amplifier, and the 3 dB bandwidth is limited by the pole at the high impedance node:

$$\omega_0 = \frac{1}{r_{o,1}\left((1 + |A_2|)C_{gs,2} + C_{d1,tot}\right)} \tag{2.8}$$

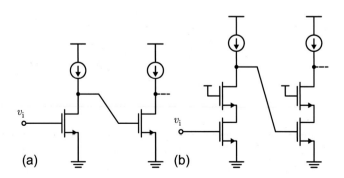

FIG. 2.11

(A) A cascade of common-source amplifiers. (B) A cascade of cascode amplifiers.

Here, $C_{d1,tot}$ is the total drain capacitance ($C_{gd} + C_{ds} + C_{wire} + \cdots$), and the effect of Miller multiplication is captured by boosting the second-stage input capacitance by the gain A_2. If the second stage is a cascode with a small voltage gain between the gate and the drain, then A_2 can be made smaller than unity to minimize its impact. So if we disregard the Miller effect, we have

$$\omega_0 = \frac{1}{r_o \left(C_{gs} + C_{d,tot}\right)} \tag{2.9}$$

as an upper bound. In most applications, we intend to embed this amplifier in a feedback loop, so the gain-bandwidth product of interest is

$$A_0\omega_0 < g_m r_o \frac{1}{r_o \left(C_{gs} + C_{d,tot}\right)} = \frac{g_m}{C_{gs} + C_{d,tot}} \approx \omega_T \tag{2.10}$$

which bears a resemblance to the unity gain of the transistor itself. So, in an approximate fashion, we can see that the maximum gain-bandwidth product of an amplifier is in fact limited by the device unity gain frequency. Even in a digital circuit, we find that f_T plays a key role. Consider the time constant of a simple gate, such as an inverter. If the fan out of an inverter is unity, in other words the inverter drives an identical copy of itself, then the load of the inverter is approximately $C_{gs} + C_{d,tot}$. The discharge current takes a complicated form, but to first order the transistor acts like a switch with on-conductance $g_{ds} = g_m$ (in the triode region), so the discharge time is given by

$$\tau_T = \frac{C_{gs}}{g_m} = \frac{1}{\omega_T} \tag{2.11}$$

A plot of f_T as a function of V_{gs} for different channel lengths is shown in Fig. 2.12. The bias and geometry dependency on g_m have already been discussed. The remaining dependency is on the device capacitance, including the gate-to-source capacitance C_{gs} and the gate-to-drain capacitance. It is important to note that f_T is usually measured using scattering parameters, and so using the "DC" values of g_m and C will fail to capture f_T at high frequency. We will return to this point in Section 2.4. In a long-channel device, C_{gs} is usually the dominant capacitor, since C_{gd} occurs only from the small overlap between the gate terminal and the drain region. In a real device, a considerable portion of the capacitance may be due to extrinsic capacitors in the wiring. In modern devices, the gate-drain overlap region is a significant fraction and C_{gd} can be comparable to C_{gs}. In a FinFET, owing to the three-dimensional structure, a considerable amount of fringing capacitance increases C_{gd} compared with a bulk device. For these reasons, FinFET f_T suffers compared with that of an equivalent bulk device. In reality, bulk devices cannot scale to FinFET dimensions, so the comparison is not fair. Nevertheless, this requires a very good extrinsic FET model to accompany the intrinsic model.

2.3.6 Noise

Thermal noise is relatively well understood in FETs and should form the core of any model. When drift current dominates, the thermal noise is a function of the channel conductance, whereas in moderate and weak inversion, the diffusion component

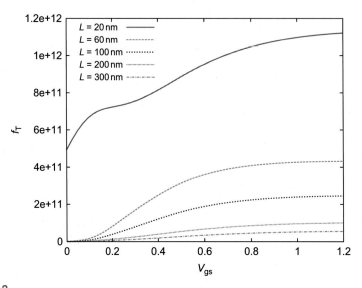

FIG. 2.12

The device unity gain frequency f_T as a function of V_{gs} for a family of channel lengths L.

gives rise to shot noise. A good model should capture the drain noise current accurately in all regions in a smooth and uniform manner. Capturing noise in saturation is complicated by the excess noise generated by the FET, captured by the parameter γ in the spot noise variance of the drain current:

$$\overline{i_{d,n}^2} = 4kT g_{ds,ch} \gamma \delta f \tag{2.12}$$

where $g_{ds,ch}$ is the channel conductance. For a long-channel device, it is well known that $\gamma = 2/3$, which is derived by considering the thermal noise contribution of an incremental region of the channel to the drain. The coupling to the gate through the gate-oxide capacitance also gives rise to a correlated gate noise current. For short-channel devices, it has been observed that γ is larger than 1, which is sometimes attributed to the hot electron effects occurring on the drain end of the transistor. A good test for a compact model is a plot of γ versus the drain and gate bias voltage, as shown in Fig. 2.13. It is important to define γ in terms of $g_{ds,ch}$ (the drain-source conductance in the "linear region" of the device) and not g_m, which leads to the erroneous conclusion that γ is much larger than 1. The reason is that noise arises from the physical resistance of the channel, which originates from the inversion layer at the source to a region near the drain. As noted before, $g_m = g_{ds}$ in the linear region of the transistor operation.

One would expect that the thermal noise of a FET varies smoothly as the device bias is varied from strong inversion to weak inversion, with the limits of weak-inversion noise given by the well-known shot noise current $2qI_{ds}$. For strong-inversion operation, a plot of γ versus V_{ds} should have physical meaning in the saturation regime. To understand the importance of γ, let us derive the noise

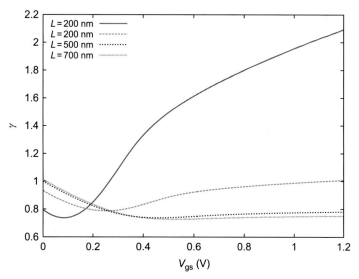

FIG. 2.13

Excess noise factor γ versus gate bias V_{gs}.

figure of a common-source and a common-gate amplifier. For the common-source case, at low frequencies, the total output noise is given by

$$\overline{i_o^2} = g_m^2\,\overline{v_{R_s}^2} + \overline{i_d^2} = g_m^2\,4kTR_s\delta f + 4kTg_{\text{ds,ch}}\gamma\delta f \tag{2.13}$$

from which we calculate the noise figure

$$F = 1 + \frac{g_{\text{ds,ch}}\gamma}{g_m^2 R_s} = 1 + \frac{\gamma/\alpha}{g_m R_s} \tag{2.14}$$

For many applications, R_s is fixed and g_m is limited by power consumption, which shows that the noise limit is ultimately determined by γ. The common-gate amplifier is even more sensitive to γ since it has no current gain,

$$\overline{i_o^2} = \overline{i_{R_s}^2} + \overline{i_d^2} = 4kTG_s\delta f + 4kTg_{\text{ds,ch}}\gamma\delta f \tag{2.15}$$

giving the well-known lower bound on the obtainable noise figure

$$F = 1 + \frac{(g_m/\alpha)\gamma}{G_s} \tag{2.16}$$

The reason the common-gate amplifier is favored over the common source amplifier is its ability to obtain a wideband impedance match by setting $g_m = R_s$, which results in

$$F = 1 + \frac{\gamma}{\alpha} \tag{2.17}$$

Later, we shall also see that the ultimate lower limit on phase noise is similarly related to γ.

In addition to thermal noise, flicker noise plays a big role in analog circuits, especially as devices are scaled to smaller dimensions. The spectrum of flicker noise in a FET overlaps with the frequency of interest in many analog signal processing applications (DC to 10 s or MHz), and a designer will often size devices in order to minimize the impact of flicker noise (Fig. 2.14). The flicker noise model should therefore be very physical in scaling limits, particularly when the designer selects different channel lengths L to minimize noise. Naturally, the bias dependence of flicker noise should be correctly captured.

In a FET, all of the intrinsic noise originates from the channel and traps in the gate-oxide interface. Extrinsic parasitic elements, such as physical resistance in the gate, source, and drain, also contribute to noise. Plotting noise parameters versus bias is a good way to tease out the various sources of noise: ones that depend on bias are generally due to intrinsic device noise, and fixed noise contributions are due to extrinsic sources. An alternative representation of the noise of a FET is to define an input pair of correlated noise voltage and current sources. This is useful for circuit design and also a good metric for testing the model. Some circuit simulators can simulate and plot these parameters for a two-port circuit, and these plots are a good way to test the physical variation in the input noise sources, which depend on both intrinsic and extrinsic noise sources.

2.3.7 Linearity and symmetry

2.3.7.1 Harmonic distortion

Linearity, or the ability of an analog circuit to faithfully reproduce the input signal at the output (amplified or buffered), is an important metric, although it receives less attention than gain and speed. The reason is simple: most analog circuits employ

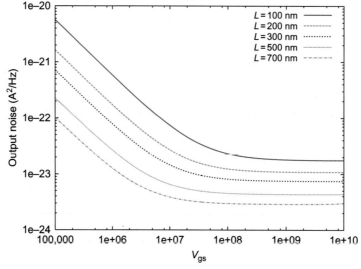

FIG. 2.14

The drain noise current spectral density versus frequency for a family of channel lengths L biased at the same bias V_{gs}.

feedback, which has the ability to reject the distortion produced by devices by the loop gain of the circuit. Open-loop amplifiers are more common in RF circuits, so intrinsic device nonlinearity is still an important metric. Even in modern nanoscale devices, linearity is playing a bigger role as loop gains drop (owing to the drop in intrinsic gain). Furthermore, when FETs are used as switches, particularly in sampled time systems, the distortion produced by the device is sampled and processed along with the signal.

The source of distortion is easy to understand since the *I-V* characteristics of a FET are described by nonlinear equations of terminal voltages:

$$I_{ds} = f\left(V_{gs}, V_{ds}, V_{bs}\right) \tag{2.18}$$

Expanding the above relation into a Taylor series of V_{gs}, while holding V_{ds} and V_{bs} constant, and disregarding the DC, we have the following "G_m" nonlinearity power series:

$$i_{ds} = g_{m1} v_{gs} + g_{m2} v_{gs}^2 + g_{m3} v_{gs}^3 + \cdots \tag{2.19}$$

where lowercase symbols denote AC quantities. The first term, g_{m1}, is nothing but the transconductance of the device. The curvature of the *I-V* curve gives rise to higher-order coefficients that describe the deviation of AC from the quiescent operating point for a small deviation of the gate-source voltage about the bias point. In a similar way, we can derive a power series in terms of the drain-source voltage,

$$i_{ds} = g_{ds1} v_{ds} + g_{ds2} v_{ds}^2 + g_{ds3} v_{ds}^3 + \cdots \tag{2.20}$$

and, in general, a two-dimensional power series can be derived when both the gate-source voltage and the drain-source voltage vary. In practice, to keep things simple, it is common to deal only with one-dimensional power series relations.

The commonest way to measure the nonlinearity of a block is to find the harmonic distortion (HD), which is defined by driving an amplifier with a pure sinusoid and observing the harmonics at the output:

$$i_{ds} = g_{m1} A_{gs} \cos(\omega t) + g_{m2} A_{gs}^2 \cos^2(\omega t) + g_{m3}^3 A_{gs}^3 \cos^3(\omega t) + \cdots \tag{2.21}$$

For example, $\cos^2(x) = 0.5(1 + \cos(2x))$, so the square term produces DC and the second harmonic. Likewise, cubing the signal gives

$$\cos^3(x) = \cos(x)\,\cos^2(x) = 0.5(\cos(x) + \cos(x)\,\cos(2x)) \tag{2.22}$$

$$= \frac{1}{2}\left(\cos(x) + \frac{1}{2}(\cos(x) + \cos(3x))\right) \tag{2.23}$$

$$= \frac{3}{4}\cos(x) + \frac{1}{4}\cos(3x) \tag{2.24}$$

which contains the fundamental signal and the third harmonic.

In general, with use of a binomial expansion on the sum of complex exponentials, it is not very difficult to show that the *n*th odd power produces harmonics starting from the fundamental and every odd harmonic (first, third, fifth, …, nth), whereas an nth even power produces all even harmonics starting from DC (zero, second, fourth, sixth, …, nth). In other words, even harmonics generate DC and every even harmonic up to n ($0 \times \omega$, $2 \times \omega$, …, $n \times \omega$). Odd harmonics generate the fundamental and every odd harmonic up to n ($1 \times \omega$, $3 \times \omega$, …, $n \times \omega$). This allows us to write the following equation for the harmonics of the output drain current when the input is excited by a sinusoid:

$$i_{ds} = \left(g_{m1} A_{gs} + \frac{3}{4} g_{m3} A_{gs}^3 + \cdots \right) \cos(\omega t)$$

$$+ \left(\frac{1}{2} g_{m2} A_{gs}^2 + \cdots \right) \cos(2\omega t) \qquad (2.25)$$

$$+ \left(\frac{1}{4} g_{m3} A_{gs}^3 + \cdots \right) \cos(3\omega t) + \cdots$$

The important point is that each harmonic arises from a potentially infinite set of powers, but for small excursions from the bias point, the most dominant terms are the smallest powers. This last point is important, and may seem like a contraction, but in fact it follows from the assumption of a "small-signal" or "weak" distortion. For example, for a practical device, when the equation is put into a power series, one finds that the relevant quantity is a normalized voltage v_{gs}/V^* (overdrive voltage), or in weak inversion, $v_{gs}q/kT$, so as long as the normalized voltage excursion is small, raising this to an increasingly larger power produces an increasingly smaller quantity in the power series. Under these conditions, the lowest-order powers completely dominate the behavior of the device distortion generation.

Given these conditions for weak distortion, the HD is defined as the ratio of the amplitude of the harmonic to the fundamental. For example,

$$HD_2 = \frac{V_{2\omega}}{V_\omega} = \frac{g_{m2}\left(A_{gs}^2/2 \right)}{g_{m1} A_{gs}} = \frac{1}{2} \frac{g_{m2}}{g_{m1}} A_{gs} \qquad (2.26)$$

$$HD_3 = \frac{V_{3\omega}}{V_\omega} = \frac{g_{m3}\left(A_{gs}^3/4 \right)}{g_{m1} A_{gs}} = \frac{1}{4} \frac{g_{m3}}{g_{m1}} A_{gs}^2 \qquad (2.27)$$

As shown in Fig. 2.15, as we sweep the amplitude of the input sinusoid, the magnitude of the second harmonic increases quadratically, whereas the third harmonic grows cubically, so their HD ratios exhibit a power law with one order less (linear and quadratic). As the amplitude grows beyond the small-signal distortion regime, eventually higher-order terms dominate and the slopes change.

To show the importance of overdrive voltage, plots of HD_2 and HD_3 versus V_{gs} are shown in Fig. 2.16. Physically, the improvement in HD with V_{gs} can be attributed to the fact that electric fields in the device are related to the overdrive voltage, and

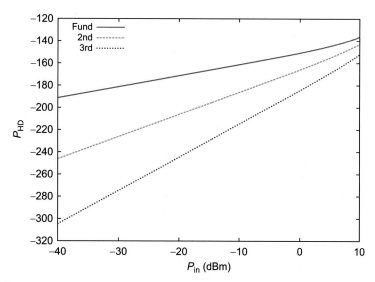

FIG. 2.15

The HD currents at 2ω and 3ω versus the input amplitude P_{in} (dBm) applied at the gate/source showing the characteristic slope of 2 and 3 on a log scale. Note the bias is fixed at V_{gs}.

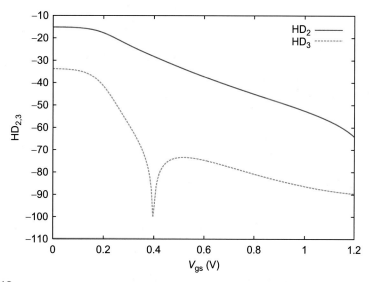

FIG. 2.16

HD_2 and HD_3 versus the bias V_{gs} for a fixed swing v_{gs} (-30 dBm). Note the "sweet spot" for HD_3.

lower fields produce comparatively linear transfer relations, as expected. The dip in HD_3 is a consequence of $g_{m3} = 0$ for a particular bias point, sometimes called the "sweet spot" of a MOSFET.

While we have defined HD for G_m nonlinearity, the same metrics can be defined for any other pair of currents/voltages in the device. While G_m nonlinearity is the commonest type of nonlinearity, output conductance nonlinearity is perhaps a close contender. Since HD is a strong function of the device bias point, it is hard to define a universal metric. The modeling engineer should ensure that the model reproduces the distortion at a range of different bias voltages (both V_{gs} and V_{ds}), and, most importantly, it is important to ensure that the slope of the HD products follows the above relations for weak inputs. For example, on a log scale, the third harmonic signal should have a slope of 3 when the input is very weak, and any deviation from this slope indicates a problem in the compact model.

This is because we expect all the compact model curves to be smooth and infinitely differentiable, which means that we can always define a power series and prove that the slope is indeed 3. But if the compact model has a source of an abrupt nonlinearity, then a nonsmooth curve will not follow our predictions. A classic example is an "if" statement in the compact model that changes the device I-V behavior (e.g., a diode model that has a hard saturation). If the signal is biased near this "if" statement, then even a small deviation produces a kink in the I-V transfer (a point where the slope is not defined). Keep in mind that for a strong enough signal, this assumption breaks down since the fundamental power is affected by all odd harmonics, which causes the slope to change dramatically, but this occurs only when a device is driven very hard (deep compression).

While it is not possible to check every possible HD product, it is good practice to check HD_2 through HD_5 to make sure all curves follow the correct slopes as predicted by small-signal distortion. Since distortion arises from the large signal behavior of a transistor, it has to be simulated using transient or steady-state simulation, such as periodic steady-state analysis using shooting in SpectreRF or harmonic balance simulation. The same can be obtained from a transient simulation by calculating the fast Fourier transform of the output waveform and plotting harmonic powers. Classic SPICE also includes a disto statement that is similar to an AC analysis, but includes the effect of higher-order harmonics, as we have done.

2.3.7.2 Gain compression

A very important metric is the gain compression that naturally occurs in all devices owing to nonlinearity. For example, if we plot the effective transconductance G_m of a device versus the input swing, we obtain a curve as shown in Fig. 2.17. As expected, for small input voltages v_{gs}, $G_m = g_m$, but as we increase the swing, the transconductance either increases or drops. The reason for this change in G_m is that all odd harmonics of the I-V relation generate fundamental current ($g_{m,2k-1}$ for $k > 1$ in our power series), and these currents can be in phase or out of phase with the fundamental generated by g_{m1}. If we examine Eq. (2.25), we find that the fundamental output current amplitude is given by

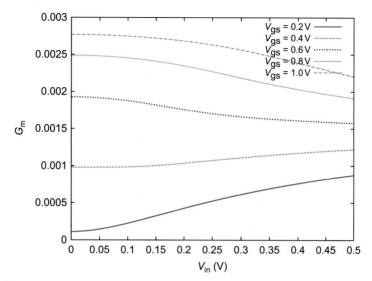

FIG. 2.17

The transconductance gain compression versus the signal swing v_{gs}.

$$i_{ds,\omega} = g_{m1} v_{gs} + \frac{3}{4} g_{m3} v_{gs}^3 + \cdots \tag{2.28}$$

If we normalize this current by the gate drive amplitude v_{gs}, we obtain the effective transconductance

$$G_m = \frac{i_{ds,\omega}}{v_{gs}} = g_{m1} + \frac{3}{4} g_{m3} v_{gs}^2 + \cdots \tag{2.29}$$

So, we see that the effective transconductance is related to the gate-source voltage excursion v_{gs}. For very small swings, in fact, the small-signal limit for $v_{gs} \to 0$, then $G_m = g_{m1}$ as expected. But for larger swings, depending on the sign of g_{m3}, the transconductance can increase or decrease. The increase in transconductance, or more generally gain, is called gain expansion. Likewise, a decrease in gain is called gain compression. In the plot in Fig. 2.17, when the device is biased in weak inversion, the gain is expansive owing to the exponential nature of the device I-V curve. For higher bias points, the gain is compressive owing to the high-field mobility effects in the FET.

Note that we have explicitly separated the gain into its components G_m and R_o, since gain expansion also occurs owing to the limited output swing (headroom) and output resistance R_o of a device. In practice, both act in conjunction to determine the overall gain compression characteristics.

Since all transistors operate with a fixed headroom, eventually the voltage swing must be clipped by the rails, and then the gain therefore drops, leading to an eventual compressive characteristic (Fig. 2.18). But the shape of the compression, in

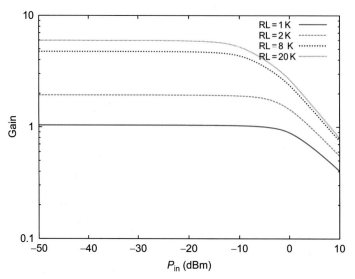

FIG. 2.18

The voltage gain compression versus the signal swing P_{in} (dBm). The drain is biased at 1.2 V using an ideal inductor and the load resistor R_L is varied. Note that high values of R_L have higher gain, but compress earlier owing to the drain voltage swing (rather than the G_m compression).

particular regions of expansion and compression, is related to the device nonlinearity and is a good way to gauge the accuracy of a compact model. For example, G_m compression occurs due to high-field effects and mobility degradation. While these effects are visible on the *I-V* curve as well, here the effects are measured by the slopes of the *I-V* curve, which may not be as apparent to the naked eye.

2.3.7.3 Memory effects

To keep things simple, we have explicitly disregarded all capacitive nonlinearity in the FET. This means that we have assumed that the drain current is an instantaneous function of the gate-source and drain-source voltages, which is not true in practice. Owing to the presence of both linear and nonlinear capacitance at the terminals of a FET, the situation becomes much more complicated, requiring a Volterra series description instead of a power series relation [1]. In practice, in most analog circuits, these so-called memory effects are disregarded. This is justified in practice if the device is operated well below the cutoff frequency of the transistor, allowing the quasi-static assumption to be valid, and well below the poles of the amplifier, so that the amplitudes of signals are not attenuated or signals are not phase shifted as the signals travel through a transistor. These conditions are often met for the desired signals that are processed by analog circuits, but not for the harmonics.

2.3.7.4 Intermodulation distortion

In practice, it is rare to amplify a single tone, and, in general, a complex signal is processed by a device. For linear time-invariant circuits, we know that we only need to characterize the frequency response of the system (or equivalently the impulse response in the time domain), but for nonlinear circuits, even memoryless ones, we have to characterize the impact of not only every tone, but also every two tones, three tones, and, in general, N tones, to completely understand the response of the system to a general signal. Owing to the complexity of doing this, a more common metric is to characterize the intermodulation between only a few tones. The commonest metrics are the two-tone response and the associated intermodulation signals (in general not harmonically related to either tone). We will return to intermodulation distortion in Section 2.4.

2.3.8 Symmetry

The symmetry of a compact model is also an important metric in analog circuits that use device bias around $V_{ds} = 0\,V$, which occurs when the device is used as a switch in sampling circuits and in commutating circuits (such as RF mixers). The origin of the asymmetry in the compact model is not the drain-source swapping in the model, as was once believed, nor is it due to source referencing, rather asymmetry can arise from both of these mechanisms if they are not handled properly. For example, an "if" statement is often used to swap the internal source/drain of a physically symmetric FET, since the label "source" and "drain" are defined only by the bias points ($V_{ds} > 0\,V$ for an n-type FET), which may change sign during the course of a circuit simulation. The model typically assigns one terminal as the drain and detects a change in sign, which can be done only with a finite precision. In other words, when $V_{ds} < E$, for some small E, the drain-source swapping will occur. This will produce a small kink in the transfer curve about $V_{ds} = 0\,V$, which requires smoothing.

The Gummel symmetry test is used to detect potential symmetry problems in a FET model. The circuit shown in Fig. 2.19 is used to excite the circuit about $V_{ds} = 0\,V$. For a uniformly doped FET, we expect the drain current to be an odd function of v_x,

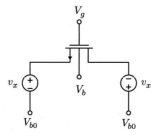

FIG. 2.19

The circuit used to characterize the Gummel symmetry test.

$$I_d(v_x) = -I_d(-v_x) \tag{2.30}$$

so

$$i_x(v_x) = \frac{1}{2}(I_d(v_x) - I_s(v_x)) \tag{2.31}$$

For a properly designed compact model, we expect i_x is a smooth function of v_x with continuity at $v_x = 0$ V, implying that $d^n i_x / dv_x^n$ for all n. From physical considerations, we expect $i_x = 0$ A at $v_x = 0$ V, and also that $d^2 i_x / dv_x^2 = 0$ due to the symmetry of waveform. For example, the first- and second-order derivatives for the FinFET common multigate (CMG) model are plotted, as shown in Fig. 2.20, clearly demonstrating smooth and consistent behavior.

In addition to the current behavior, the terminal charge behavior of the Gummel symmetry circuit should also satisfy physical conditions on symmetry and smoothness. Sweeping v_x from a negative value to an equal positive value, one can plot the gate, source, drain, and bulk charges of the MOSFET (Q_g, Q_s, Q_d, and Q_b), respectively. The derivatives of charges versus v_x should be continuous around $v_x = 0$ V, and the derivatives of charges should pass smoothly through $v_x = 0$ V and have no discontinuities. Furthermore, the derivatives of Q_s and Q_d should be equal and opposite at $v_x = 0$ V because the device is assumed to be symmetric.

An ideal symmetric device should also exhibit symmetry in the capacitances C_{gs} and C_{gd}, with equal values at $v_x = 0$ V. Similar comments apply to C_{bs} and C_{bd}, and C_{dd} and C_{ss}.

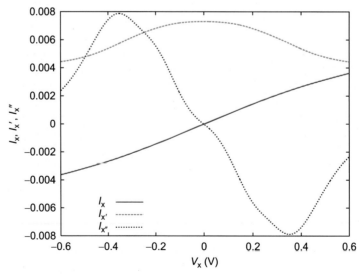

FIG. 2.20

The Gummel symmetry test for the BSIM-CMG FinFET model.

2.4 **RF metrics**

A good analog compact model is the foundation for a good RF compact model. In addition to the already mentioned important metrics for analog circuits, we find that RF circuits require new modules in the compact model to capture gate resistance, substrate resistance, and nonquasi-static (NQS) effects. Since many RF circuits do not use feedback, they are more sensitive to distortion produced by the transistors, placing stronger demands on the accuracy of the model. Finally, since inductors can be used to tune out capacitive parasitic elements, there are also lower tolerances for error in the capacitance of the device. Furthermore, we find that f_T is not the best metric for speed, although it continues to play an important role.

2.4.1 **Two-port parameters**

At higher frequencies, it is more common to rely on two-port parameters of a transistor. The reasons for the importance of two-port parameters, as opposed to say a hybrid-pi model or a full compact model, are both historical and practical. In the past, compact models were relatively simple and did not capture the detailed device behavior and extrinsic parasitic elements, particularly for operation near f_T of the device when NQS effects are important to model. An RF model that includes the intrinsic device and packaging parasitic elements can become extremely complex for analysis by hand (Fig. 2.21). To circumvent this issue, designers often rely on measured two-port parameters at high frequency rather than the extrapolated model behavior from *I-V/C-V* curves, in order to understand the device behavior.

Two-port parameters have many limitations, most importantly arising from their small-signal assumption and their frequency dependence. But many RF circuits

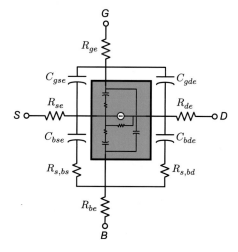

FIG. 2.21

A full FET model includes the intrinsic transistor in addition to external parasitic elements.

(power amplifiers are a good exception to this) operate with weak signals (signals captured off the air) over a narrow range of frequencies (so-called narrowband communication) and the small-signal narrowband assumption is valid in many circumstances, especially if there are no interfering signals. Moreover, by using measured two-port (or three-/four-port) parameters over a range of bias points, one can construct a table-lookup model to capture the dynamics for modulating signals. The benefit of two-port parameters is that they are completely general and capture all of the transistor dynamics (by definition) at a particular frequency of interest. Moreover, they can be used to answer some very general questions about a transistor device, such as "What is the maximum possible power gain for a device?" For these reasons, we will make extensive use of two-port parameters and the rich theory to delve into these questions.

Two-port parameters from a measured device are often compared to a compact model as a metric, but often many details are hard to decipher directly from these parameters as both circuit designers and device engineers have more intuition for circuits and small-signal equivalent circuits. A good solution is to convert general two-port parameters into an equivalent circuit representation, such as the popular hybrid-pi model, in order to gain intuition about how well the model matches measurements. Keep in mind that the two-port parameters vary with frequency, so the equivalent circuit will also vary, but if the two-port is chosen wisely, then many of the parameters remain fairly constant. For example, if we consider the two-port parameters of a transistor in the common-source configuration, as shown in Fig. 2.22, then it is easy to show that the equivalent pi circuit is related to the Y parameters as follows:

$$Y_{11} = Y_\pi + Y_\mu \tag{2.32}$$

$$Y_{22} = Y_o + Y_\mu \tag{2.33}$$

$$Y_{21} = g_m - Y_\mu \tag{2.34}$$

$$Y_{12} = -Y_\mu \tag{2.35}$$

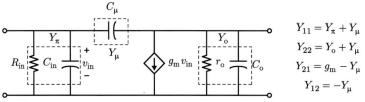

FIG. 2.22

The measured Y parameters can be converted directly into circuit pi parameters using the relations shown. Since this is an oversimplified model, the parameters are frequency dependent.

We can solve these equations for four pi-circuit parameters from measured results:

$$Y_\pi = Y_{11} + Y_{12} \tag{2.36}$$

$$Y_\mu = -Y_{12} \tag{2.37}$$

$$Y_o = Y_{22} + Y_{12} \tag{2.38}$$

$$G_m = Y_{21} - Y_{12} \tag{2.39}$$

It is important to note that Y_μ and G_m are, in general, complex. The real part of C_μ reflects that the feedback in a transistor can experience a phase shift at high frequencies, and the imaginary part of G_m reflects the same in the forward direction. It is convenient to think of G_m in magnitude/phase terms,

$$G_m = |G_m| \angle G_m \tag{2.40}$$

and it is also convenient to decompose the input network Y_π from a shunt representation to a series one, since in practice the gate resistance is fairly constant with frequency:

$$Y_\pi = j\omega C_\pi + G_\pi = \frac{1}{\left(1/(j\omega C_{gs})\right) + R_g} \tag{2.41}$$

The output network, on the other hand, is already in a convenient form:

$$Y_o = j\omega C_{ds} + g_o \tag{2.42}$$

Since these capacitors and resistors represent the entire transistor, they are, in fact, lumped representations of many sources of capacitance arising from both the intrinsic device and the extrinsic parasitic elements, including the substrate network. Note that this common-source representation cannot capture the model accuracy for source side parasitic elements, which may be different owing to layout differences. For this reason, a two-port representation is not complete and a three-port or four-port characterization is preferred (for the back gate or second gate of the device). For a typical device, a plot of the pi parameters is shown in Fig. 2.23. Over a narrow range of frequencies, these parameters are relatively constant and a good indication of the device behavior for a fixed operating point.

2.4.2 The need for speed

RF circuits work from low gigahertz to several thousand gigahertz, from the so-called RF spectrum to the millimeter-wave spectrum, and even at "terahertz" frequencies. There is a quest to run circuits as fast as possible to exploit new bandwidth and to maximize communication rates. For analog circuits, we found that f_T was a good substitute for estimating the gain-bandwidth product of a circuit. But in RF circuits, capacitive parasitic elements can be tuned out and the transistor can be operated over a narrow bandwidth at frequencies in excess of f_T. So what is the right metric for a transistor?

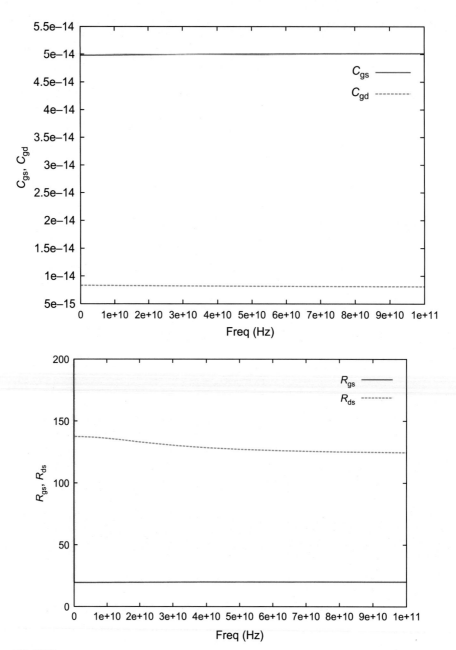

FIG. 2.23

The capacitance and resistance pi parameters of a core device versus frequency. Over a wide range of frequency, these parameters are constant and correspond closely to the often-used hybrid-pi equivalent circuit model.

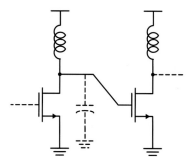

FIG. 2.24

The cascade of tuned amplifier stages. L is chosen to tune out the capacitive parasitic elements at the drain node of each transistor at the desired operating frequency.

Let us start with a simple exercise and derive the gain and bandwidth for the cascade of tuned amplifiers shown in Fig. 2.24. Here, the voltage gain per stage is limited not by the output resistance, but rather by the inductor quality factor Q_L:

$$A_v = g_m r_o \big\| R_L \approx g_m Q_L \omega L \qquad (2.43)$$

For a fixed device transconductance g_m, we wish to maximize the product of Q_L and the inductance, L. We wish to use the smallest possible inductors in order to minimize their physical area. The value of L is not arbitrary, though, and at minimum it must tune out the next-stage capacitance. So for simplicity, let us assume no explicit capacitance is added to the circuit so that $C_L = C_{gs} + C_{gd}$. To minimize C_{gd}, we can use a cascode stage as shown. So to first order,

$$A_v = g_m Q_L \omega L = g_m Q_L \frac{1}{\omega C_{gs}(1 + \mu)} = \frac{g_m}{C_{gs}(1 + \mu)} \frac{Q_L}{\omega} \qquad (2.44)$$

Interestingly, we see that f_T has found its way into our equation,

$$A_v \approx \frac{f_T}{f} Q_L \qquad (2.45)$$

which states that our center frequency f times the gain is limited by the product of f_T and Q_L. This drives home the well-known point of the important role played by passive devices at RF frequencies. But is this equation fundamental? No, since we just took a fixed topology common-source amplifier and found the voltage gain per stage in a cascade of identical stages.

A more fundamental question is given a transistor, what is the maximum gain that one can obtain under the optimal conditions? A related question is, what the maximum frequency where one can in fact get power gain from a transistor?

A unique aspect of RF circuits is that power gain is often more important than voltage gain, since RF circuits must drive (or are driven by) fixed impedances, often with a low value of 50 Ω. In this case, a more relevant question is what is the most gain we can get out of a transistor? The well-known result from two-port theory is that the gain is maximized by biconjugate matching, or in other words by designing the source and load admittance to be conjugately matched with the input and output admittance of the two-port:

$$Y_s = Y_{in}^* = \left(y_{11} - \frac{y_{12}Z_{21}}{y_{22} + Y_L}\right)^* \tag{2.46}$$

$$Y_L = Y_{out}^* = \left(y_{22} - \frac{y_{12}y_{21}}{y_{11} + Y_S}\right)^* \tag{2.47}$$

where y_{ij} are the two-port Y parameters of the two-port (transistor), and Y_s and Y_L are the source and load admittance. If these values are fixed (e.g., by the antenna port), then an impedance transformation network can be used to obtain the optimal Y_s and Y_L given above. A very interesting result is that if the above source/load impedance exists, then the maximum gain obtainable is given by

$$G_{max} = \frac{Y_{21}}{Y_{12}}\left(K - \sqrt{K^2 - 1}\right) \tag{2.48}$$

where $K \geq 1$ is the stability factor of the two-port:

$$K = \frac{2\Re(Y_{11})\Re(Y_{22}) - \Re(Y_{12}Y_{21})}{|Y_{12}Y_{21}|} \tag{2.49}$$

A stable two-port has $K \geq 1$, which means that it will not oscillate for any physical load or source admittance ($\Re(Y_{S|L}) > 0$). By definition, an unstable two-port has infinite gain, since it generates an output signal for a zero input, which is why we restrict our choice to $K \geq 1$. Since the gain increases as we reduce the stability factor, a common metric is the maximum stable gain (MSG):

$$\text{MSG} = G_{max|K=1} = \frac{Y_{21}}{Y_{12}} \tag{2.50}$$

The way to understand the MSG metric is to imagine that you have a two-port that is conditionally stable, so that $K < 1$. Instability implies that the input and/or output admittance of the two-port has a negative real part, $\Re(Y_{S|L}) < 0$. We can make this two-port stable by intentionally adding loss to its ports until the net real part is zero, $\Re(Y_{S|L}) = 0$. At this point, the two-port is conditionally stable and its gain is given by MSG. While adding any more loss makes the two-port more stable, it also reduces the gain.

An example may shed some light on this. Consider a FET transistor modeled by the simple hybrid-pi model shown in Fig. 2.25. At very low frequency, the FET is unconditionally stable, since C_μ provides negligible feedback. But at higher frequencies, for an inductive load, it is easy to see that the input admittance has a negative real part, since the current flowing through the feedback capacitor has a quadrature phase relation with the voltage at the drain:

$$i_\mu = \left(v_g - v_d\right)j\omega C_\mu \approx \left(v_g - v_g(-g_m j\omega L)\right)j\omega C_\mu \tag{2.51}$$

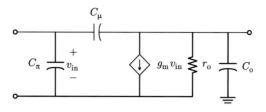

FIG. 2.25

The hybrid-pi model for a simple FET transistor amplifier.

$$= v_g \left(j\omega C_\mu - g_m \omega^2 L C_\mu \right) \tag{2.52}$$

$$\Re(Y_{in}) = \Re\left(\frac{i_\pi + i_\mu}{v_g} \right) = -g_m \omega^2 L C_\mu \tag{2.53}$$

A more exact relation can be derived directly from the complete two-port parameters:

$$Y_{in} = Y_{11} - \frac{Y_{12} Y_{21}}{j\omega L + Y_{22}} \tag{2.54}$$

These relations can be simplified to

$$Y_{in} = j\omega \left(C_\pi + C_\mu \right) - \frac{-j\omega C_\mu g_m}{g_o + j\omega C_o + (1/(j\omega L))} \tag{2.55}$$

$$\approx j\omega \left(C_\pi + C_\mu \right) - \frac{-j\omega C_\mu g_m}{1/(j\omega L)} \tag{2.56}$$

$$\Re(Y_{in}) \approx -g_m \omega^2 L C_\mu \tag{2.57}$$

which matches our intuitive calculation described above. The key point is that the approximation is valid in the frequency range where the inductive load presents a higher susceptance compared to the device's output conductance and susceptance. This can easily happen in a real device. To make the device stable, one can either add gate resistance to the device or add shunt resistance (or series) in a manner that increases the real component of the load voltage compared with the imaginary part. By doing so, we can improve the stability factor to the point $K = 1$.

The maximum gain of a device is dependent on the value of K. For $K < 1$, the maximum gain is obtained by stabilizing the device and is given by the MSG. For $K > 1$, the maximum gain is given by Eq. (2.48). A plot of the maximum gain for a typical device, then, follows the curve shown in Fig. 2.26. There is a kink in the curve at the point where K crosses unity as described.

For the example device, we can derive K from Eq. (2.49)

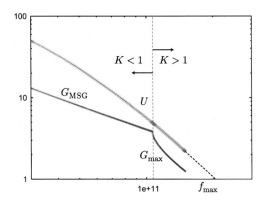

FIG. 2.26

The maximum gain available from a transistor two-port has two regions: when $K < 1$, it is given by MSG, and for higher frequencies when $K > 1$, it is given by G_{max} or the available gain.

$$K = \frac{\Re\left(j\omega C_\mu\left(g_m - j\omega C_\mu\right)\right)}{\omega C_\mu\sqrt{g_m^2 + \omega^2 C_\mu^2}} \tag{2.58}$$

$$= \frac{\omega C_\mu}{\sqrt{g_m^2 + \omega^2 C_\mu^2}} \tag{2.59}$$

For frequencies of interest, $\omega C_\mu \ll g_m$,

$$K \approx \frac{\omega C_\mu}{g_m} < 1 \tag{2.60}$$

which shows that the transistor is conditionally stable as predicted before. Note that we have disregarded the physical gate resistance in a real device, which changes the above calculation since we assumed $\Re(Y_{11}) = 0$.

2.4.2.1 The maximum unity power gain frequency (f_{max})

The commonest figure of merit for the high-frequency performance of a transistor is the maximum frequency of oscillation, f_{max}, or equivalently, the highest frequency at which we can obtain power gain from a transistor. Let us work with a simple hybrid-pi model as before, but introduce gate resistance and disregard C_μ (for simplicity), as shown in Fig. 2.27. In this case, owing to the unilateral nature ($Y_{12} = 0$), we can simply find the maximum gain by impedance matching the source and load directly:

$$Z_s = Z_{in}^* = \left(R_g + \frac{1}{j\omega C_{gs}}\right)^* = R_g + j\omega L_s \tag{2.61}$$

$$Y_L = Y_{out}^* = \left(G_o + j\omega C_{ds}\right)^* = G_o + j\omega L_d \tag{2.62}$$

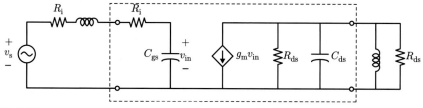

FIG. 2.27

A simple model for a FET used to calculate the power gain. The model does not include feedback capacitance but includes gate resistance.

Under these matched conditions, the power gain can be easily calculated without recourse to the full two-port parameters, since the reactances are absorbed by the source/load, as shown in Fig. 2.27:

$$P_{in} = \frac{1}{2} V_{in} I_{in} = \frac{1}{2} \cdot \frac{\nu_s}{2} \cdot \frac{\nu_s}{2R_i} = \frac{1}{8} \frac{\nu_s^2}{R_i} \tag{2.63}$$

$$P_L = \frac{\nu_o^2}{2r_o} = \frac{1}{2r_o} \left(\frac{g_m r_o}{2} \cdot \nu_\pi \right)^2 = \frac{1}{2r_o} \left(\frac{g_m r_o}{2} \cdot \frac{\nu_s}{2R_i} \frac{1}{\omega C_\pi} \right)^2 \tag{2.64}$$

Taking the ratio, and noting that $g_m/C_\pi \approx \omega_T$, we have a simple expression for the power gain:

$$P_L = \left(\frac{\omega_T}{\omega} \right)^2 \frac{r_o}{R_i^2} \frac{1}{32} \nu_s^2 \tag{2.65}$$

$$G_p = \frac{P_L}{P_{in}} = \frac{1}{4} \left(\frac{r_o}{R_i} \right) \left(\frac{\omega_T}{\omega} \right)^2 \tag{2.66}$$

Now we are in a position to find the maximum frequency f_{max} for which the transistor provides power gain:

$$G_p = 1 = \frac{1}{4} \left(\frac{r_o}{R_i} \right) \left(\frac{f_T}{f_{max}} \right)^2 \tag{2.67}$$

$$f_{max} = \frac{f_T}{2} \sqrt{\frac{r_o}{R_i}} \tag{2.68}$$

Despite the simplicity of our transistor model, this equation is very insightful as it connects the maximum power gain to the device f_T, or unity gain frequency. This affirms our faith in f_T as an important metric for RF and analog circuits, but it also shows that f_T is not the complete story. The device's f_{max} may in fact be larger than f_T if the output resistance r_o is a large factor bigger than the device's input resistance R_i. For example, it can be shown that R_i depends on the gate polysilicon resistance

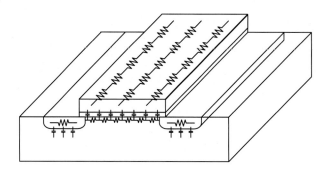

FIG. 2.28

The origin of gate resistance has two components: one from the physical gate material, which is minimized with a layout using multifinger devices, and the other from the actual channel resistance, which is unavoidable and layout invariant.

and the channel resistance as seen from the gate terminal (gate AC travels through the gate oxide and flows out through the channel and into the source and drain terminals). In the limit, if we lay out the device with many fingers to minimize the physical gate resistance (see Fig. 2.28), only the channel resistance contributes to R_i [2] (see Section 2.4.3 for more details):

$$R_i \approx \frac{1}{5g_m} \tag{2.69}$$

so

$$f_{max} = \frac{f_T}{2}\sqrt{\frac{r_o}{1/(5g_m)}} = \frac{f_T}{2}\sqrt{5g_m r_o} = \frac{f_T}{2}\sqrt{5A_o} \tag{2.70}$$

Since for any respectable transistor $A_0 > 1$, we have $f_{max} > f_T$. Although we have disregarded the feedback capacitor C_μ, which also plays a critical role, a more thorough analysis yields the following expression for f_{max} [3], which is very similar to the one adopted by the International Technology Roadmap for Semiconductors [4]:

$$f_{max} \approx \frac{f_T}{2\sqrt{R_g\left(g_m C_{gd}/C_{gg}\right) + \left(R_g + r_{ch} + R_s\right)g_{ds}}} \tag{2.71}$$

which highlights the importance of minimizing the loss in the device, such as the drain resistance R_d, source resistance R_s, and gate resistance R_g (part of R_i in our hybrid-pi model, the rest coming from r_{ch}). These parasitic elements are in large part determined by the layout and the process technology, which is a good metric for testing both the intrinsic and the extrinsic transistor model.

In RF circuits, we can neutralize this capacitance by connecting an inductor L_μ in parallel to resonate it out at any given frequency and over a narrow bandwidth. In differential circuits, we can neutralize the capacitance with a cross-coupled matching capacitor, as shown in Fig. 2.29. In fact, is the optimal L_μ the value that completely cancels the capacitance?

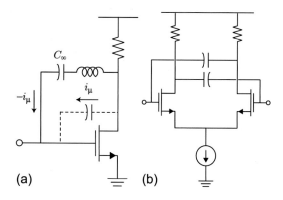

FIG. 2.29

Two techniques to neutralize the feedback capacitance (C_{gd}) in a FET include (A) adding an inductor in shunt to resonate out the capacitance and (B) alternatively in a fully differential or balanced amplifier, cross-coupling can be added to produce a negative capacitance in shunt, which works over a broad frequency range.

For a FET, we know that $Y_{21} \approx g_m$ and $Y_{12} \approx j\omega C_{gd}$, so

$$\text{MSG} = \left|\frac{Y_{21}}{Y_{12}}\right| \approx \frac{g_m}{\omega C_{gd}} \tag{2.72}$$

This implies that as we reduce the effective C_{gd} by neutralization, the gain increases:

$$\text{MSG} \approx \frac{g_m}{\omega(C_{gd} - C_n)} \tag{2.73}$$

where C_n is the amount of neutralization. It can be shown that under these conditions

$$\text{MSG} = \sqrt{\frac{g_m^2}{\omega^2(C_{gd} - C_n)} + 1} \tag{2.74}$$

with a corresponding stability factor given by

$$k = \left(1 + \frac{2g_g g_{ds}}{\omega^2(C_{gd} - C_n)^2}\right)\text{MSG}^{-1} \tag{2.75}$$

Stability requires $K > 1$, which occurs for a range of neutralization values n:

$$n_1 \leq n \leq n_2 \tag{2.76}$$

where n is defined as the relative amount of neutralization

$$n = \frac{C_n}{C_{gd}} \tag{2.77}$$

and the critical values of n occur at

$$n_1 = 1 - \frac{1}{\omega C_{gd}}\sqrt{\frac{g_g g_{ds}}{U - 1}} \tag{2.78}$$

$$n_2 = 1 + \frac{1}{\omega C_{gd}} \sqrt{\frac{g_g g_{ds}}{U-1}} \tag{2.79}$$

where U is Mason's unilateral gain for the two-port, which will be discussed next. In fact, the maximum gain occurs at these critical points, rather than at $n = 1$, and the maximum gain can be computed to be [5]

$$G_{max} = 2U - 1 \tag{2.80}$$

So what is U and how does it relate to the power gain of the two-port?

2.4.2.2 Mason's unilateral gain U

One of the most important metrics in high-frequency transistor characterization is U, or Mason's unilateral gain [6]. The idea behind U has an interesting history, recounted in Ref. [7]. The definition of U is the power gain for a two-port under the general four-port lossless embedding shown in Fig. 2.30. The idea is to provide lossless "feedback" to unilaterize the two-port, and then under these conditions U is the gain we obtain from the two-port:

$$U = \frac{|Y_{21} - Y_{12}|^2}{4(\mathfrak{R}(Y_{11})\mathfrak{R}(Y_{22}) - \mathfrak{R}(Y_{12})\mathfrak{R}(Y_{21}))} \tag{2.81}$$

The U function has several important properties:

1. If $U > 1$, the two-port is active; if $U \leq 1$, the two-port is passive.
2. U is the maximum unilateral power gain of a device under a lossless reciprocal embedding.
3. U is the maximum gain of a three-terminal device regardless of the common terminal.

These properties have contributed to the widespread use of U as a metric to test a transistor power gain, and it is a good metric to test a compact model for the same reasons. Unlike MSG, U is extremely sensitive to the loss of a device. This is because, even if $K < 1$, we can add a network around the device to make it unilateral, and hence stable, and then we can extract the MSG from the device in this

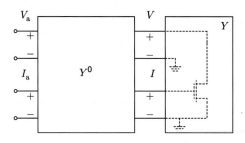

FIG. 2.30

A two-port network Y is embedded into a four-port network Y^0 providing lossless feedback to unilaterize the device.

configuration. We find that the less loss there is in a device, the more gain we can extract, in stark contrast to MSG, which in fact requires us to *add* loss to stabilize the device when $K < 1$. For these reasons, we find it much easier to match a model to data for MSG rather than U. U is a function of what really matters to obtain power gain, and it is therefore the metric of choice.

Is U the maximum possible power gain we can obtain? As we have already seen in the case of neutralization, it is not in fact the maximum. While neutralization provides a gain of $2U - 1$, it can be shown that the maximum possible gain is as high as

$$G_{\max} = 2U - 1 + 2\sqrt{U(U - 1)} \approx 4U \tag{2.82}$$

2.4.3 NQS model

In the derivation of most compact models, the *I-V* relations and *C-V* relations are derived independently, with the implicit assumption that charges respond instantaneously to external voltages. This assumption is known as the quasi-static assumption, which states that if the voltages at the terminals of a device are varied slowly enough, then we would expect the *C-V* relations and *I-V* relations to accurately describe the current flow into the transistor. But how fast can we go and still consider this assumption valid? How long does it take charge to flow into the channel? A very naive calculation would say that the *RC* time constant associated with charging the channel is approximately the resistance of the channel r_{ds} times the gate-oxide capacitance C_{gs}. As we have seen, the channel resistance r_{ds} in the triode region, or equivalently the channel conductance g_{ds} in the triode region, is equal to g_m of the device in saturation. So with this approximation, the time constant is

$$\tau = \frac{1}{g_m} \times C_{gs} = \frac{C_{gs}}{g_m} = \frac{1}{\omega_T} \tag{2.83}$$

which shows that the device f_T is again playing an important role. It seems that if we approach the device f_T, our quasi-static assumption begins to fail.

To see this in a compact model, we can plot the real and imaginary parts of Y_{11}, where port 1 is the gate and port 2 is the drain. For a real device, we would expect to see a real component of Y_{11} owing to the channel conductance and the distributed nature of the gate-to-channel capacitance. In a quasi-static model, though, the gate-source impedance is purely reactive and the real part of Y_{11} can arise only from feedback from C_{gd}. In Fig. 2.31A, we demonstrate this by plotting $\Re(Y_{11})$ and the component R_{gs} of the pi model for a model with the NQS effect turned on and off (using NQSMOD parameter). It is clear that the NQS resistor has a significant impact on the overall device. As we have seen, loss in the device plays an important role in determining the maximum available gain from the device. Without this resistor, we have the nonphysical result that a device has infinite available gain (disregarding feedback), since the input would dissipate no power in a reactive load, whereas the output would generate power.

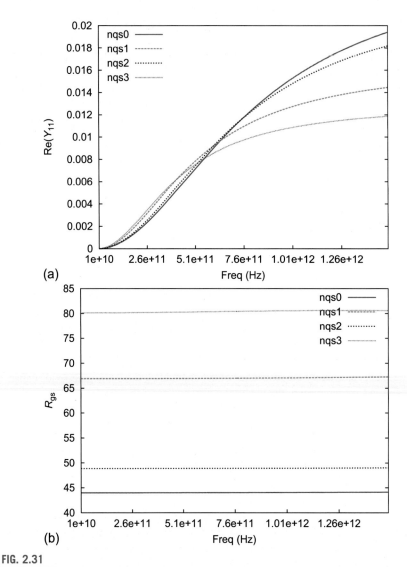

FIG. 2.31

The plot of the (A) real part of Y_{11} demonstrates the impact of the NQS effect, which (B) is evident as an extra component of gate resistance R_{gs}.

We can also see the frequency dependence of the NQS effect by observing the effective transconductance G_m versus frequency. We define the effective transconductance from the Y parameters as the forward minus the reverse short-circuit transconductance:

$$G_m = |Y_{21} - Y_{12}| \tag{2.84}$$

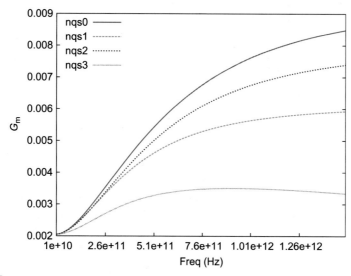

FIG. 2.32

A comparison of various NQS models built into the BSIM-CMG model and their impact on the effective G_m.

For a physical device, we would expect that G_m should drop with frequency, but a simulation of a model without the NQS effect, shown in Fig. 2.32, shows the opposite trend, predicting that the device transconductance gets better at high frequency. With the NQS model, the correct behavior is observed, which can be verified with measurements. Note, the full segmented model described in Ref. [8] is needed to capture this effect fully.

Various techniques have been developed to model NQS effect with a SPICE model. Since SPICE is inherently an "ordinary differential equation" solver, it cannot intrinsically handle distributed components. To model distributed effects, we need to approximate the distributed nature using lumped circuits. For example, a relaxation time approximation models the charge deficit or surplus in the channel [9], which decays to its nominal quasi-static value with a time constant similar to $1/\omega_T$. The NQS model can also be lumped into a gate resistance model (and added to the physical gate resistance) [10]. The most accurate models use charge segmentation and the continuity relation to model the nonuniform charge distribution in the channel [11,12]. These various models are compared in Fig. 2.32, showing that each model has a frequency range over which it performs quite well. Only when devices are pushed beyond f_T is it necessary to invoke the NQS model.

2.4.4 Noise

The noise performance of a two-port can be analyzed in general in terms of the equivalent input noise voltage/current sources and their correlation (Fig. 2.33). Partitioning the input noise current into two components, a component correlated

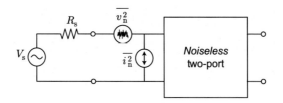

FIG. 2.33

A general two-port noisy network can be partitioned into a noiseless two-port and two correlated noise sources at the inputs and/or outputs. An input-referred two-port is shown here.

(parallel) to the noise voltage and a component uncorrelated (perpendicular) to the noise voltage, we have

$$i_n = i_c + i_u \tag{2.85}$$

where we assume that $<i_u, v_n> = 0$ and

$$i_c = Y_c v_n \tag{2.86}$$

We can therefore write

$$v_{eq} = v_n(1 + Y_c Z_s) + Z_s i_u \tag{2.87}$$

2.4.4.1 Minimum achievable noise figure (F_{min})

With this model, we can calculate the noise figure and minimize it by finding the optimal source impedance. Let $\overline{v_n^2} = 4kTBR_n$, $\overline{i_u^2} = 4kTBG_u$, and $\overline{v_s^2} = 4kTBR_s$. Then

$$F = 1 + \frac{R_n|1 + Y_c Z_s|^2 + |Z_s|^2 G_u}{R_s} \tag{2.88}$$

If we let $Y_c = G_c + jB_c$ and $Y_s = Z_s^{-1} = G_s + jB_s$, it is not too difficult to show that the optimum source impedance to minimize F is given by

$$B_{opt} = B_s = -B_c \tag{2.89}$$

$$G_{opt} = G_s = \sqrt{\frac{G_u}{R_n} + G_c^2} \tag{2.90}$$

The minimum achievable noise figure is

$$F_{min} = 1 + 2G_c R_n + 2\sqrt{R_n G_u + G_c^2 R_n^2} \tag{2.91}$$

Through some algebraic manipulations, one can derive [13]

$$F = F_{min} + R_n R_s |G_{opt} - G_s|^2 \tag{2.92}$$

where R_n is called the noise sensitivity parameter. This terminology is clear since the rate of deviation from the optimal noise figure is determined by R_n. If a two-port has a

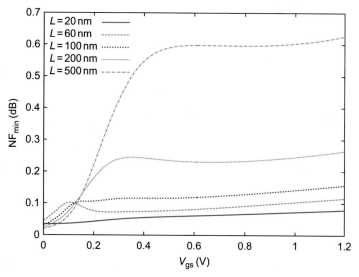

FIG. 2.34

A plot of the minimum achievable FET noise figure versus bias (V_{gs}) for a family of channel lengths (L).

small value of R_n, then we can be sloppy and sacrifice the noise match for gain. If R_n is large, though, we have to pay careful attention to the noise match. A plot of NF_{min} (logarithm of F_{min}) versus the device bias V_{gs} is shown in Fig. 2.34. These plots were generated for a fixed frequency of 20 GHz and a family of channel lengths L is used to clearly show the trend that increasing the device f_T (minimize L) improves the achievable noise figure.

2.4.4.2 Simple model for FET noise

Consider the following noise sources associated with the source R_s, the gate resistance R_g, the channel resistance R_{ch}, and the load R_L:

$$\overline{v_s^2} = 4kT\,BR_s$$
$$\overline{v_g^2} = 4kT\,BR_g$$
$$\overline{i_d^2} = 4kT\,Bg_{d0}\gamma B$$
$$\overline{i_L^2} = 4kT\,BG_L$$

The gate noise can be used to model the channel noise seen on the gate side and the correlation is disregarded (in addition to the physical thermal noise on the gate side). This is known as the Pospieszalski noise model, and it matches measurements pretty well [14].

$$F = 1 + \frac{\overline{v_g^2}}{\overline{v_s^2}} + \frac{\overline{i_d^2} + \overline{i_L^2}}{g_m^2\,\overline{v_s^2}} \qquad (2.93)$$

$$= 1 + \frac{R_g}{R_s} + \frac{g_{d0}\gamma + G_L}{R_s g_m^2} \tag{2.94}$$

If we assume $g_m = g_{d0}$ (long channel),

$$F = 1 + \frac{R_g}{R_s} = \frac{\gamma}{g_m R_s} + \frac{G_L G_s}{g_m^2} \tag{2.95}$$

If we make g_m sufficiently large, the gate resistance will dominate the noise. The gate resistance has two components, the physical gate resistance and the induced channel resistance:

$$R_g = R_{poly} + \delta R_{ch} = \frac{1}{3} \frac{W}{L} R_\square + \frac{1}{5} \frac{1}{g_m} \tag{2.96}$$

For the optimal source impedance, one can show that

$$F_{min} = 1 + 2 \left(\frac{\omega}{\omega_T} \right) \sqrt{g_m R_g \gamma} \tag{2.97}$$

Doing the best layout possible, we obtain the limit that $R_g = 1/5 g_m$ and

$$F_{min} = 1 + 2 \left(\frac{\omega}{\omega_T} \right) \sqrt{\gamma/5} \tag{2.98}$$

Again we see our old friend f_T and γ playing a central role.

2.4.4.3 Phase noise

Owing to random noise from devices, an oscillator does not have a delta-function power spectrum, but rather has a very sharp peak at the oscillation frequency (Fig. 2.35). The amplitude of the noise power spectrum drops very quickly, though, as one moves away from the center frequency. For example, a cell phone oscillator has a phase noise that is 100 dB down at an offset of only 0.01% from the carrier! Note that the noise shaping is not due to the LC tank or crystal resonator used to build an oscillator, which has a much wider bandwidth.

The importance of phase noise is that it places limits on RF communication. For example, phase noise in a transmit chain will "leak" power into adjacent channels as shown in Fig. 2.36. Since the power transmitted is large—say, about 30 dBm—an adjacent channel in a narrowband system may reside only about 200 kHz away,[b] placing a stringent specification on the transmitter spectrum.

In a receive chain, the fact that the local oscillator is not a perfect delta function means that there is a continuum of local oscillators that can mix with interfering signals and produce energy at the same intermediate frequency. Here, we observe an adjacent channel signal mixing with the "skirt" of the local oscillator and falling on top of the a weak intermediate-frequency signal from the desired channel.

---

[b]In fact, these are the specifications for the popular 2G digital mobile standard known as GSM.

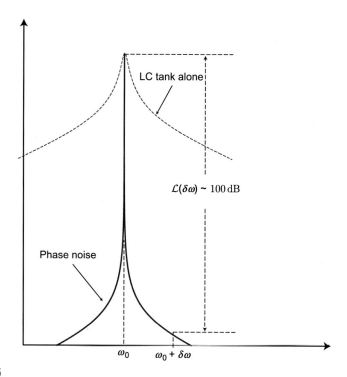

$\mathcal{L}(\delta\omega) \sim 100\,\text{dB}$

FIG. 2.35

The power spectrum of an oscillator has a finite width in the frequency domain owing to the noise in electronic devices, which alters the period of oscillation (jitter).

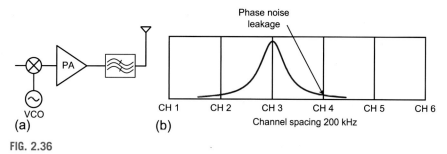

FIG. 2.36

(A) In a typical up-conversion transmitter, the "local oscillator" has phase noise and this is amplified by the power amplifier driving the antenna. (B) This phase noise leaks into adjacent channels, making it impossible or difficult for receivers to distinguish the information from the noise.

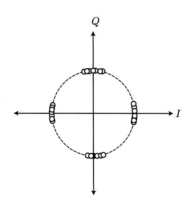

FIG. 2.37

The ideal constellation of a transmitter and receiver is degraded by the "jitter" or phase noise in the RF carriers and clocks used to modulate and sample the data.

In a digital communication system, phase noise can lead to a lower noise margin. In Fig. 2.37, we see that the phase noise causes the constellation of a quadrature phase-shift keying modulation to spread out. In orthogonal frequency-division multiplexing systems, a wide bandwidth is split into subchannels. The phase noise leads to intercarrier interference and a degradation in the digital communication bit error rate.

2.4.4.4 Phase noise derivation: Lorentzian spectrum

While the full derivation of phase noise requires the solving of stochastic differential equations, we can gain insight with a very simple linear time-invariant model. It can be readily shown that the power spectrum of v_o is given by [15,16]

$$v_o^2 = \overline{i_n^2} \frac{R_1^2}{(1 - (g_m R_1 / n))^2 + 4Q^2 \left(\delta\omega^2 / \omega_0^2\right)} \tag{2.99}$$

Thus, the noise has a Lorentzian shape for white noise. For offset frequencies of interest,

$$4Q^2 \frac{\delta\omega^2}{\omega_0^2} \gg \left(1 - \frac{g_m R_1}{n}\right)^2 \tag{2.100}$$

The spectrum normalized to the peak is given by

$$\left(\frac{v_o}{V_o}\right)^2 \approx \frac{\overline{i_n^2} R_1^2}{V_o^2} \left(\frac{\omega_0}{\delta\omega}\right)^2 \frac{1}{4Q^2} \tag{2.101}$$

The above equation is in the form of Leeson's equation. It compactly expresses the oscillator noise as noise power over signal power (N/S), divided by Q^2 and drops like $1/\delta\omega^2$.

2.4.4.5 Phase noise and flicker noise

As we have seen, phase noise depends on the device noise, which is similar to our analysis of linear gain. Namely, phase noise at some frequency offset is generated by device noise at the same frequency. But in our simple linear time-invariant model, we disregarded the important quasi-periodically time-varying nature of the transistor operating point. As shown in Fig. 2.38, this time-varying behavior, similar to that of a mixer, allows noise from low frequency to mix up and appear around the RF carrier. Thus, in RF circuits, noise at low frequency (due to flicker) is also important and cannot be disregarded.

2.4.5 Linearity

At RF frequencies, in addition to HD metrics, there is also considerable interest in intermodulation distortion, especially odd-order intermodulation. The reason for this is that many RF systems are narrowband and thus a bandpass network can attenuate distortion products that fall out of the desired band, such as harmonics, but odd-order intermodulation products fall into the band and cannot be filtered. Moreover, even-order intermodulation products have DC components that are important in direct-conversion applications, since most of the signal processing is at the baseband (DC).

To derive the intermodulation of a general nonlinearity, we can follow the same approach as before. Consider applying a two-tone signal to a memoryless nonlinearity described by a power series:

$$i_{ds} = g_{m1}\left(V_{gs1}\cos(\omega_1 t) + V_{gs2}\cos(\omega_2 t)\right) + g_{m2}\left(V_{gs1}\cos(\omega_1 t) + V_{gs2}\cos(\omega_2 t)\right)^2 + \cdots$$

$$(2.102)$$

If we focus only on the second power term, we see that in addition to harmonics at $2\omega_{1,2}$ we also generate a new second-order intermodulation (IM$_2$) term,

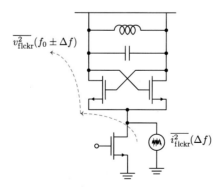

FIG. 2.38

The noise from low frequencies, such as flicker noise in the current source, is upconverted to RF through the time-varying "mixer" action of the oscillator.

$$g_{m2}\left(V_{gs1}\cos(\omega_1 t) + V_{gs2}\cos(\omega_2 t)\right)^2 = g_{m2}V_{gs,1,2}^2 \frac{1 + \cos(2\omega_{1,2})}{2}$$
$$+ 2V_{gs1}\cos(\omega_1 t)V_{gs2}\cos(\omega_2 t) \tag{2.103}$$

where the latter term can be simplified to

$$= \cdots + g_{m2}V_{gs1}V_{gs2}(\cos(\omega_1 + \omega_2)t + \cos(\omega_1 - \omega_2)t) \tag{2.104}$$

These two "sum and difference" frequency terms are the second-order intermodulation terms. The ratio of their power relative to the fundamental is used to define the IM_2 metric (by applying equal voltages for both tones):

$$IM_2 = \frac{g_{m2}V_{gs,1}^2}{g_{m1}V_{gs}} = \frac{g_{m2}}{g_{m1}}V_{gs} \tag{2.105}$$

We see that for a memoryless system, IM_2 is in fact related to HD_2 (see Eq. 2.26) as follows:

$$IM_2 = 2HD_2 \tag{2.106}$$

This relationship clearly breaks down if there is memory in the system. A simple low-pass filter will attenuate HD_2 but leave the low-frequency difference tone at $\omega_1 - \omega_2$ intact.

In a similar way, we can proceed to analyze the third-order two-tone response, building from here

$$= \cdots + g_{m3}\left(V_{gs1}\cos(\omega_1 t) + V_{gs2}\cos(\omega_2 t)\right)^3 \tag{2.107}$$

$$= \cdots + g_{m3}\left(V_{gs1}\cos(\omega_1 t) + V_{gs2}\cos(\omega_2 t)\right)^2\left(V_{gs1}\cos(\omega_1 t) + V_{gs2}\cos(\omega_2 t)\right) \tag{2.108}$$

$$= \cdots + g_{m3}V_{gs1}V_{gs2}\left[\frac{1}{2}(\cos(2\omega_{1,2}) + 1) + \cos(\omega_1 + \omega_2)t + \cos(\omega_1 - \omega_2)t\right]$$
$$\left(V_{gs1}\cos(\omega_1 t) + V_{gs2}\cos(\omega_2 t)\right) \tag{2.109}$$

Expanding the products, we see terms that now have terms at $2\omega_2 + \omega_1$ and $2\omega_1 + \omega_2$, generated by

$$\left(\frac{1}{2} + \frac{1}{4}\right)\cos(\omega_1 + \omega_2)t \times \cos(\omega_{1,2}t) \tag{2.110}$$

and also terms at $2\omega_2 - \omega_1$ and $2\omega_1 - \omega_2$, generated by

$$\left(\frac{1}{2} + \frac{1}{4}\right)\cos(\omega_1 + \omega_2)t \times \cos(\omega_{1,2}t) \tag{2.111}$$

These are called the third-order intermodulation terms, or IM_3. Computing the ratio of the third-order intermodulation to the fundamental (with equal powers in each tone), we have

$$IM_3 = \frac{g_{m3}V_{gs}^3(3/4)}{g_{m1}V_{gs}} = \frac{3}{4}\frac{g_{m3}}{g_{m1}}V_{gs}^2 \tag{2.112}$$

We see that IM_3 also has many of the same features as HD_3; in fact, for a memoryless system they are related by a factor of 3. What is notably different about IM_3 (and in fact all odd-order intermodulation products) is that there are spectral tones created in band, or near the original two-tone frequencies. In particular, these tones are at $2\omega_2 - \omega_1$ and $2\omega_1 - \omega_2$, and as $\omega_1 \approx \omega_2$ in a narrowband system, these distortion products fall in band and can do the most harm. It is for this reason that IM_3 plays such an important role in RF systems.

Fig. 2.39 displays the output signal at the fundamental versus the distortion products at frequencies $\omega_2 - \omega_1$ (second order) and $2\omega_2 - \omega_1$ (third order). The distance between the fundamental curve and the second- and third-order curves is the intermodulation on a log scale. The intercept (extrapolated) is another metric that is often used to describe the third- and second-order intermodulation with a single number. By definition, at the intercept, V_{IIP} is the input signal level such that the distortion product and the fundamental output have equal magnitude.

IM_3 and IM_2 are plotted versus the signal power in Fig. 2.40. These plots are very similar to the HD plots and have a slope of 2 and 1, respectively, on a decibel scale. Compared with the previous plots, the slopes are 1 minus the slope of the power that generates the nonlinearity, since IM_2 and IM_3 are ratios normalized by the fundamental.

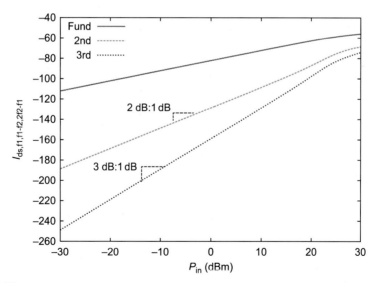

FIG. 2.39

The power of fundamental and the second- and third-order intermodulation products as a function of signal power. The intermodulation products have a slope of 2 and 3 dB per decibel increase in signal power and the distance to the fundamental is the intermodulation ratio on a log scale.

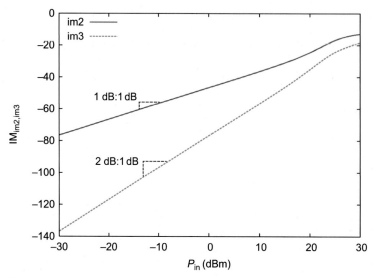

FIG. 2.40

IM_2 and IM_3 as a function of signal power have a slope of 1 and 2 dB per decibel increase in signal power, respectively.

For the second-order distortion products,

$$IM_2 = 1 = \frac{g_{m2}}{g_{m1}} V_{IIP,2} \qquad (2.113)$$

$$V_{IIP,2} = \frac{g_{m1}}{g_{m3}} \qquad (2.114)$$

For the third-order distortion products,

$$IM_3 = 1 = \frac{3}{4} \frac{g_{m3}}{g_{m1}} V_{IIP,3}^2 \qquad (2.115)$$

$$V_{IIP,3} = \sqrt{\frac{4}{3} \frac{g_{m1}}{g_{m3}}} \qquad (2.116)$$

The advantage of the intercept is that we can calculate (or measure) it once, and then for any given input signal level power, we can simply apply the rule that the IM_3 relation grows by 2 dB for every 1 dB of back-off from the intercept, whereas the IM_2 relation grows by a decibel for every decibel.

2.4.5.1 Memory effects

As previously mentioned, for RF systems we expect that any reactance in the circuit, such as device capacitors, will interact with the distortion products generated by the transistor nonlinearity and attenuate (and perhaps magnify) the distortion products.

This extra complication is handled by using a Volterra series [1] in place of a power series. Moreover, any nonlinearity in the capacitors (such as the variation of C_{gs} with signal power) will also create distortion products. This means that the *I-V* and *C-V* relations both need to be extracted accurately for RF applications. From a compact modeling perspective, a simple solution is to focus on the *C-V* curve of the device and ensure a good fit. Similar to the case of the *I-V* curve, the fit needs to be very good in the sense that not only must the absolute values fit the measurements, but the curves must also be smooth and symmetric since distortion products are created by the derivatives of the *C-V* curve. Owing to space limitations, the details of this are omitted here.

2.4.5.2 Other distortion metrics

While IM_2 and IM_3 are excellent vehicles for understanding the distortion generation in narrowband systems, when an actual modulation is applied to a transistor, all intermodulation products produce distortion, causing the spectrum of the signal to expand. This is called spectral regrowth, and in addition to corruption of the modulation pattern, it also introduces unwanted out-of-band interference, in particular to neighboring channels (in a frequency-division multiplexed system). The distortion in the modulation pattern is measured by the error vector magnitude, which measures the deviation from the ideal constellation points due to the nonlinearity (and noise). As these metrics are more applicable at the system level, rather than at the device level, they are hard to use directly in compact modeling. In practice, if intermodulation products match up to several orders, this is a good sign that the compact model is representing the distortion products well. In particular, the intermodulation products must also match for different tone spacings (even though simple theory predicts the intermodulation products do not depend on tone spacing, the more detailed Volterra series analysis will show otherwise). Furthermore, three or more tones should be used to emulate the effect of a broadband modulation, such as an orthogonal frequency-division multiplexing or multitone signal.

2.5 **Conclusion**

Today, compact models play a key binding role between technology and circuits, for diverse applications including precision analog, digital, mixed signal, and RF and millimeter-wave applications. These vast applications place stringent demands on the accuracy of the compact model, requiring extensive testing and validation. In this chapter, we have examined important properties of a compact model for both analog and RF applications. We have seen that common metrics such as threshold voltage variation, intrinsic gain, device unity gain f_T, maximum oscillation frequency f_{max}, minimum achievable noise figure F_{min}, and Mason's unilateral gain U all play an important role in evaluating the accuracy of a compact model. We have also seen that other metrics and properties that capture linearity, device variability, symmetry, and other subtler factors are important. Many subsequent chapters will focus on compact model modules that directly impact these metrics.

References

[1] W.J. Rugh, Nonlinear System Theory: The Volterra-Wiener Approach, Johns Hopkins University Press, Baltimore, 1981.

[2] A.M. Niknejad, Electromagnetics for High-Speed Analog and Digital Communication Circuits, Cambridge University Press, Cambridge, 2007.

[3] C. Doan, S. Emami, A. Niknejad, R. Brodersen, Millimeter-wave CMOS design, IEEE J. Solid State Circuits 40 (1) (2005) 144–155.

[4] International Technology Roadmap for Semiconductor, RF and Analog/Mixed-Signal Technologies (RFAMS), 2011, Available from: https://www.semiconductors.org/wp-content/uploads/2018/08/2011RFAMS.pdf.

[5] Z. Deng, A. Niknejad, A layout-based optimal neutralization technique for mm-wave differential amplifiers, in: IEEE Radio Frequency Integrated Circuits Symposium (RFIC), May 23–25, 2010, 2010, pp. 355–358.

[6] S.J. Mason, Power gain in feedback amplifiers, Trans. IRE Professional Group Circuit Theory CT-1 (2) (1954) 20–25.

[7] M.S. Gupta, Power gain in feedback amplifiers, a classic revisited, IEEE Trans. Microw. Theory Tech. 40 (5) (1992) 864–879.

[8] S. Venugopalan, From Poisson to Silicon—Advancing Compact SPICE Models for IC Design (Ph.D. dissertation), Electrical Engineering and Computer Sciences, University of California at Berkeley, Technical Report No. UCB/EECS-2013-166, 2013, http://www.eecs.berkeley.edu/Pubs/TechRpts/2013/EECS-2013-166.html.

[9] M. Chan, K. Hui, R. Neff, C. Hu, P.K. Ko, A relaxation time approach to model the nonquasi-static transient effects in MOSFETs, in: International Electron Devices Meeting, Technical Digest, December 1994, 1994, pp. 169–172.

[10] X. Jin, J.-J. Ou, C.-H. Chen, W. Liu, M.J. Deen, P.R. Gray, C. Hu, An effective gate resistance model for CMOS RF and noise modeling, in: International Electron Devices Meeting (IEDM), Technical Digest, December 1998, 1998. pp. 961, 964.

[11] A.J. Scholten, R. van Langevelde, L.F. Tiemeijer, R.J. Havens, D.B.M. Klaassen, Compact MOS modelling for RF CMOS circuit simulation, in: Simulation of Semiconductor Processes and Devices, Springer, Vienna, 2001, pp. 194–201.

[12] H. Wang, T.-L. Chen, G. Gildenblat, Quasi-static and nonquasi-static compact MOSFET models based on symmetric linearization of the bulk and inversion charges, IEEE Trans. Electron Devices 50 (11) (2003) 2262–2272.

[13] D.M. Pozar, Microwave Engineering, second ed., Wiley, New York, NY, 1997.

[14] M.W. Pospieszalski, On the measurement of noise parameters of microwave two-ports, IEEE MTT-S Int. Microw. Symp. Digest 34 (4) (1986) 456–458.

[15] D. Leeson, A simple model of feedback oscillator noise spectrum, Proc. IEEE 54 (2) (1966) 329–330.

[16] A.M. Niknejad, R.G. Meyer, Design, Simulation and Applications of Inductors and Transformers for Si RF ICs, Kluwer Academic Publishers, Boston, 2000.

Core model for FinFETs

In the implementation of circuit simulators, compact models are preferred over other numerical approaches because the former can offer, in addition to good computational efficiency, good accuracy [1]. The good accuracy of a compact model mainly relies on the amount of physics and the assumptions behind its mathematical derivation. Indeed, FinFET/GAA devices are constructed in the nanoscale regime; therefore, accurate compact models must include several physical effects: charge quantization, gate oxide tunneling, gate capacitance degradation, SCEs, etc.

Most of the compact models are based on a "core model," which is a model obtained using a long-channel assumption, so-called gradual-channel-approximation (GCA) [2], and simplifying other physical effects like charge quantization or gate oxide tunneling. A core model is crucial for the completed compact model because it gives the basis of a mathematical framework that is continuous and smooth, a key requirement for a robust model. In this context, core models are further improved by the inclusion of correction terms that represent advanced physical effects [1, 3]. Regularly, core models are obtained by solving Poisson's equation under the GCA condition and assuming Boltzmann's statistics for the carriers. Even though the use of GCA condition and Boltzmann's statistics alleviates the difficulty in obtaining a solution from Poisson's equation, a direct analytical solution is only available for the cases of undoped double-gate (DG) [4] and cylindrical gate-all-around (Cy-GAA) FETs [5], where the three-dimensional (3D) Poisson's equation can be reduced to a one-dimensional (1D) form. If depletion charges arisen from dopants are included, Poisson's equation becomes highly nonlinear. It is then more challenging to obtain a direct analytical solution [3,6–9]. However, in realistic FinFET/GAA devices, doping is needed to be used for multiple threshold voltage devices that are required in contemporary system-on-chip technologies for better power-performance-area trade-off [10]; therefore, core models including doping effects must be developed. Finding a direct analytical solution becomes even more difficult to obtain for complex FinFET geometries because they lack structural symmetry [11]. Indeed, compact models for asymmetric geometries, such as triple-gate (TG), rectangular (Re) GAA, or Pi-gate FETs, are rarely found in the literature and are only accomplished by the extensive use of fitting parameters or numerical techniques [11–14]. However, some of these asymmetric geometries offer simpler fabrication processes than other symmetric geometries [15]. Therefore, it is important to develop physical-based core model for FinFET/GAA devices with complex geometry, for comprehensive understanding and circuit design.

FinFET/GAA Modeling for IC Simulation and Design. https://doi.org/10.1016/B978-0-323-95729-8.00004-0

Various compact models have been proposed for several FinFET/GAA device geometries, such as DG [3,4,6–8,16–19], TG [11,12,20], Re-GAA [11,13,14], or Cy-GAA [5,9,21,22]. The core model of BSIM-CMG is a surface potential model, which is based on the solution of the Poisson's equation for a doped DG FinFET. The surface potential obtained at the channel is utilized to obtain the mobile charge in the channel that eventually leads to the computation of the drain current of the device. A description of this model is shown in Section 3.1. Recently, a new core model has been added to BSIM-CMG, titled Unified FinFET Compact Model, due to its capability to describe the electrical behavior of several FinFETs with complex cross sections [23]. This new feature of BSIM-CMG is presented in Section 3.2.

Although the core model for BSIM-CMG is based on the double-gate FinFET structure, the same expressions can be extended for modeling GAA devices such as that in nanosheet FETs, expected to be used for 2 nm technology and beyond. Exact surface potential and drain current expression for GAA devices with a cylindrical channel geometry had be developed, implemented in BSIM-CMG, and tested [22]. However, real GAA technology does not have cylindrical channel geometry, but likely one with a rectangular cross section (with process-induced corner rounding). Therefore it is a closer approximation to utilize double-gate FinFET core model as the basic model, with corrections in effective channel width and subband effects [24] to consider GAA. The latest BSIM-CMG model offers a model switch (GEOMOD=5) to activate models for special considerations of GAA FETs.

The Unified FinFET Compact Model is extended to model GAA devices, as we will describe in detail in Chapter 4. In this chapter, we focus on the core model for the FinFET geometry.

3.1 Core model for double-gate FinFETs

The core model used in BSIM-CMG is based on a solution of Poisson's equation for a long-channel DG FinFET, assuming a finite doping in the channel to mimic the doped channels currently used in FinFET fabrication [10]. It is challenging to obtain a direct analytical solution of the Poisson's equation of doped FinFETs due to the high nonlinearity of the equation; therefore, to overcome this limitation, a perturbation approach is used to approximately solve the Poisson's equation in the presence of body doping [3,6].

Fig. 3.1 shows a two-dimensional cross section of a DG FinFET, which is being used as a reference for the model derivation. Poisson's equation, assuming GCA, Boltzmann's distribution for the inversion carriers, and considering only mobile carriers (e.g., electrons in an NMOS FinFET), can be expressed as

$$\frac{\partial^2 \psi(x, y)}{\partial x^2} = \frac{q}{\varepsilon_{ch}} \left(n_i e^{\frac{\psi(x,y) - \psi_B - V_{ch}(y)}{V_{tm}}} + N_{ch} \right) \tag{3.1}$$

where $\psi(x, y)$ is the electrostatic potential in the channel, q is the magnitude of the electronic charge, n_i is the intrinsic carrier concentration, ε_{ch} is the dielectric constant of the channel (fin), V_{tm} is the thermal voltage given by $k_B T/q$, where k_B and T are the Boltzmann constant and the temperature, respectively, V_{ch} is the quasi-Fermi potential of the channel ($V_{ch}(0) = V_s$ and $V_{ch}(L) = V_d$) which only has a y spatial

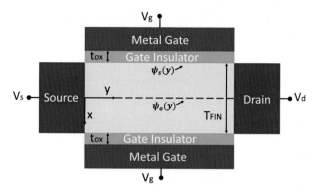

FIG. 3.1

Schematic representation of a symmetric double-gate FinFET.

dependence, N_{ch} is the channel doping, and $\psi_B = V_{tm} \ln(N_{ch}/n_i)$. Note that Eq. (3.1) is a 1D Poisson's equation, the other two dimensions have been neglected due to the long length of the channel (GCA condition) and the symmetry of the DG structure in the z-direction. In other FinFET geometries, where channel length is short and fin cross sections are complex, full 3D Poisson's equation must be solved.

Figs. 3.2–3.4 show fin potential versus fin position obtained from Eq. (3.1) for different conditions. Eq. (3.1) has been solved numerically using the finite element method to generate the fin potential data. Using these results, there are several points that must be contrasted with conventional planar MOSFETs. For example, as shown

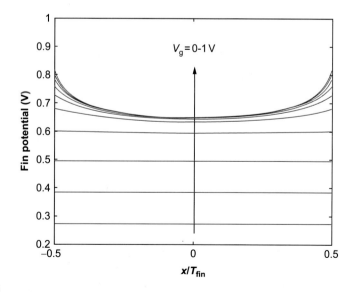

FIG. 3.2

Fin potential versus position obtained from the numerical solution of Eq. (3.1) for a lightly doped fin. $N_{ch} = 1 \times 10^{15}$ cm^{-3}, $TFIN = 20$ nm, $t_{ox} = 1$ nm, and $V_{ch} = 0$ V have been used for simulation.

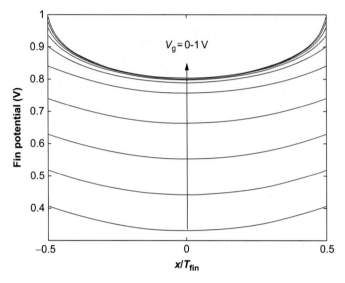

FIG. 3.3

Fin potential versus position obtained from the numerical solution of Eq. (3.1) for a doped fin. $N_{ch} = 5 \times 10^{18}$ cm^{-3}, $TFIN = 20$ nm, $t_{ox} = 1$ nm, and $V_{ch} = 0$ V have been used for simulation.

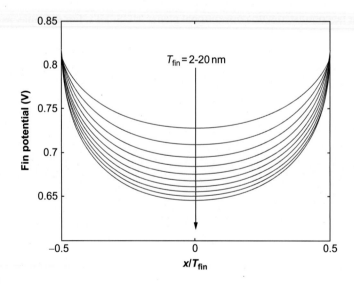

FIG. 3.4

Fin potential versus position obtained from the numerical solution of Eq. (3.1) for different fin thicknesses at strong inversion bias. $N_{ch} = 1 \times 10^{15}$ cm^{-3}, $V_{ch} = 0$ V, $TFIN = 20$ nm, $t_{ox} = 1$ nm, and $V_g = 1$ V have been used for simulation. As TFIN decreases, center potential increases, which increases the number of mobile carriers in the center of the fin.

in Figs. 3.2 and 3.3, the center potential is not fixed as in the case of the bulk potential in planar MOSFETs. Indeed, center potential has a different value depending on the bias condition and it tends to a fixed value in the strong operation regime. In addition, FinFETs may not require the use of high doping concentration for the channel to counter short-channel effects. In this context, lightly doped fins consequently increase carrier mobility and reduce device variability coming from random dopant fluctuations. The use of lightly doped channels implies that potential in the subthreshold region is mostly flat, as shown in Fig. 3.2, which makes the mobile carriers and subthreshold current to be proportional to the fin thickness. Thus, in order to decrease leakage, fin thickness must be scaled down. Including dopants in the channel can also be a good option for multiple threshold FinFETs [10]. Fig. 3.3 shows how the potential changes when a heavily doped fin is used. The potential in the subthreshold region is bent due to the ionized dopants, which change the threshold of the device depending on the amount of dopants and thickness of the fin. Finally, another important point to notice is that as the thickness of the fin is decreased, the potential in the center of the channel increases as shown in Fig. 3.4. An increment of the center potential increases the amount of mobile carriers in the center of the fin where mobility is larger than the surface, thus higher mobility could be obtained for those carriers. Figs. 3.2–3.4 were obtained by a numerical solution which is not suitable for a compact modeling perspective; therefore, a compact model would be presented starting from Eq. (3.1).

In order to obtain the potential in the channel for Eq. (3.1), ψ is written following the perturbation approach [3]:

$$\psi(x, y) \cong \psi_1(x, y) + \psi_2(x, y) \tag{3.2}$$

Here, ψ_1 is the potential contribution due to the inversion carriers and without the effect of the ionized dopants, N_{ch}, and is given by

$$\frac{\partial^2 \psi_1(x, y)}{\partial x^2} = \frac{q n_i}{\varepsilon_{ch}} e^{\frac{\psi_1(x,y) - \psi_B - V_{ch}(y)}{V_{tm}}} \tag{3.3}$$

and ψ_2 is the potential contribution due to the presence of the ionized dopants N_{ch}, and without the effect of the inversion carriers. It is the perturbation potential and is given by

$$\frac{\partial^2 \psi_2(x, y)}{\partial x^2} = \frac{q N_{ch}}{\varepsilon_{ch}} \tag{3.4}$$

The geometrical symmetry of a DG FinFET leads to the fact that the vertical component of the electric field $\mathcal{E}_x$ at the center of the channel is zero; then, it is possible to integrate Eq. (3.3) twice to obtain $\psi_1(x, y)$ as a function of the potential in the center of the body $\psi_0(y)$:

$$\psi_1(x, y) = \psi_0(y) - 2 V_{tm} \ln \left[\cos \left(\sqrt{\frac{q}{2\varepsilon_{ch} V_{tm}} \frac{n_i^2}{N_{ch}} e^{\frac{\psi_0(y) - V_{ch}(y)}{V_{tm}}}} \times \frac{x}{2} \right) \right] \tag{3.5}$$

In order to find ψ_2, it is also possible to apply $\mathcal{E}_x = 0$ at the center of the channel and setting $\psi_2(x = 0, y) = 0$. Then, integrating Eq. (3.4) twice, it is possible to obtain

$$\psi_2(x, y) = \frac{q N_{ch} x^2}{2\varepsilon_{ch}} \tag{3.6}$$

The surface potential ψ_s at any point y along the surface is obtained by evaluating the sum of ψ_1 and ψ_2 at the surface of the fin:

$$\psi_s(y) \cong \psi_1(-T_{fin}/2, y) + \psi_2(-T_{fin}/2, y) \tag{3.7}$$

Gauss's law and the boundary conditions at the channel-insulator interface lead to a second important equation:

$$V_{gs} = V_{fb} + \psi_s(y) + \varepsilon_{ch}\mathcal{E}_{xs}/C_{ox} \tag{3.8}$$

where V_{gs} is the gate voltage, V_{fb} is the flat-band voltage, C_{ox} is the gate oxide capacitance per unit area, given by ε_{ox}/T_{ox}, where ε_{ox} and T_{ox} are the oxide dielectric constant and oxide thickness, respectively, and $\mathcal{E}_{xs}$ is the vertical component of the electric field at the surface, which can be obtained by integrating Eq. (3.1):

$$\mathcal{E}_{xs} = \sqrt{\frac{2qn_i}{\varepsilon_{ch}}\left[V_{tm}\left(e^{\frac{\psi_s(y)}{V_{tm}}} - e^{\frac{\psi_0(y)}{V_{tm}}}\right)e^{\frac{-\psi_B - V_{ch}(y)}{V_{tm}}} + e^{\frac{\psi_B}{V_{tm}}}(\psi_s(y) - \psi_0(y))\right]} \tag{3.9}$$

Eqs. (3.7), (3.8) represent a self-consistent system of equations that can be used to obtain ψ_0 and ψ_s. However, through a change of variable, given by

$$\beta = \sqrt{\frac{q}{2\varepsilon_{ch}V_{tm}}\frac{n_i^2}{N_{ch}}e^{\frac{\psi_0 - V_{ch}}{V_{tm}}}}\frac{T_{FIN}}{2} \tag{3.10}$$

Eqs. (3.7), (3.8) can be written as a single equation:

$$f(\beta) \equiv \ln(\beta) - \ln(\cos(\beta)) - \frac{V_{gs} - V_{fb} - V_{ch}}{2V_{tm}} + \ln\left(\frac{2}{T_{fin}}\sqrt{\frac{2\varepsilon_{ch}V_{tm}N_{ch}}{qn_i^2}}\right)$$

$$+ \frac{2\varepsilon_{ch}}{T_{fin}C_{ox}}\sqrt{\beta^2\left(\frac{e^{\frac{\psi_{pert}}{V_{tm}}}}{\cos^2(\beta)} - 1\right) + \frac{\psi_{pert}}{V_{tm}^2}[\psi_{pert} - 2V_{tm}\ln(\cos(\beta))]} = 0 \tag{3.11}$$

where ψ_{pert} is given by ψ_2 evaluated at $x = T_{fin}/2$. Eq. (3.11) is an implicit equation in β which must be solved using numerical methods, then, once β is calculated, the surface potential and the charge in the channel can be obtained. Fig. 3.5 shows the surface potential obtained from Eq. (3.11) and the numerical solution of Eq. (3.1) for different doping concentrations. The amount of doping in the channel determines the threshold voltage of the device as shown in Fig. 3.6, which represents the mobile charge density obtained from the proposed compact model and the numerical solution of Eq. (3.1) for different doping concentrations. In the case of lightly doped DG FinFETs, the thickness of the channel determines the amount of mobile carrier charge density in the channel in a linear manner, as shown in Fig. 3.7.

Solving Eq. (3.11) using numerical methods is not practical for compact modeling applications because their use increases the computation time and may cause divergence problems [25]. Therefore, Eq. (3.11) is solved by first using an analytical approximation for the initial guess, followed by two quartic modified iterations [26]. This approach makes the model numerically robust and accurate (see Appendix Appendix). The surface potentials at the source end ψ_s and drain end ψ_d are calculated by setting $V_{ch} = V_s$ and $V_{ch} = V_d$, respectively. For a lightly doped body,

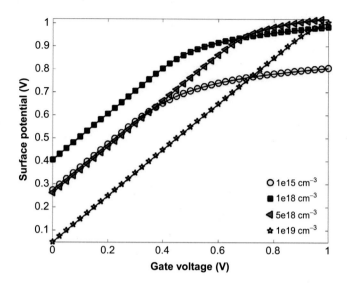

FIG. 3.5

Surface potential versus V_G of DG FinFETs ($T_{Fin} = 20$ nm) at $V_{DS} = 0.0$ V obtained from the proposed model (*lines*) and numerical simulations (*symbols*) for different channel dopings.

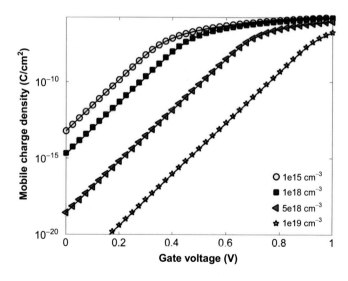

FIG. 3.6

Mobile electron charge density versus V_G of DG FinFETs ($T_{Fin} = 20$ nm) at $V_{DS} = 0.0$ V obtained from the proposed model (*lines*) and numerical simulations (*symbols*) for different channel dopings.

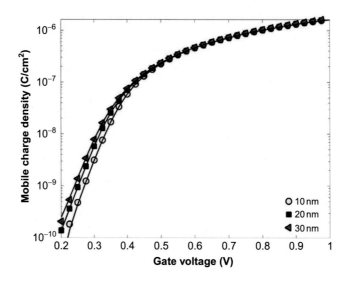

FIG. 3.7

Mobile electron charge density versus V_G of DG FinFETs (using intrinsic Fin channel) at $V_{DS} = 0.0$ V obtained from the proposed model (*lines*) and numerical simulations (*symbols*) for different Fin thicknesses.

Eq. (3.11) can be simplified further [3] to speed up the simulation. This option can be selected in the BSIM-CMG model by setting the parameter COREMOD. A separate model has been derived for the Cy gate geometry, which has been discussed in detail in Ref. [22].

In order to complete the core model, the drain to source current I_{ds} model in BSIM-CMG is obtained from a solution of the drift-diffusion equation, assuming a long-channel DG FinFET:

$$I_{ds}(y) = \mu(T)W Q_{inv}(y)\frac{dV_{ch}}{dy} \tag{3.12}$$

where $\mu(T)$ is the low-field and temperature-dependent mobility, W is the total effective width, and Q_{inv} is the inversion charge per unit area in the body. Eq. (3.12) includes drift and diffusion transport mechanisms through the use of the quasi-Fermi potential.

Integrating both sides of Eq. (3.12), and considering the fact that under quasistatic operation I_{ds} is constant along the channel, it is possible to express Eq. (3.12) in its integral form:

$$I_{ds} = \frac{W}{L}\mu(T)\int_{Q_{invs}}^{Q_{invd}} Q_{inv}\left(\frac{dV_{ch}}{dQ_{Inv}}\right)dQ_{inv} \tag{3.13}$$

where L is the effective channel length, and Q_{invs} and Q_{invd} are the inversion charge densities at the source and drain ends, respectively, given by

$$Q_{inv,d/s} = C_{ox}(V_{gs} - V_{fb} - \psi_{d/s}) - Q_{bulk} \tag{3.14}$$

Here, Q_{bulk} is the fixed depletion charge and is given by $qN_{ch}T_{fin}$ and $\psi_{d/s}$ are obtained by solving Eq. (3.11). The term dV_{ch}/dQ_{inv} in Eq. (3.13) can be calculated as a function of Q_{inv} using a simple, but accurate, implicit equation for Q_{inv} [3]:

$$Q_{inv}(y) \approx \sqrt{2qn_i\varepsilon_{ch}V_{tm}}\; e^{\frac{\psi_s(y)-\psi_B-V_{ch}(y)}{2V_{tm}}} \sqrt{\frac{Q_{inv}(y)}{Q_{inv}(y)+Q_0}} \tag{3.15}$$

where $Q_0 = Q_{bulk} + 5C_{fin}V_{tm}$, with $C_{fin} = \varepsilon_{ch}/T_{fin}$. Using this approximation, Eq. (3.13) can be integrated analytically, leading to the following basic equation for I_{ds}:

$$I_{ds} = \mu(T) \cdot \frac{W}{L} \cdot \left[\frac{Q_{inv,s}^2 - Q_{inv,d}^2}{2C_{ox}} + 2V_{tm}(Q_{inv,s} - Q_{inv,d}) - V_{tm}Q_0 \ln\left(\frac{Q_0 + Q_{inv,s}}{Q_0 + Q_{inv,d}}\right)\right] \tag{3.16}$$

Note that Q_{inv} charges are calculated using Eqs. (3.11), (3.13). Fig. 3.8 shows an example drain current obtained from the proposed model and numerical simulations.

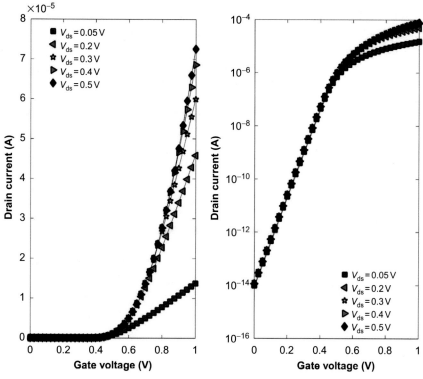

FIG. 3.8

Drain current versus V_G of DG FinFETs (using intrinsic Fin channel) at different V_{DS} values obtained from the proposed model (*lines*) and numerical simulations (*symbols*). Linear (*left*) and logarithm (*right*) scales are shown. $HFIN = L = 1\ \mu m$, $t_{ox} = 1\ nm$, $N_{ch} = 1 \times 10^{18}\ cm^{-3}$, $\mu_e = 100\ cm^2V^{-1}s^{-1}$, and a gate work-function equal to 4.6 eV have been used for the simulations.

It shows that BSIM-CMG model captures the behavior of FinFET under all bias conditions: subthreshold, triode, and saturation conditions.

3.2 Unified FinFET compact model

The typical rectangular cross section of FinFETs is hardly found on industry FinFETs. Indeed, whether intentional or due to manufacturing variation, industry FinFET cross sections are nonuniform and similar to rounded trapezoidal shapes [27,28], as shown in Fig. 3.9. In order to capture fin shape effects on device performance, a compact model for FinFETs with complex cross sections is important. In this section, the unified FinFET compact model is presented for devices with complex fin cross sections. FinFET structures, such as DG, Cy-GAA, rectangular GAA, or trapezoidal triple-gate FinFETs, are all modeled under the same framework. A single unified core model is used for different FinFET structures as those shown in Fig. 3.10, and only model parameters are different for each FinFET structure, which are precalculated for each device type and dimension. The proposed core model can be used with short-channel effects' submodels in a manner similar to what is done in previous versions of BSIM-CMG models [29]. The model presented in this section has been recently incorporated into BSIM-CMG [23].

Several compact models have been proposed for FinFETs with complex cross-sectional shapes. The work presented in Ref. [11] developed compact models for different undoped or lightly doped FinFET shapes utilizing a combination of the compact models for DG [4] and Cy-GAA FinFETs [5]. In Ref. [30], a compact model

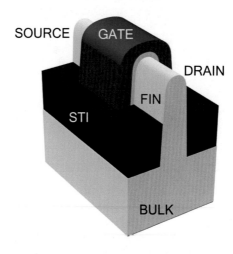

FIG. 3.9

A three-dimensional schematic representation of a FinFET with a complex fin cross section. The fin shape is similar to industry FinFETs reported in Refs. [27,28].

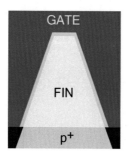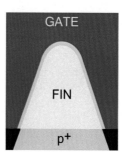

FIG. 3.10

Cross-sectional schematics of FinFETs with complex fin shapes. The *left* structure represents a perfectly rectangular fin shape; the two subsequently structures represent trapezoidal fin shape with and without rounded corners. These structures may differ from the typical rectangular FinFET shape due intentional manufacturing process or random structure variation. The electrical behavior of each one of these devices can be obtained with the proposed unified compact model.

for undoped or lightly doped FinFETs was extended to model FinFETs with different cross-sectional shapes by obtaining an equivalent channel thickness for each structure. Another compact model has been recently proposed for FinFET devices with different cross-sectional shapes [31,32], where new models for doped FinFETs were developed in a universal model framework. In this section, based on the approach presented in Ref. [23], a new normalized unified FinFET core model is presented [23]. The new normalized charge model is obtained from the solutions of the Poisson's equation for DG and Cy-GAA FinFETs, which leads to a single closed-form relationship between the mobile charge and the applied terminal voltages given as follows [23]:

$$v_G - v_o - v_{ch} = -q_m + \ln(-q_m) + \ln\left(\frac{q_t^2}{e^{q_t} - q_t - 1}\right) \tag{3.17}$$

where v_o and q_t are represented by

$$v_o = v_{FB} - q_{dep} - \ln\left(\frac{2qn_i^2 A_{ch}}{v_T C_{ins} N_{ch}}\right) \tag{3.18}$$

$$q_t = (q_m + q_{dep})r_N \tag{3.19}$$

In the previous equations, v_G and v_{ch} are the normalized gate and channel potentials expressed by

$$v_G = \frac{V_G}{v_T} \tag{3.20}$$

$$v_{ch} = \frac{V_{ch}}{v_T} \tag{3.21}$$

q_m and q_{dep} are the normalized mobile and depletion charges:

$$q_m = \frac{Q_m}{v_T C_{ins}} \tag{3.22}$$

$$q_{dep} = \frac{-qN_{ch}A_{ch}}{v_T C_{ins}} \tag{3.23}$$

r_N is given by

$$r_N = \frac{A_{Fin}C_{ins}}{\varepsilon_{ch}W^2} \tag{3.24}$$

A_{ch} is the area of the channel, N_{ch} is the doping in the channel, W is the channel width, and C_{ins} is the insulator capacitance per unit length. It is important to notice that on the right side of Eq. (3.17) there are three terms that determine the behavior of the charge in the channel. A linear term, which is important in the strong inversion, the first logarithm term, which is important in the subthreshold region, and the last logarithm term, which is important in the moderate inversion. Therefore, Eq. (3.17) can represent the mobile carrier concentration in the channel for all bias conditions in a continuous and smooth manner, crucial for circuit simulation success.

The normalized drain current is obtained from the solution of the Poisson-carrier transport equation [23] and it is represented by

$$i_{DS} = \left[\frac{q_m^2}{2} - 2q_m - q_H \ln\left(1 - \frac{q_m}{q_H} \right) \right] \Bigg|_{q_{m,S}}^{q_{m,D}} \tag{3.25}$$

where q_H is equal to:

$$q_H = \frac{1}{r_N} - q_{dep} \tag{3.26}$$

The drain current normalization is given by

$$i_{DS} = \frac{-I_{DS}L}{\mu_m v_T^2 C_{ins}} \tag{3.27}$$

Eq. (3.25) has three terms that determine the behavior of the current under different bias conditions. The first term, quadratic, is important in the saturation and triode conditions, the second term, linear, is important in the triode and sub-threshold conditions, and finally, the last term, logarithm, is important in the subthreshold and moderate inversion conditions. This shows that Eq. (3.25) represents the drain current characteristics in all regions of device operation, that is, subthreshold, linear, and saturation regions, in a continuous and explicit expression. Considering each region of operation, the proposed model can be reduced to simple expressions. In the subthreshold region, the drain current is approximately given by

$$I_{DS} \approx \frac{\mu}{L} v_T^2 C_{ins} \exp\left(\frac{V_G - V_{TH}}{v_T} \right) \times \left(1 - \exp\frac{-V_{DS}}{v_T} \right) \tag{3.28}$$

where V_{TH} is the threshold voltage of FinFETs [32]. Note that Eq. (3.28) is independent of C_{ins} for undoped devices, thus Eq. (3.28) can further be simplified to

$$I_{DS} \approx A_{ch}\frac{\mu}{L}v_Tq\frac{n_i^2}{N_{ch}}\exp\left(\frac{V_G - V_{FB}}{v_T}\right) \times \left(1 - \exp\frac{-V_{DS}}{v_T}\right) \qquad (3.29)$$

Using Eq. (3.29), it is possible to conclude that to decrease the leakage in lightly doped FinFETs, A_{ch} must be scaled down. The drain currents in the linear and saturation regions are approximately given by

$$I_{DS} \approx \frac{\mu}{L}C_{ins}(V_G - V_{TH} - V_{DS}/2)V_{DS} \qquad (3.30)$$

$$I_{DS} \approx \frac{\mu}{2L}C_{ins}(V_G - V_{TH})^2 \qquad (3.31)$$

Eqs. (3.30), (3.31) are the very well-known equations from the quadratic model for conventional long-channel CMOS MOSFETs.

Using the presented results, it should be note that only four different model parameters are needed for the modeling of FinFET devices: A_{ch}, N_{ch}, W, and C_{ins}. Using these parameters, a FinFET with simple cross section, such as DG FinFET, can be accurately modeled for different channel doping concentrations, as shown in Fig. 3.11. The model parameters used for DG FinFETs are given as follows [23]:

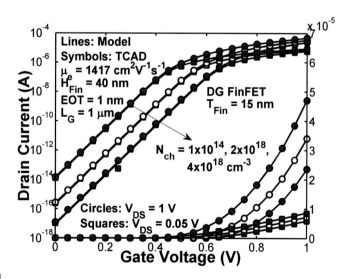

FIG. 3.11

Drain current versus gate voltage of DG FinFETs at $V_{DS} = 0.05$ V (*squares*) and $V_{DS} = 1$ V (*circles*) obtained from the proposed model (*lines*) and numerical simulations (*symbols*) for three different channel dopings: $N_{ch} = 1 \times 10^{14}$ cm^{-3} (*red symbols*), $N_{ch} = 2 \times 10^{18}$ cm^{-3} (*empty symbols*), and $N_{ch} = 4 \times 10^{18}$ cm^{-3} (*blue symbols*). $\mu = 1470$ cm^2/Vs, $L_G = 1$ μm, $T_{CH} = 15$ nm, $H_{CH} = 40$ nm, $EOT = 1$ nm, and metal gate work-function equal to 4.6 eV have been used.

$$A_{ch} = H_{Fin} T_{Fin} \tag{3.32}$$

$$W = 2H_{Fin} \tag{3.33}$$

$$C_{ins} = \frac{\varepsilon_{ins}}{EOT} W \tag{3.34}$$

$$N_{ch} \tag{3.35}$$

A trapezoidal triple-gate FinFET is a good example of a FinFET with a complex cross section. Indeed, the industry transistors reported in Refs. [27,28] have fin cross sections similar to trapezoidal shapes. The proposed model can be used to model these type of devices through the use of the following four model parameters:

$$A_{ch} = H_{Fin} \frac{(T_{Fin,top} + T_{Fin,base})}{2} \tag{3.36}$$

$$W = 2\sqrt{\frac{(T_{Fin,base} - T_{Fin,top})^2}{4} + H_{Fin}^2} + T_{Fin,top} \tag{3.37}$$

$$C_{ins} = \frac{\varepsilon_{ins}}{EOT} W \tag{3.38}$$

$$N_{ch} \tag{3.39}$$

Using these parameters, trapezoidal triple-gate FinFETs can be accurately modeled as shown in Fig. 3.12, where a T-TG FinFET has been doped at different doping concentrations as it is used in chips with multithreshold voltage levels.

In the case of a channel dimension variation, the model can accurately predict the trend of current changes, as shown in Fig. 3.13. Note that a $T_{Fin,top}$ variation is more important for the on-current than a $T_{Fin, base}$ variation. In addition, the off-current linearly varies as a function of $T_{Fin,top}$ or $T_{Fin,base}$, as expected.

The proposed model accurately models experimental long-channel FinFETs without the use of fitting parameters, as shown in Figs. 3.14 and 3.15, which compare the proposed core compact model and the data from a fabricated long-channel Fin-FET. Note that only mobility model has been additionally included to the core model in the figures. The accuracy of the proposed model is tested by the drain current and its derivatives agreement between the data and model. This is a very important test because g_m and g_{DS} are crucial quantities in the ultimate performance of circuits constructed from these FinFETs.

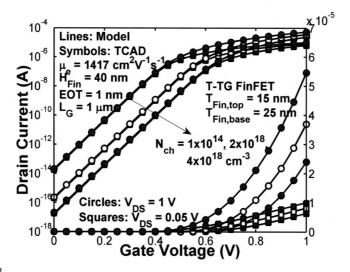

FIG. 3.12

I_D versus V_G of trapezoidal triple-gate FinFETs ($T_{Fin,Top}$ = 15 nm and $T_{Fin,Base}$ = 25 nm) at V_{DS} = 0.05 V (*squares*) and V_{DS} = 1 V (*circles*) obtained from the proposed model (*lines*) and numerical simulations (*symbols*) for three different channel dopings: N_{ch} = 1 × 10^{14} cm^{-3} (*red symbols*), N_{ch} = 2 × 10^{18} cm^{-3} (*empty symbols*), and N_{ch} = 4 × 10^{18} cm^{-3} (*blue symbols*).

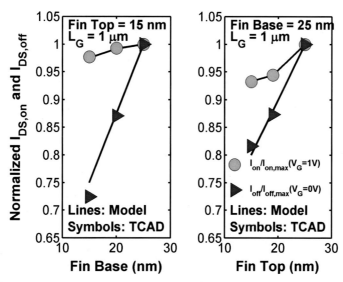

FIG. 3.13

Normalized ($I_{DS}/I_{DS,MAX}$) on (*circles*) and off (*triangles*) drain currents.

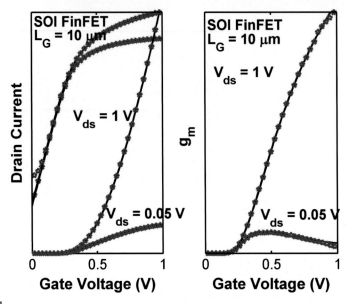

FIG. 3.14

Drain current versus gate voltage (*left figure*) of a long-channel SOI FinFET obtained from experimental data and from the proposed model only using the ideal unified long-channel core model. Transconductance (*right figure*) also shows good agreement between model and experimental data. *Symbols* are the experimental data and *lines* the proposed model.

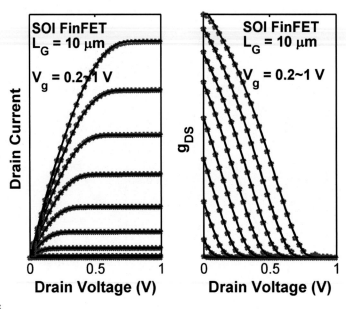

FIG. 3.15

Drain current versus drain voltage (*left figure*) of a long-channel SOI FinFET obtained from experimental data and from the proposed model only using the ideal unified long-channel core model. Output conductance (*right figure*) also shows good agreement between model and experimental data. *Symbols* are the experimental data and *lines* the proposed model.

Appendix **Explicit surface potential model**

The surface potential model used in BSIM-CMG was initially derived in an implicit form in Section 3.1, which relies on numerical calculations to be solved, such as the Newton-Raphson method [26,33]. The Newton-Raphson method is an iterative algorithm that starts from an initial guess to then keep refining it until the method arrives at a final solution. This refinement is done by approximating the function to be solved at the given guess by a linear system, which is then used to obtain the new guess for the solution [34]. Although the Newton-Raphson method could potentially have a rapid convergence (e.g., quadratic [26,34]), its convergence is no guarantee. Indeed, it could potentially fail by, for example, getting into an infinite loop. This is not practical for compact models, where convergence and speed are key requirements. However, by making sure that the initial guess is close enough to the final solution and that the solution is not singular, that is, the derivative of the function is nonzero at the solution, it could be assured that the Newton-Raphson method will converge [26]. Therefore, in order to implement the model proposed in Section 3.1 in a circuit simulator, a good initial guess formula, so-called continuous starting function (CSF) [25], for Eq. (3.11) is presented in this section. In addition, the Newton-Raphson method could be replaced for a higher-order method such as the quartic modified iteration [26]. The main difference between quartic modified and the Newton-Raphson iterations is that first refine the initial guess by approximating the function to be solved using a high-order approximation, instead of the linear one used in the Newton-Raphson iteration. Therefore, quartic modified iteration could give a faster convergence rate, which would reduce the iterations needed to solve Eq. (3.11), making the proposed compact model fast.

A.1 **Continuous starting function**

An initial guess for Eq. (3.11) has been already proposed in Ref. [3]. It was derived by considering the asymptotic behavior of β in Eq. (3.11) under the two main regimes of operation: subthreshold regime and strong inversion regime. Although the initial guess proposed in Ref. [3] works well for most conditions, it has few important issues that must be solved. First, it is not a single continuous and smooth starting function because it was constructed by taking the minimum between two different functions. The lack of smoothness implies a higher number of iterations needed to arrive to a final solution. The second issue is that the initial guess does not work very well for devices with highly doped or wide channels, where β can get values as low as 1×10^{-15}. This makes the model unstable, depending on device geometry and bias conditions. This is a critical point, especially when a compact model is used for circuit-device optimization where the optimization algorithm could bring the evaluation of the model to extreme conditions. Due to the mentioned issues, in this section a new CSF is proposed to solve Eq. (3.11). In a similar manner to the initial guess proposed in Ref. [3], the new CSF will be obtained by considering the asymptotic behavior of Eq. (3.11); however, highly doped conditions would be consider first to later update the derived CSF to lightly doped channels.

In the case of a FinFET device with a highly doped channel, ψ_{pert} is large and β tends to 0, which makes $\cos(\beta)$ to tend to 1 and $\ln(\cos(\beta))$ to 0. Using these assumptions and neglecting $\ln\beta$ term, which is close to be constant in the strong inversion condition, a β function for highly doped channel FinFETs can be obtained in the strong inversion by solving Eq. (3.11) for the β^2 term inside the square root:

$$\beta_{\text{SI}} = e^{\frac{-\psi_{\text{pert}}}{2V_{\text{tm}}}} \frac{A_{g0}}{r} \sqrt{\left(\frac{F - F_{\text{th,SI}}}{A_{g0}} + 1\right)^2 - 1} \tag{A.1}$$

where A_{g0}, r, F, and $F_{\text{th,SI}}$ are defined as follows:

$$A_{g0} = \frac{r\psi_{\text{pert}}}{V_{\text{tm}}} \tag{A.2}$$

$$r = \frac{2\epsilon_{\text{si}}t_{\text{ox}}}{\epsilon_{\text{ox}}T_{\text{FIN}}} \tag{A.3}$$

$$F = \frac{V_G - V_{\text{FB}} - V_{\text{CH}} - \psi_{\text{pert}}}{2V_{\text{tm}}} + \ln\left(\frac{T_{\text{FIN}}}{2} \sqrt{\frac{qn_i^2}{2\epsilon_{\text{si}}N_{\text{ch}}V_{\text{tm}}}}\right) \tag{A.4}$$

$$F_{\text{th,SI}} = (2r - 1)\frac{\psi_{\text{pert}}}{2V_{\text{tm}}} \tag{A.5}$$

Eq. (A.1) is only valid when $(F - F_{\text{th,SI}})/A_{g0} > -1$. In the subthreshold region, if β is even smaller than that in the previous assumption, then β^2 can be neglected in Eq. (3.11). The $\ln\beta$ is not any more constant in the subthreshold region; from it, a β function for the subthreshold can be obtained as follows [3]:

$$\beta_{\text{ST}} = e^{F - F_{\text{th}}} e^{\frac{-\psi_{\text{pert}}}{2V_{\text{tm}}}} \tag{A.6}$$

Note that Eq. (A.6) is a good approximation for the subthreshold region of lightly and heavily doped channels as well. Eqs. (A.1), (A.6) are good approximations for β; however, a single CSF has not been constructed yet. In order to obtain a single CSF, Eq. (A.1) can be modified to

$$\beta_{\text{doped}} = e^{\frac{-\psi_{\text{pert}}}{2V_{\text{tm}}}} \frac{A_{g0}}{r} \sqrt{\left(\frac{\ln(1 + e^{2(F - F_{\text{th,SI}})})}{2A_{g0}} + 1\right)^2 - 1} \tag{A.7}$$

The earlier expression can be evaluated in all bias conditions and it tends to Eq. (A.1) when $F > F_{\text{th}}$. In the case of $F \ll F_{\text{th, SI}}$, Eq. (A.7) tends to an expression similar but not exactly equal to Eq. (A.6). In order to make Eq. (A.7) to reduce to Eq. (A.6) for $F \ll F_{\text{th,SI}}$, $F_{\text{th,SI}}$ can be updated by equating Eqs. (A.6), (A.7) at the subthreshold, given:

$$F_{\text{th}} = (2r - 1)\frac{\psi_{\text{pert}}}{2V_{\text{tm}}} + \ln\left(\frac{\sqrt{A_{g0}}}{(2r - 1)\frac{\psi_{\text{pert}}}{2V_{\text{tm}}}}\right) \tag{A.8}$$

Figs. A.1 and A.2 show β obtained from Eq. (3.11) solved using the Newton-Raphson iteration and from the initial guess expressed by Eq. (A.7) using Eq. (A.8), as expected,

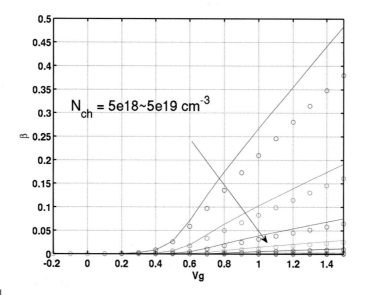

FIG. A.1

β obtained from Eq. (3.11) solved using the Newton-Raphson iteration (*symbols*) and from the initial guess expressed by Eq. (A.7) (*lines*) using Eq. (A.8) in a linear scale, for different doping concentrations. Eq. (A.7) should be extended to cover lightly doped devices. $T_{FIN} = 20$ nm, $t_{ox} = 1$ nm, and a gate work-function equal to 4.4 eV have been used for simulations.

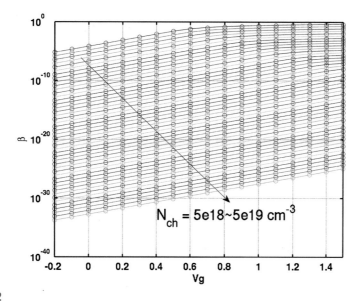

FIG. A.2

β obtained from Eq. (3.11) solved using the Newton-Raphson iteration (*symbols*) and from the initial guess expressed by Eq. (A.7) (*lines*) using Eq. (A.8) in a logarithm scale, for different doping concentrations. Eq. (A.7) should be extended to cover lightly doped devices. $T_{FIN} = 20$ nm, $t_{ox} = 1$ nm, and a gate work-function equal to 4.4 eV have been used for simulations.

the proposed guess is close to the final solution in all bias regimes for highly doped channels. In order to extend Eq. (A.7) to devices with lightly doped channels, only two majors changes are needed. First, it must be noticed that as channel doping decreases, A_{g0} tends to 0; this would give invalid results. Therefore, analyzing the asymptotic behavior of Eq. (3.11), it can be noticed that, as doping decreases, a valid value for A_{g0} should be 1; thus A_{g0} can be updated to the following expression:

$$A_{\mathrm{g}} = \frac{\dfrac{r\psi_{\mathrm{pert}}}{V_{\mathrm{tm}}}}{1 - e^{\frac{-r\psi_{\mathrm{pert}}}{V_{\mathrm{tm}}}}} \tag{A.9}$$

The second change that must be done is to limit the value of β obtained from Eq. (A.7) to $\pi/2$. This limit is obtained analyzing the behavior of Eq. (3.11) in strong inversion for lightly doped devices, as also explained in Ref. [3]. Therefore, the final CSF proposed here is given as follows:

$$\beta_0 = \frac{1}{\dfrac{1}{\beta_{\mathrm{doped}}} + \dfrac{2}{\pi}} \tag{A.10}$$

where β_0 is obtained from Eq. (A.7) using F_{th} and A_{g} from Eqs. (A.8), (A.9), respectively. Figs. A.3 and A.4 show β obtained from Eq. (3.11) is solved using the

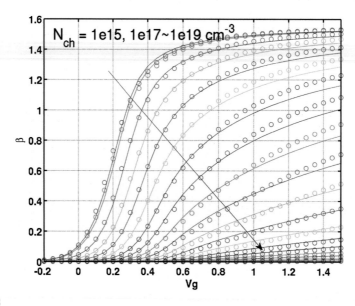

FIG. A.3

β obtained from Eq. (3.11) solved using the Newton-Raphson iteration (*symbols*) and from the proposed CSF expressed by Eq. (A.10) (*lines*) in a linear scale, for different doping concentrations. Eq. (A.10) is valid for devices with lightly or heavily doped channels. Note that for lightly doped devices β has a similar behavior, which changes as doping is increased. $T_{\mathrm{FIN}} = 20$ nm, $t_{\mathrm{ox}} = 1$ nm, and a gate work-function equal to 4.4 eV have been used for simulations.

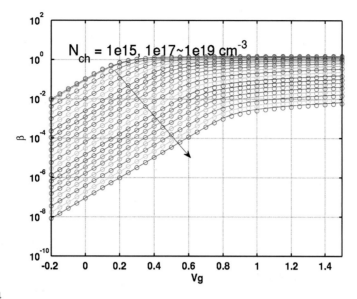

FIG. A.4

β obtained from Eq. (3.11) solved using the Newton-Raphson iteration (*symbols*) and from the proposed CSF expressed by Eq. (A.10) (*lines*) in a logarithm scale, for different doping concentrations. Eq. (A.10) is valid for devices with lightly or heavily doped channels. Note that for lightly doped devices β has a similar behavior, which changes as doping is increased. $T_{FIN} = 20$ nm, $t_{ox} = 1$ nm, and a gate work-function equal to 4.4 eV have been used for simulations.

Newton-Raphson iteration and from the initial guess expressed by Eq. (A.10). As expected, Eq. (A.10) is valid for devices with lightly or heavily doped channels.

A.2 Quartic modified iteration: Implementation and evaluation

In order to complete the explicit surface potential model, a quartic modified iteration has to be derived. The quartic modified iteration updates the initial guess expressed by Eq. (A.10) using a high-order correction:

$$\beta_1 = \beta_0 - \frac{f_0}{f_1}\left(1 + \frac{f_0 f_2}{2f_1^2} + \frac{f_0^2(3f_2^2 - f_1 f_3)}{6f_1^4}\right) \tag{A.11}$$

where f_n is the nth derivative of Eq. (3.11) with respect to β, that is, $f_n = \left.\frac{\partial^n f}{\partial \beta^n}\right|_{\beta=\beta_0}$.

Using this algorithm, an explicit surface potential model can be calculated using the following steps:

1. Calculate the continuous starting function in the following order:

$$T_a = \psi_{pert}/V_{tm}^2 \tag{A.12a}$$

$$T_b = e^{\psi_{pert}/V_{tm}} \tag{A.12b}$$

$$T_c = 2V_{tm} \tag{A.12c}$$

$$r = \frac{2\epsilon_{si}t_{ox}}{\epsilon_{ox}T_{FIN}} \tag{A.12d}$$

$$A_g = \frac{\dfrac{r\psi_{pert}}{V_{tm}}}{1 - e^{\frac{-r\psi_{pert}}{V_{tm}}}} \tag{A.12e}$$

$$F_{th} = (2r - 1)\frac{\psi_{pert}}{2V_{tm}} + \ln\left(\frac{\sqrt{A_g}}{(2r - 1)\dfrac{\psi_{pert}}{2V_{tm}}}\right) \tag{A.12f}$$

$$\beta_{doped} = e^{\frac{-\psi_{pert}}{2V_{tm}}}\frac{A_g}{r}\sqrt{\left(\frac{\ln(1 + e^{2(F - F_{th})})}{2A_g} + 1\right)^2 - 1} \tag{A.12g}$$

$$\beta_0 = \frac{1}{\dfrac{1}{\beta_{doped}} + \dfrac{2}{\pi}} \tag{A.12h}$$

2. Compute:

$$\tan g0 = \tan(\beta_0) \tag{A.13a}$$

$$\cos g0 = \cos(\beta_0) \tag{A.13b}$$

$$\sec g0 = \cos g0^{-1} \tag{A.13c}$$

$$\sec g0sq = \sec g0^2 \tag{A.13d}$$

$$\ln g0 = \ln(\beta_0) \tag{A.13e}$$

3. Compute:

$$T_0 = 1 + \beta_0 \tan g0 \tag{A.14a}$$

$$T_1 = \beta_0^2(T_b\sec g0sq - 1) + T_{abb}(\psi_{pert} - T_c\ln(\cos g0)) \tag{A.14b}$$

$$T_2 = \sqrt{T_1} \tag{A.14c}$$

$$T_3 = -2\beta + T_{abb}T_c\tan g0 + 2T_b\beta\sec g0sqT_0 \tag{A.14d}$$

$$T_4 = -2 + 2T_b\beta_0^2\sec g0sq^2 + \sec g0sq(2T_b + T_aT_c + 8T_b\beta\tan g0 + 4T_b\beta_0^2\tan g0^2) \tag{A.14e}$$

$$T_5 = 2\tan g0T_4 + 4(3T_b\beta\sec g0sq^2T_0 + \tan g0 + 2T_b\sec g0sqT_0\tan g0) \tag{A.14f}$$

4. Compute derivatives:

$$f_0 = \ln g0 - lln(\cos g0) + rT_2 - F \tag{A.15a}$$

$$f_1 = \beta^{-1} + \tan g0 + \frac{rT_3}{2T_2} \tag{A.15b}$$

$$f_2 = -\beta_0^{-2} + \sec g0sq - \frac{rT_3^2}{4T_2^3} + \frac{rT_4}{2T_2} \tag{A.15c}$$

$$f_0 = 2\beta_0^{-3} + 2 \sec g0 sq \tan g0 + \frac{3rT_3}{4T_2^3}\left(\frac{T_3^2}{2T_2^2} - T_4\right) + \frac{rT_5}{2T_2} \tag{A.15d}$$

5. Update β_0 to β_1 using a quartic modified iteration:

$$\beta_1 = \beta_0 - \frac{f_0}{f_1}\left(1 + \frac{f_0 f_2}{2f_1^2} + \frac{f_0^2(3f_2^2 - f_1 f_3)}{6f_1^4}\right) \tag{A.16a}$$

Steps 2–5 can be repeated more than once. The number of iterations can be obtained by doing an extensive analysis of the proposed model accuracy under difference device and bias conditions. Fig. A.5 shows the error of β obtained from the proposed explicit surface potential model with one and two iterations with respect to β obtained using the Newton-Raphson method under all different combinations (>52,500 simulations) of doping concentration (1×10^{15} to 1×10^{19} cm^{-3}), channel width (1–30 nm), dielectric thickness (0.5–5 nm), gate voltage (−0.2 to 1.5 V), and temperature (−100°C to 100°C). Using one Householder's iteration, the RMS error is 0.28% with a peak error of 2.47%. Using two Householder's iterations, the RMS error is 1.58×10^{-6}% with a peak error of 3.31×10^{-5}%. Therefore, only two iterations are needed to obtain an accurate solution of β, as shown in Fig. A.6.

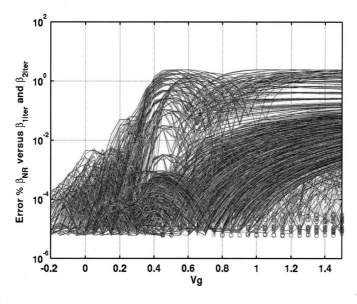

FIG. A.5

Error of β obtained from the proposed explicit surface potential model with one (*lines*) and two (*symbols*) iterations with respect to β obtained using the Newton-Raphson method under all different combinations of doping concentration (1×10^{15} to 1×10^{19} cm^{-3}), channel width (1–30 nm), dielectric thickness (0.5–5 nm), gate voltage (−0.2 to 1.5 V), and temperature (−100°C to 100°C).

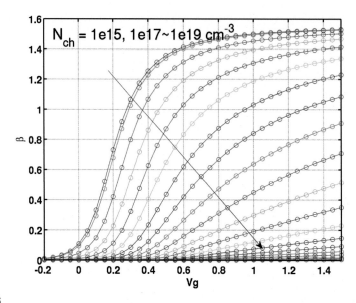

FIG. A.6

β obtained from Eq. (3.11) solved using the Newton-Raphson iteration (*symbols*) and from the proposed explicit surface potential model using two Householder's iterations (*lines*) in a linear scale, for different doping concentrations. $T_{FIN} = 20$ nm, $t_{ox} = 1$ nm, and a gate work-function equal to 4.4 eV have been used for simulations.

References

[1] A. Ortiz-Conde, F.J. Garcia-Sanchez, J. Muci, S. Malobabic, J.J. Liou, A review of core compact models for undoped double-gate SOI MOSFETs, IEEE Trans. Electron Devices 54 (1) (2007) 131–140.

[2] H.C. Pao, C.T. Sah, Effects of diffusion current on characteristics of metal-oxide (insulator)-semiconductor transistors, Solid-State Electron. 9 (10) (1966) 927–937.

[3] M.V. Dunga, Nanoscale CMOS Modeling, EECS Department, University of California, Berkeley, 2008.

[4] Y. Taur, An analytical solution to a double-gate MOSFET with undoped body, IEEE Electron Device Lett. 21 (5) (2000) 245–247.

[5] Y. Chen, J. Luo, A comparative study of double-gate and surrounding-gate MOSFETs in strong inversion and accumulation using an analytical model, Integrion 1 (2) (2001) 6.

[6] M.V. Dunga, C.H. Lin, X. Xi, D.D. Lu, A.M. Niknejad, C. Hu, Modeling advanced FET technology in a compact model, IEEE Trans. Electron Devices 53 (9) (2006) 1971–1978.

[7] O. Moldovan, A. Cerdeira, D. Jimenez, J.P. Raskin, V. Kilchytska, D. Flandre, N. Collaert, B. Iñiguez, Compact model for highly-doped double-gate SOI MOSFETs targeting baseband analog applications, Solid-State Electron. 51 (5) (2007) 655–661.

[8] F. Liu, L. Zhang, J. Zhang, J. He, M. Chan, Effects of body doping on threshold voltage and channel potential of symmetric DG MOSFETs with continuous solution from accumulation to strong-inversion regions, Semicond. Sci. Technol. 24 (2009) 085005.

[9] F. Liu, J. He, L. Zhang, J. Zhang, J. Hu, C. Ma, M. Chan, A charge-based model for long-channel cylindrical surrounding-gate MOSFETs from intrinsic channel to heavily doped body, IEEE Trans. Electron Devices 55 (8) (2008) 2187–2194.

[10] C.-H. Lin, R. Kambhampati, R.J. Miller, T.B. Hook, A. Bryant, W. Haensch, P. Oldiges, I. Lauer, T. Yamashita, V. Basker, et al., Channel doping impact on FinFETs for 22 nm and beyond, in: 2012 Symposium on VLSI Technology (VLSIT), IEEE, 2012, pp. 15–16.

[11] B. Yu, J. Song, Y. Yuan, W.Y. Lu, Y. Taur, A unified analytic drain-current model for multiple-gate MOSFETs, IEEE Trans. Electron Devices 55 (8) (2008) 2157–2163.

[12] A. Tsormpatzoglou, C.A. Dimitriadis, R. Clerc, G. Pananakakis, G. Ghibaudo, Semianalytical modeling of short-channel effects in lightly doped silicon trigate MOSFETs, IEEE Trans. Electron Devices 55 (10) (2008) 2623–2631.

[13] E. Moreno, J.B. Roldán, F.G. Ruiz, D. Barrera, A. Godoy, F. Gámiz, An analytical model for square GAA MOSFETs including quantum effects, Solid-State Electron. 54 (11) (2010) 1463–1469.

[14] E. Moreno Perez, J.B. Roldan Aranda, F.J. Garcia Ruiz, D. Barrera Rosillo, M.J. Ibanez Perez, A. Godoy, F. Gamiz, An inversion-charge analytical model for square gate-all-around MOSFETs, IEEE Trans. Electron Devices 58 (9) (2011) 2854–2861.

[15] J.T. Park, J.P. Colinge, C.H. Diaz, Pi-gate SOI MOSFET, IEEE Electron Device Lett. 22 (8) (2001) 405–406.

[16] Y. Taur, X. Liang, W. Wang, H. Lu, A continuous, analytic drain-current model for DG MOSFETs, IEEE Electron Device Lett. 25 (2) (2004) 107–109.

[17] J.M. Sallese, F. Krummenacher, F. Prégaldiny, C. Lallement, A. Roy, C. Enz, A design oriented charge-based current model for symmetric DG MOSFET and its correlation with the EKV formalism, Solid-State Electron. 49 (3) (2005) 485–489.

[18] A. Ortiz-Conde, F.J.G. Sánchez, J. Muci, Rigorous analytic solution for the drain current of undoped symmetric dual-gate MOSFETs, Solid-State Electron. 49 (4) (2005) 640–647.

[19] G.D.J. Smit, A.J. Scholten, G. Curatola, R. van Langevelde, G. Gildenblat, D.B.M. Klaassen, PSP-based scalable compact FinFET model, Proc. Nanotech. 3 (2007) 520–525.

[20] H. Abd El Hamid, J.R. Guitart, V. Kilchytska, D. Flandre, B. Iniguez, A 3-D analytical physically based model for the subthreshold swing in undoped trigate FinFETs, IEEE Trans. Electron Devices 54 (9) (2007) 2487–2496.

[21] D. Jiménez, B. Iniguez, J. Sune, L.F. Marsal, J. Pallares, J. Roig, D. Flores, Continuous analytic IV model for surrounding-gate MOSFETs, IEEE Electron Device Lett. 25 (8) (2004) 571–573.

[22] S. Venugopalan, D.D. Lu, Y. Kawakami, P.M. Lee, A.M. Niknejad, C. Hu, BSIM-CG: a compact model of cylindrical/surround gate MOSFET for circuit simulations, Solid-State Electron. 67 (1) (2012) 79–89.

[23] J.P. Duarte, N. Paydavosi, S. Venugopalan, A. Sachid, C. Hu, Unified FinFET Compact Model: Modelling Trapezoidal Triple-Gate FinFETs, SISPAD, 2013.

[24] B. Yu, H. Lu, M. Liu, Y. Taur, Explicit continuous models for double-gate and surrounding-gate MOSFETs, IEEE Trans. Electron Devices 54 (10) (2007) 2715–2722.

[25] P. Sebah, X. Gourdon, Newton's method and high order iterations, 2001. http://numbers.computation.free.fr/Constants/Algorithms/newton.html. Technical report.

[26] C.-Y. Chang, T.-L. Lee, C. Wann, L.-S. Lai, H.-M. Chen, C.-C. Yeh, C.-S. Chang, C.-C. Ho, J.-C. Sheu, T.-M. Kwok, et al., A 25-nm gate-length FinFET transistor module for 32 nm node, in: 2009 IEEE International Electron Devices Meeting (IEDM), IEEE, 2009, pp. 1–4.

[27] C. Auth, C. Allen, A. Blattner, D. Bergstrom, M. Brazier, M. Bost, M. Buehler, V. Chikarmane, T. Ghani, T. Glassman, et al., A 22 nm high performance and low-power CMOS technology featuring fully-depleted tri-gate transistors, self-aligned contacts and high density MIM capacitors, in: 2012 Symposium on VLSI Technology (VLSIT), IEEE, 2012, pp. 131–132.

[28] N. Paydavosi, S. Venugopalan, Y.S. Chauhan, J.P. Duarte, S. Jandhyala, A.M. Niknejad, C. Hu, BSIM-SPICE models enable FinFET and UTB IC designs, IEEE Access 1 (2013) 201.

[29] N. Chevillon, J.-M. Sallese, C. Lallement, F. Prégaldiny, M. Madec, J. Sedlmeir, J. Aghassi, Generalization of the concept of equivalent thickness and capacitance to multigate MOSFETs modeling, IEEE Trans. Electron Devices 59 (1) (2012) 60–71.

[30] J.P. Duarte, S.-J. Choi, D.-I. Moon, J.-H. Ahn, J.-Y. Kim, S. Kim, Y.-K. Choi, A universal core model for multiple-gate field-effect transistors. Part I: charge model, IEEE Trans. Electron Devices 60 (2) (2013) 840–847.

[31] J.P. Duarte, S.-J. Choi, D.-I. Moon, J.-H. Ahn, J.-Y. Kim, S. Kim, Y.-K. Choi, A universal core model for multiple-gate field-effect transistors. Part II: drain current model, IEEE Trans. Electron Devices 60 (2) (2013) 848–855.

[32] W.H. Press, B.P. Flannery, S.A. Teukolsky, W.T. Vetterling, B.P. Flannery, Numerical Recipes—The Art of Scientific Computing (FORTRAN), Cambridge University Press, Cambridge, 1989.

[33] J. Roychowdhury, Numerical Simulation and Modelling of Electronic and Biochemical Systems, Now Publishers Inc., 2009.

[34] A. Dasgupta et al., BSIM compact model of quantum confinement in advanced nanosheet FETs, IEEE Trans. Electron Devices 67 (2) (2020) 730–737, https://doi.org/10.1109/TED.2019.2960269.

Gate-all-around FETs

4.1 Introduction

The semiconductor landscape has seen continued scaling of devices for improved performance at reduced costs [1–6]. The newest version of the heart of all semiconductor electronics is the Gate-All-Around Field Effect Transistor (GAAFET), also known as nanosheet FETs, RibbonFETs, or multibridge channel FETs [7–12]. These are the logical successors to the tri-gate FinFETs and have gates encompassing the channel on all four sides to improve electrostatic control while scaling down device dimensions further. Fig. 4.1 shows the structure of stacked nanosheet GAA-FETs. These devices have already been announced for the next few technology nodes [11, 12].

The existing BSIM-CMG model is well equipped to capture the behavior of GAAFETs. However, there are certain advanced physical aspects that have recently been incorporated into this industry standard model to better predict the terminal characteristics. Stacked nanosheet GAAFETs have a completely different geometry with stacked layers of Silicon channels in every fin. A new enhanced core model has been developed specifically for this. Also, these devices have very confined cross-sections resulting in significant quantum mechanical effects. The quantum effects take the form of geometry-dependent subband separation as well as geometry dependence of electrical dimensions, density of states (DOS), and effective mass. All of these directly affect the terminal characteristics. Moreover, the structure results in a geometry-dependent mobility, which further complicates current calculations. All of these have been discussed in detail in this chapter along with model enhancements to capture the mentioned effects.

4.2 Core model

As described in Chapter 3, the BSIM-CMG core model is able to capture the electrostatics of FinFETs with nonuniform trapezoidal FinFETs through the unified FinFET compact model. This core model is obtained from the solution of Poisson's equation for DG and Cy-GAA FinFETs, leading to a single closed-form relationship between the mobile charge and the applied terminal voltages given as follows (see Chapter 3):

FinFET/GAA Modeling for IC Simulation and Design. https://doi.org/10.1016/B978-0-323-95729-8.00003-9

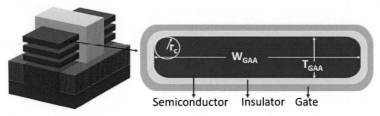

FIG. 4.1

Pictorial representation of a nanosheet structure showing corner rounding. The stacked nanosheet GAAFET structure is shown alongside for reference.

$$v_g - v_0 - v_{ch} = -q_m + \ln(-q_m) + \ln\left(\frac{q_t^2}{e^{q_t} - q_t - 1}\right) \tag{4.1}$$

where

$$v_0 = v_{fb} - q_{dep} - \ln\left(\frac{2qn_i^2 A_{ch}}{v_T C_{ins} N_{ch}}\right) \tag{4.2}$$

and

$$q_t = (q_m + q_{dep})r_N \tag{4.3}$$

These equations are written in terms of normalized quantities: gate potential (v_g), channel potential (v_{ch}), mobile charges (q_m), and depletion charges (q_{dep}). The normalization factors are given in Section 3.1. Also, A_{ch} is the area of the channel cross-section, N_{ch} is the doping in the channel, W is the effective electrical width of the channel, and C_{ins} is the insulator capacitance.

The fact that the unified core model has roots in DG and Cy-GAA FETs allows us to use it for GAAFETs. Based on the cross-sectional shape of a GAA channel, they can be classified as

(a) nanowires if the physical width (W_{GAA}) and thickness (T_{GAA}) of the channel are roughly equal, that is, $W_{GAA} \approx T_{GAA}$ or
(b) nanosheets if the physical width of the channel is much larger than the physical thickness of it, that is, $W_{GAA} \gg T_{GAA}$.

Since the electrostatics of the nanowire case, that is, Cyl-GAA, is already built into the unified core model, we simply need to provide appropriate values of W, A_{ch}, and C_{ins} as given in the following equations, to capture the behavior of this type of device.

$$W = 2\pi R \tag{4.4}$$

$$A_{ch} = \pi R^2 \tag{4.5}$$

and

$$C_{ins} = \frac{2\pi \epsilon_{ins}}{\ln\left(1 + \dfrac{T_{ins,eff}}{R}\right)} \tag{4.6}$$

where $R = W_{Si}/2 = T_{Si}/2$ is the radius of the circular cross section, $T_{ins,eff}$ is the effective insulator thickness, and ϵ_{ins} is the insulator permittivity.

On the other hand, the nanosheet structures can be thought of as nonuniform FinFETs lying on their sides. The unified model is able to capture this behavior in the same way as it does for nonuniform FinFETs with appropriate W, A_{ch}, and C_{ins}, similar to FinFETs, as given follows:

$$W = 2W_{GAA} + 2T_{GAA} \tag{4.7}$$

$$A_{ch} = W_{GAA}T_{GAA} \tag{4.8}$$

and

$$C_{ins} = \frac{\epsilon_{ins}}{T_{ins,eff}} \tag{4.9}$$

In the extreme case of a very wide and thin sheet, the situation is electrically similar to that of a DG-FET.

4.2.1 Core enhancement: Geometry

Although the existing core can capture the fundamental behavior of GAA devices, it needs to be enhanced to be able to capture all practical effects seen in real devices. First, note that there can be devices at the transition between the wide and thin nanosheets and nanowires where the physical width is larger than the physical thickness, but has comparable values. The electrostatics of these devices will lie somewhere in between the two extremes of nanowires and nanosheets. This brings in a complicated geometry dependence.

Second, the effects of corner rounding for nanosheets become a significant contributor to the electrostatics. A pictorial representation of corner rounding is shown in Fig. 4.1. This results in complicated field, potential, and charge variation at the corners, which often result in significant effects on terminal characteristics. This is especially true if the corner rounding is comparable to the width or thickness of the nanosheet. The impact of corner rounding on device characteristics is shown in Fig. 4.2.

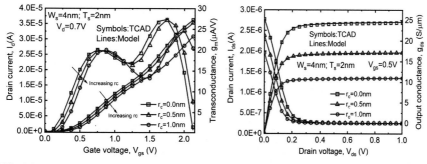

FIG. 4.2

Impact of corner rounding on device characteristics. Increasing corner rounding results in more quantum confinement.

Third, practical nanosheet GAAFETs have multiple channels stacked in a single fin, as shown in Fig. 4.1. This is done to improve the current output per unit footprint area.

To account for all of these nonidealities, we have a separate geometry module, GEOMOD = 5, that accounts for GAAFETs. This module uses the same core as described earlier, but with W and A_{ch} calculation as follows [13]:

$$W = \sum_{i=1}^{N_{GAA}} 2(W_{GAA} + T_{GAA}) - \Delta_{W,i} \tag{4.10}$$

and

$$A_{ch} = \sum_{i=1}^{N_{GAA}} W_{GAA} T_{GAA} - \Delta_{A,i} \tag{4.11}$$

where N_{GAA} is the number of stacked GAA structures per fin while $\Delta_{W,i}$ and $\Delta_{A,i}$ are correction factors for electrical width and area calculation for the ith GAA structure. The correction factors are particularly helpful to account for any nonidealities in the structures, including corner rounding. Also, the model allows different correction factors for each of the stacked GAA sheets. This allows us to capture sheet-to-sheet variation within the same fin.

As an example, let us consider a structure with rounded corners, as shown in Fig. 4.1. Let the corner-rounding radii be r_1, r_2, r_3, and r_4 for the four corners. The effective electrical width can be calculated as

$$W = \left(2W_{GAA} - \sum_{k=1}^{4} r_k\right) + \left(2T_{GAA} - \sum_{k=1}^{4} r_k\right) + \frac{\pi}{2} \sum_{k=1}^{4} r_k \tag{4.12}$$

which gives us the width correction factor as

$$\Delta_W = \left(\frac{\pi}{2} - 2\right) \sum_{k=1}^{4} r_k \tag{4.13}$$

Similarly, the cross-sectional area of the channel can be calculated as

$$A_{ch} = \left(W_{GAA} T_{GAA} - \sum_{k=1}^{4} r_k^2\right) + \frac{\pi}{4} \sum_{k=1}^{4} r_k^2 \tag{4.14}$$

which gives us the area correction factor as

$$\Delta_A = \left(\frac{\pi}{4} - 1\right) \sum_{k=1}^{4} r_k^2 \tag{4.15}$$

We can account for any other geometrical nonideality, including variations between stacked sheets, using the correction factors in a similar way.

4.2.2 Core enhancement: Electrostatic quantum effects

GAAFET channels have higher confinement compared to FinFETs due to the gate-all-around structure. Moreover, at advanced technology nodes, the device geometries

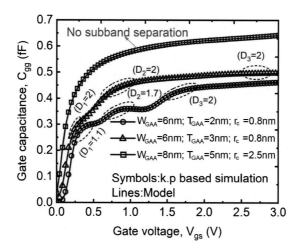

FIG. 4.3

Impact of quantum effects on device electrostatics. Higher confinement leads to a change in the density of states, subband energies, and subband separation.

are so scaled that the quantum mechanical confinement effects become significant. For confined GAAFETs, the quantum mechanical effects result in the conduction band splitting into multiple subbands and changes in the DOS as well as the effective mass (m^*). These not only affect the device electrostatics but also the transport, resulting in a significant impact on terminal characteristics. Fig. 4.3 shows the impact on gate capacitance due to these effects for a confined structure. The well-known effect of centroid shift, as has been discussed in the previous chapter, is also present in GAAFETs.

To understand and model these effects, consider the charge in a semiconductor, which can be calculated as a sum of the charge contributions from individual subbands and is given as [13–15]

$$Q = \sum_i q \int_{E_i}^{\infty} g_i(E) \frac{1}{1 + \exp\left(\dfrac{E - E_f}{kT}\right)} dE \tag{4.16}$$

where E_i denotes the ith subband energy, g_i is the DOS corresponding to the ith subband, q is the electronic charge, E_f is the Fermi level, k is the Boltzmann constant, and T is the temperature. The integration in Eq. (4.16) can be solved to get the total charge as [13–15]

$$Q = \sum_i N_{Ci} F_{\frac{D_i}{2} - 1}\left(\frac{E_f - E_i}{kT}\right) \tag{4.17}$$

where D_i is the electrostatic dimension for the ith subband and F_j is the Fermi integral of the jth order. N_{Ci} is the effective DOS for the ith subband and is given as [13–15]

$$N_{Ci} = q \frac{D_i(m_i^*)^{D_i/2}}{(2\pi)^{D_i/2} \Gamma\left(1 + \frac{D_i}{2}\right) \hbar^{D_i}} \qquad (4.18)$$

where m_i^* is the effective mass corresponding to the ith subband, $\hbar = \frac{h}{2\pi}$ with h being the Plank's constant, and $\Gamma()$ is the gamma function.

Note that this formulation has two bottle necks to be implemented as a compact model: the Fermi integral and the gamma function, both of which do not have closed-form analytical formulations. To solve this issue, the Fermi function for any order j can be approximated as [13]

$$F_j(x) = a_j H(x + 1.5, 2, \delta_j)^j e^{x+1.5-H(x+1.5, 2, \delta_j)} \qquad (4.19)$$

where $a_j = 1.16 + 0.07(j + 1) - 0.23(j + 1)^2$ and $\delta_j = 4 + 2j$. Also, $H(x, x_0, \delta)$ is a smoothing function given as

$$H(x, x_0, \delta) = \frac{x + x_0 + \sqrt{(x - x_0)^2 + \delta}}{2} \qquad (4.20)$$

The accuracy of this approximation is demonstrated in Fig. 4.4. The exponential term is responsible for accurate behavior in the negative x, which maps to the subthreshold region of operation for FETs. Moreover, this approximation also preserves the temperature dependence of the Fermi function, a critical consideration to be able to capture the temperature dependence of charges and terminal characteristics accurately. Finally, this approximation is infinitely differentiable, thus getting rid of any convergence issues for higher-order derivatives of charges and current and ensuring high accuracy for radio frequency (RF) and harmonics simulations.

The gamma function is approximated as [13]

$$\Gamma(x) = a_e e^{-4.6(x-1)} + a_7(2x - 3)^8 + \sum_{p=0}^{6} (-1)^p a_p (2x - 2)^p \qquad (4.21)$$

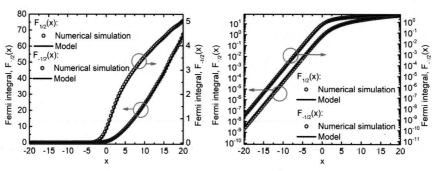

FIG. 4.4

Validation of the analytical model for the Fermi function.

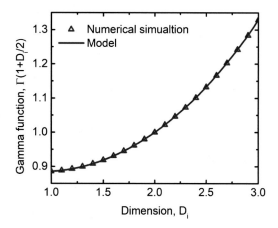

FIG. 4.5

Validation of the analytical model for the Gamma function.

where $a_e = 0.0385$, $a_7 = 7.5893 \times 10^{-7}$, $a_6 = 6.9583 \times 10^{-5}$, $a_5 = 6.583 \times 10^{-4}$, $a_4 = 0.0065$, $a_3 = 0.026$, $a_2 = 0.1371$, $a_1 = 0.194$, and $a_0 = 0.959$. Fig. 4.5 shows the accuracy of this gamma function approximation in our range of interest dictated by the variation in the electrostatic dimension, which lies between one and three for FETs. Note that the approximation has been formulated to not have any convergence issues for arbitrarily high orders of derivatives and to retain good accuracy at least up to the ninth-order derivative of charge.

4.2.3 Width scaling

It is important that our charge model accurately captures the channel geometry dependence. It is especially important to capture the dependence on nanosheet width (W_{GAA}) since this is a design parameter [12,13]. Note that, reducing the width increases the quantum confinement, which affects the subband energies and their separation. It also affects the DOS for each subband through a change in the electrostatic dimension corresponding and the effective mass. The effect of width scaling is captured through geometry-dependent models for dimension, effective mass, and subband energy in Eq. (4.16).

4.2.3.1 Dimension

Decreasing the nanosheet width results in an increase in the quantum confinement, which in turn reduces the electrostatic dimension (D_i). The electrostatic dimension can actually be extracted from the band structure of the semiconductor channel. Note that, for practical devices, as width scales continuously, there is a continuous variation in the band structure, which results in a continuous variation in the electrostatic dimension. Unlike existing notions, there is no sharp transition from 1D/2D to 2D/3D. The rate of change depends on various factors such as material, other confining

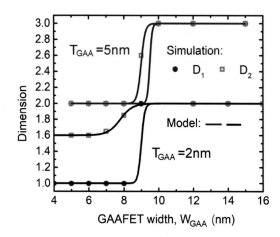

Validation of model for continuously variable dimension.

dimensions, etc. The variation in the first-principle-based band-structure-extracted electrostatic dimension is shown in Fig. 4.6. This variation is modeled as

$$D_i = D_{iL} + \frac{D_{iH} - D_{iL}}{1 + \exp\left(\dfrac{W_{GAA0,i} - W_{GAA}}{D_{r,i}}\right)} \tag{4.22}$$

where D_{iL} and D_{iH} represent the dimensions for the ith subband at small and large nanosheet widths, respectively. $W_{GAA0,i}$ is the critical nanosheet width at which the dimension change happens for the ith subband and $D_{r,i}$ captures the rate of change with width scaling for this subband. The validation of this model with first-principle based simulations is shown in Fig. 4.6.

4.2.3.2 Effective mass

Changing the quantum confinement, by changing the channel geometry, affects the band structure which results in a change of the effective mass and the band gap. It is well known that the effective mass increases with increasing confinement [13–16]. The dependence of effective mass for the ith subband on the width can be modeled as

$$m_i^* = m^* + \frac{\Delta_{m,i}}{1 + \kappa_m W_{GAA}^{\gamma_m}} \tag{4.23}$$

where m^* is the effective mass for the bulk channel material.

4.2.3.3 Subband energies

The subband energies and their separation depend on the quantum confinement. For GAAFETs with small channel cross-sections, the quantum confinement is primarily dominated by geometry confinement. Electrical confinement only comes into play when geometry confinement is much smaller, or in other words, the cross-sectional geometry is large. For nanosheet GAAFETs which have cross-sections small enough to show subband separation, electrical confinement has a minor role and the subband energies are mostly dominated by geometry confinement. The subband energies can be modeled as a function of the cross-sectional width as

$$E_i = G_{c,i} \frac{i^2 h^2}{m_i^* W_{GAA}^{\gamma_G}} \tag{4.24}$$

This formulation is adopted from the solution of the Schrödinger equation for a particle in a 1D box. However, the quantum confinement in GAAFETs is 2D. Moreover, the assumptions of infinite potential boundary are invalid for practical devices. Also, effects of corner rounding, potential perturbation due to applied biases, and the resulting correlation effects are not accounted for in the simple particle-in-a-box model. To compensate for these approximations, $G_{c,i}$ and γ_G are used as parameters.

For added accuracy, the impact of electrical confinement can be included as [14]

$$E_i = H(E_{i,\text{geometry}}, E_{i,\text{electrical}}, \delta) \tag{4.25}$$

where $H()$ is defined in Eq. (4.20) and δ is a smoothing parameter also defined in Eq. (4.20). $E_{i,\text{geometry}}$ is the geometry confinement-dependent subband energy as calculated in Eq. (4.24). $E_{i,\text{electrical}}$ is the electrical confinement-dependent subband energy, which can be given as [14]

$$E_{i,\text{electrical}} = E_c \left[\frac{(\psi_s + \psi_d)^2 \hbar^2}{m_i^* W_{GAA}^{\gamma_E}} \right]^{1/3} \zeta_i \tag{4.26}$$

where ψ_s and ψ_d are the surface potentials on the source and drain sides, respectively, of the channel. ζ_i is the ith zero of the Airy function and can be approximated analytically as

$$\zeta_i = -\left[\frac{3\pi}{2} \left(i - \frac{1}{4} \right) \right]^{2/3} \tag{4.27}$$

The form of Eq. (4.26) is derived using the idea of noninteracting triangular potential wells, which is not strictly applicable for practical GAAFETs. We have also assumed an infinite potential boundary at the semiconductor insulator interface, which is also not accurate. To compensate for the errors due to these approximations, the terms E_c and γ_E are introduced as parameters. Fig. 4.7 shows the variation in subband energy from first-principle simulations and the model.

4.2.3.4 Threshold voltage

With a reduction in nanosheet width, the threshold voltage also increases. This is due to the increase in band gap as well as the increase in the first subband energy due to increasing quantum confinement. The model for a quantum mechanical shift in threshold voltage is described in Section 5.4. Fig. 4.8 shows the variation in the threshold voltage with change in cross-sectional geometry.

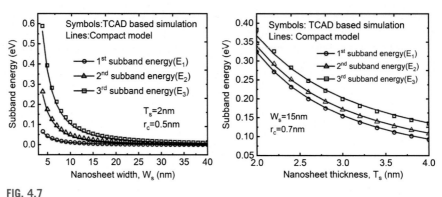

FIG. 4.7

Variation of subband energy with changing cross-sectional dimensions.

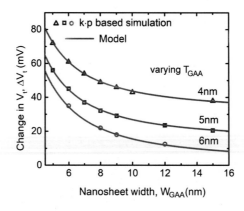

FIG. 4.8

Variation of threshold voltage with cross-sectional dimension. Smaller width implies larger quantum confinement and higher threshold shift.

4.2.4 Thickness dependence

For nanosheet GAAFETs, the channel thickness (T_{GAA}) also contributes to the quantum confinement. As a result, the subband energies, the effective masses of the subbands, and the threshold voltage are thickness dependent. It is important that the compact model captures these dependencies. Even though the channel thickness is supposed to be a constant for a given technology, there are often minor thickness variations from sheet to sheet due to fabrication variability. For efficient circuit design, the compact model needs to capture this variability information and accurately predict its effect on terminal characteristics.

4.2.4.1 Effective mass

To model the thickness variability in effective mass, the terms $\Delta_{m,i}$ and γ_m in Eq. (4.23) are made functions of thickness as

$$\Delta_{m,i} = \frac{\Delta_{0},i}{T_{\mathrm{GAA}}^{\alpha_m}} \tag{4.28}$$

$$\gamma_m = \frac{\gamma_0}{T_{\mathrm{GAA}}^{\beta_m}} \tag{4.29}$$

where $\Delta_{0,i}$, γ_0, α_m, and β_m are parameters, which have a fixed value for a given technology node.

4.2.4.2 Subband energies

Decreasing the thickness reduces the overall cross-sectional area of the channel which, in turn, increases quantum confinement. This results in an increase in subband energies and subband separation with a reduction in channel thickness. Moreover, the width dependence of the subband energies also depends on the thickness. This is because, for a practical nanosheet, the width and thickness direction are coupled to create an overall 2D confinement.

The thickness dependence of subband energies is modeled by making the terms $G_{\mathrm{c},i}$, E_{c}, γ_{G}, and γ_{E} in Eqs. (4.24) and (4.26) functions of channel thickness as

$$G_{\mathrm{c},i} = \frac{G_{\mathrm{c}0,i}}{T_{\mathrm{GAA}}^{\alpha_{\mathrm{G},i}}} \tag{4.30}$$

$$E_{\mathrm{c}} = \frac{E_{\mathrm{c}0}}{T_{\mathrm{GAA}}^{\alpha_{\mathrm{E}}}} \tag{4.31}$$

$$\gamma_{\mathrm{G}} = \frac{\gamma_{\mathrm{G}0}}{T_{\mathrm{GAA}}^{\beta_{\mathrm{G}}}} \tag{4.32}$$

$$\gamma_{\mathrm{E}} = \frac{\gamma_{\mathrm{E}0}}{T_{\mathrm{GAA}}^{\beta_{\mathrm{E}}}} \tag{4.33}$$

where $G_{\mathrm{c}0,i}$, $E_{\mathrm{c}0}$, $\gamma_{\mathrm{G}0}$, $\gamma_{\mathrm{E}0}$, $\alpha_{\mathrm{G},i}$, α_{E}, β_{G}, and β_{E} are parameters that have a fixed value for a given technology node.

4.2.5 Implementation

These geometry-dependent quantum mechanical effects can be turned on by selecting $SUBBANDMOD = 1$. Once activated, the geometry-dependent effective mass, dimension, and subband energies are used to calculate the effective normalized charge density in the semiconductor at the source side ($q_{\mathrm{nds},i}$) and at the drain side ($q_{\mathrm{ndd},i}$) as per Eq. (4.17). The initial estimate for the source and drain side charge densities ($q_{m,\mathrm{s}}$ and $q_{m,\mathrm{d}}$, respectively) are modified using a weighted average as

$$q_{m,\mathrm{s(new)}} = N_{\mathrm{c,3D}}q_{m,\mathrm{s}} + N_{\mathrm{c},q}\sum_{i=1}^{N_{\mathrm{GAA}}} q_{\mathrm{nds},i} \tag{4.34}$$

$$q_{m,\mathrm{d(new)}} = N_{\mathrm{c,3D}}q_{m,\mathrm{s}} + N_{\mathrm{c},q}\sum_{i=1}^{N_{\mathrm{GAA}}} q_{\mathrm{ndd},i} \tag{4.35}$$

where q_m is the charge calculated as per the unified model discussed in Section 3.2. The weights are given as

$$N_{c,3D} = \max\left(\frac{1}{1 + \exp\left(\dfrac{2.75 - T_{GAA} \times 10^9}{0.78}\right)}, 0.5\right) \tag{4.36}$$

$$N_{c,q} = \frac{1}{1 + \exp\left(\dfrac{N_{c,3D} - 0.999}{10^{-4}}\right)} \tag{4.37}$$

The rest of the iterative calculation is the same as discussed in Chapter 3.

4.3 Core enhancement: Mobility

Cross-sectional geometry has a significant effect on the mobility in nanosheet GAA-FETs [11,16,17]. This is in part due to the confined nature of the gate-all-around cross-section, which results in quantum mechanical effects as well as an impact on the various scattering components and their field dependence, and in part due to the construction with gate-on-all-sides resulting in current flow along various crystal orientations. The accurate geometry-dependent mobility model includes all of these effects. The general field-dependent mobility model will be discussed in Section 5.5. The framework can be roughly described as

$$\mu_{\text{eff}} = \frac{U0}{1 + UA \cdot E_{\text{eff}}^{EU} + \dfrac{UD}{\left[\dfrac{1}{2}\left(1 + \dfrac{qia}{qb}\right)\right]^{UCS}}} \tag{4.38}$$

4.3.1 Quantum effects

The variation of mobility for a nanosheet GAAFET is shown in Fig. 4.9. From these experimental results [11], it is evident that the low-field mobility drops significantly as confinement increases. This is because increasing confinement results in an increase in the conductivity effective mass, which inversely affects the mobility. The variation in effective mass can be modeled, based on an approximate $k \cdot p$ formulation [18], as a function of the channel thickness as [16]

$$\frac{m^*}{m_0} = S_m \frac{\kappa_1 + \sqrt{\kappa_1^2 + 4A_{ch}\kappa_2}}{\kappa_2} \tag{4.39}$$

where

$$\kappa_1 = m_0 E_{g,\text{bulk}} T_{GAA}^2 - \frac{h^2}{32 m_0 a_0^2} \tag{4.40}$$

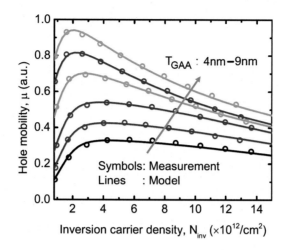

FIG. 4.9

Variation of mobility with nanosheet cross-section.

$$\kappa_2 = m_0 \left(E_{g,\text{bulk}} + \frac{2h^2}{m_0 a_0^2} \right) T_{\text{GAA}}^2 \tag{4.41}$$

Here, m_0 is the rest mass of the charge carrier, a_0 is the lattice constant, and $E_{g,\text{bulk}}$ is the band gap of the bulk channel material. S_m is a correction factor that compensates for the errors due to the approximations involved in the derivation of this effective mass variation. The low-field mobility can then be written as

$$U0_{\text{new}} = U0 \frac{m_0}{m^*} \tag{4.42}$$

The variation of the experimentally extracted peak mobility [11], along with the results from the model [16], is shown in Fig. 4.10.

4.3.2 Field dependence

Fig. 4.9 shows that not only does the low-field mobility reduce with increasing confinement, but the electric field dependence of the mobility also changes [11,16]. The field-dependent degradation is due to the combined effects of various scattering types: primarily the phonon and surface-roughness scatterings captured by the second term in the denominator of Eq. (4.38) and the Coulomb scattering captured by the third term in the denominator of Eq. (4.38). With increasing confinement, a change in the effective field distribution as well as the scattering rates is expected. Moreover, the centroid shift is a function of the geometry. For confined structures, the centroid shifts away from the interface leading to less surface roughness. All of these effects together make the bias-dependent degradation of mobility a function of the cross-sectional geometry.

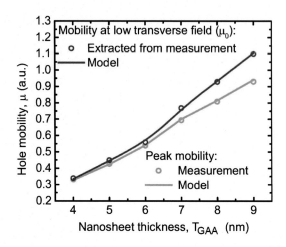

FIG. 4.10

Variation of peak mobility with changing nanosheet thickness. This is primarily due to the impact of quantum confinement.

To capture these effects, we modify UA, EU, and UD as

$$UA_{\text{new}} = UA_{\text{thin}} + \frac{UA + UA_{\text{thin}} + (T_{\text{GAA}} - UA_{\text{TSAT}})UA_{\text{RTSC}}}{1 + \exp\left(\dfrac{UA_{\text{TNI}} - T_{\text{GAA}}}{UA_{\text{IR}}}\right)} \tag{4.43}$$

$$EU_{\text{new}} = \min\left[370\frac{EU_{\text{thin}} - EU}{(T_{\text{GAA}} \cdot 10^9)^{EU_{\text{PTSC}}}} + \frac{EU_{\text{thin}} - EU}{1 + \exp\left(\dfrac{EU_{\text{thin}} - EU_{\text{TNI}}}{EU_{\text{IR}}}\right)} + EU, EU_{\text{thin}}\right] \tag{4.44}$$

$$UD_{\text{new}} = UD + (UD_{\text{thin}} - UD)[\max(UD_{\text{TSAT}} - T_{\text{GAA}}, 0)]^{UD_{\text{PTSC}}} \tag{4.45}$$

Here, UD_{thin}, EU_{thin}, and UA_{thin} refer to the limiting values of the corresponding parameters for ultrathin nanosheets. These are input parameters along with the regular UD, EU, and UA to allow users to tune the range of variation better.

Fig. 4.9 shows that this model is able to match experimental data accurately.

4.3.3 Crystal orientation

Another significant geometry-dependent impact on mobility in nanosheets is due to the different crystal orientations of the current flow paths [11,17]. This effect is specific to nanosheet GAAFETs due to the gate-all-around structure and affects mobility both with width and thickness scaling. Fig. 4.11 shows experimental data where increasing width results in higher current for n-type devices but lower current for p-type devices.

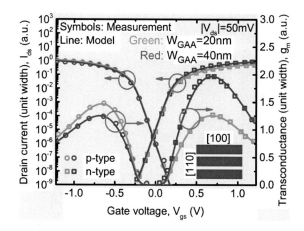

FIG. 4.11

Effect of crystal orientation on the scaling trends in nanosheet FETs.

The origin of this effect is the difference in crystal orientations for the top/bottom surfaces to the sidewalls. As a result, the effective mobility depends on the relative magnitudes of the mobilities along these different crystal directions weighted by the effective width of the current paths along those directions. This can be modeled as

$$\mu_{\text{eff}} = \mu_{\text{tb}} \left[\frac{W_{\text{GAA}}}{W_{\text{GAA}} + T_{\text{GAA}}} + \frac{\mu_{\text{sw}}}{\mu_{\text{tb}}} \frac{T_{\text{GAA}}}{W_{\text{GAA}} + T_{\text{GAA}}} \right] \qquad (4.46)$$

where μ_{tb} is the mobility along the top/bottom surface and μ_{sw} is the mobility along the sidewalls. The ratio $\frac{\mu_{\text{sw}}}{\mu_{\text{tb}}}$ is treated as a parameter η_{μ}.

Fig. 4.11 shows the validation of this model with experimental data. In this case $\mu_{\text{tb}} = \mu_{100}$, $\mu_{\text{sw}} = \mu_{110}$, and $\eta_{\mu} = 0.48$ for electrons and 1.5 for holes. The different trends in n-type and p-type devices are due to η_{μ} values being less than and greater than unity, respectively. Moreover, Fig. 4.12 shows further validation of this model with experimental results from nanosheets using reported η_{μ}.

4.3.4 Implementation

The overall updated mobility model includes all of the mentioned changes as

$$\mu_{\text{eff}} = \frac{U0_{\text{new}} \left[\frac{W_{\text{GAA}}}{W_{\text{GAA}} + T_{\text{GAA}}} + \eta_{\mu} \frac{T_{\text{GAA}}}{W_{\text{GAA}} + T_{\text{GAA}}} \right]}{1 + UA_{\text{new}} \cdot E_{\text{eff}}^{EU_{\text{new}}} + \frac{UD_{\text{new}}}{\left[\frac{1}{2} \left(1 + \frac{qia}{qb} \right) \right]^{UCS}}} \qquad (4.47)$$

This geometry-dependent mobility can be turned on by selecting *MOBSCMOD* = 1. The accuracy and utility of this model is illustrated by Fig. 4.13, where the dotted lines, which over predict the current, correspond to older versions of the model without geometry-dependent mobility (*MOBSCMOD* = 0). However, the updated model presented here (solid lines) is able to capture the experimental trends accurately.

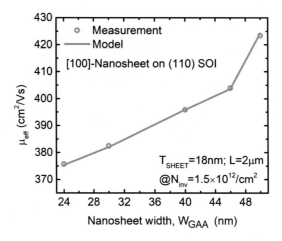

FIG. 4.12

Variation of mobility with nanosheet width due to the crystal-orientation impact.

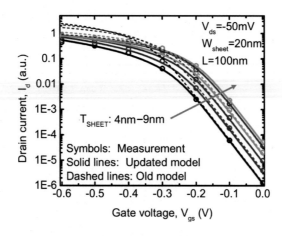

FIG. 4.13

Validation of the updated mobility model through terminal current for different nanosheet thicknesses.

References

[1] H. Mertens, et al., Vertically stacked gate-all-around Si nanowire CMOS transistors with dual work function metal gates, in: IEDM Tech. Dig., 2016, pp. 19.7.1–19.7.4, https://doi.org/10.1109/IEDM.2016.7838456.

[2] M. Karner, et al., Vertically stacked nanowire MOSFETs for sub-10 nm nodes: advanced topography, device, variability, and reliability simulations, in: IEDM Tech. Dig., 2016, pp. 30.7.1–30.7.4, https://doi.org/10.1109/IEDM.2016.7838516.

[3] N. Singh, et al., Ultra-narrow silicon nanowire gate-all-around CMOS devices: impact of diameter, channel-orientation and low temperature on device performance, in: IEDM Tech. Dig., 2006, pp. 1–4, https://doi.org/10.1109/IEDM.2006.346840.

[4] Y. Jiang, et al., Performance breakthrough in 8 nm gate length Gate-All-Around nanowire transistors using metallic nanowire contacts, in: Proc. Symp. VLSI Technol., 2008, pp. 34–38, https://doi.org/10.1109/VLSIT.2008.4588553.

[5] Y. Cui, Z. Zhong, D. Wang, W.U. Wang, C.M. Lieber, High performance silicon nanowire field effect transistors, Nano Lett. 3 (2) (2003) 149–152, https://doi.org/10.1021/nl0258751.

[6] N. Singh, et al., High-performance fully depleted silicon nanowire (diameter $\leq$ 5 nm) gate-all-around CMOS devices, IEEE Electron Device Lett. 27 (5) (2006) 383–386, https://doi.org/10.1109/LED.2006.873381.

[7] N. Loubet, et al., Stacked nanosheet gate-all-around transistor to enable scaling beyond FinFET, in: Proc. Symp. VLSI Technol., 2017, pp. T230–T231, https://doi.org/10.23919/VLSIT.2017.7998183.

[8] K.H. Yeo, et al., Gate-all-around (GAA) twin silicon nanowire MOSFET (TSNWFET) with 15 nm length gate and 4 nm radius nanowires, in: IEDM Tech. Dig., 2006, pp. 1–4, https://doi.org/10.1109/IEDM.2006.346838.

[9] S. Barraud, et al., Tunability of parasitic channel in gate-all-around stacked nanosheets, in: IEDM Tech. Dig., 2018, pp. 21.3.1–21.3.4, https://doi.org/10.1109/IEDM.2018.8614507.

[10] R. Ritzenthaler, et al., Vertically stacked gate-all-around Si nanowire CMOS transistors with reduced vertical nanowires separation, new work function metal gate solutions, and DC/AC performance optimization, in: IEDM Tech. Dig., 2018, pp. 21.5.1–21.5.4, https://doi.org/10.1109/IEDM.2018.8614528.

[11] C.W. Yeung, et al., Channel geometry impact and narrow sheet effect of stacked nanosheet, in: IEDM Tech. Dig., 2018, pp. 28.6.1–28.6.4, https://doi.org/10.1109/IEDM.2018.8614608.

[12] G. Bae, et al., 3 nm GAA technology featuring multi-bridge-channel FET for low power and high performance applications, in: IEDM Tech. Dig., 2018, pp. 28.7.1–28.7.4, https://doi.org/10.1109/IEDM.2018.8614629.

[13] A. Dasgupta, S.S. Parihar, P. Kushwaha, H. Agarwal, M.-Y. Kao, S. Salahuddin, Y.S. Chauhan, C. Hu, BSIM compact model of quantum confinement in advanced nanosheet FETs, IEEE Trans. Electron Devices 67 (2) (2020) 730–737, https://doi.org/10.1109/TED.2019.2960269.

[14] A. Dasgupta, P. Rastogi, A. Agarwal, C. Hu, Y.S. Chauhan, Compact modeling of cross-sectional scaling in gate-all-around FETs: 3-D to 1-D transition, IEEE Trans. Electron Devices 65 (3) (2018) 1094–1100, https://doi.org/10.1109/TED.2018.2797687.

[15] A. Dasgupta, A. Agarwal, Y.S. Chauhan, Unified compact model for nanowire transistors including quantum effects and quasi-ballistic transport, IEEE Trans. Electron Devices 64 (4) (2017) 1837–1845, https://doi.org/10.1109/TED.2017.2672207.

[16] A. Dasgupta, S.S. Parihar, H. Agarwal, P. Kushwaha, Y.S. Chauhan, C. Hu, Compact model for geometry dependent mobility in nanosheet FETs, IEEE Electron Device Lett. 41 (3) (2020) 313–316, https://doi.org/10.1109/LED.2020.2967782.

[17] J. Wang, A. Rahman, G. Klimeck, M. Lundstrom, Bandstructure and orientation effects in ballistic Si and Ge nanowire FETs, in: IEDM Tech. Dig., 2005, p. 533, https://doi.org/10.1109/IEDM.2005.1609399.

[18] M. Willatzen, L.C.L.Y. Voon, The K P Method, Springer, Berlin, Germany, 2009, https://doi.org/10.1007/978-3-540-92872-0.

Channel current and real device effects

5.1 Introduction

This chapter discusses channel current modeling in the BSIM-CMG model. Modeling of the core current valid for wide-long double-gate MOSFET has already been discussed in Chapter 3. Some of the important effects affecting channel current in short-/medium-channel devices are described here. Short-channel effects (SCEs) in FinFETs are the same as those present in bulk MOSFETs but less severe due to better electrostatic control on the channel from multiple gates in a FinFET. Some of the dominant effects are

- threshold voltage (V_{th}) roll-off;
- subthreshold slope (n) degradation;
- mobility degradation due to vertical electric field;
- carrier velocity saturation;
- series resistance effect (see Chapter 7);
- drain-induced barrier lowering (DIBL);
- channel length modulation (CLM); and
- quantum mechanical effects (QMEs)

5.2 Threshold voltage modulation

In a short-channel device, the drain starts affecting the potential barrier seen by carriers while entering from the source side as the source is in close proximity with the drain. Although there are numerous models for the modeling of threshold voltage roll-off in double-gate devices, BSIM-CMG uses a physical but simple model such that it is computationally efficient and implementable.

SCEs originate due to the influence of the two-dimensional (2D) field in the channel. The 2D Poisson's equation inside the body in the subthreshold regime, where inversion carriers can be ignored compared to bulk charge, can be written as

$$\frac{d^2\psi(x,y)}{dx^2} + \frac{d^2\psi(x,y)}{dy^2} = \frac{qN_A}{\epsilon_{Si}} \tag{5.1}$$

where N_A is the channel doping concentration. Here, the x-axis is perpendicular and y-axis is parallel to the channel. $\psi(x, y)$ is the electrostatic potential at any point (x, y) in the channel.

113

FinFET/GAA Modeling for IC Simulation and Design. https://doi.org/10.1016/B978-0-323-95729-8.00005-2

5.2.1 Characteristic length

The characteristic field penetration length, also known as scale length, is an important parameter, which defines the amount of SCE in the transistor and captures the variation in the amount of drain field penetrating into the silicon body. It is a function of physical parameters such as T_{Si} and T_{ox} [1,2].

To develop the characteristic length model, a parabolic potential profile in the channel perpendicular to the silicon-insulator interface (x-direction in Fig. 5.1) is assumed as follows:

$$\psi(x,y) = C_0(y) + C_1(y)x + C_2(y)x^2 \tag{5.2}$$

where C_0, C_1, and C_2 are independent of x but are functions of y. These can be obtained by applying three boundary conditions along the silicon body.

$$\psi\left(x = \frac{T_{Si}}{2}, y\right) = \psi_s(y) \tag{5.3}$$

$$\left.\frac{d\psi(x,y)}{dx}\right|_{x=0} = 0 \tag{5.4}$$

$$\left.\frac{d\psi(x,y)}{dx}\right|_{x=\frac{T_{Si}}{2}} = \frac{V_{gs} - V_{fb} - \psi_s}{T_{ox}}\frac{\epsilon_{ox}}{\epsilon_{Si}} \tag{5.5}$$

where ψ_s is the surface potential. Using the parabolic profile in the vertical direction and applying the boundary condition of $\frac{d\psi}{dx} = 0$ for $x = \frac{T_{Si}}{2}$, we obtain $\psi(x, y)$ as

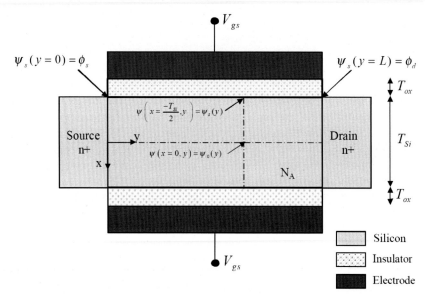

FIG. 5.1

Schematic representation of a symmetric common-gate DG-FET.

$$\psi(x,y) = \psi_s(y) + \frac{\epsilon_{ox}}{\epsilon_{Si}} \frac{\psi_s(y) - (V_{gs} - V_{fb})}{T_{ox}} x - \frac{\epsilon_{ox}}{\epsilon_{Si}} \frac{\psi_s(y) - (V_{gs} - V_{fb})}{T_{ox}T_{Si}} x^2 \tag{5.6}$$

If ψ_c is the potential distribution along the channel at the center, the relation between ψ_s and ψ_c will be obtained by substituting $x = \dfrac{T_{Si}}{2}$

$$\psi_s(y) = \frac{1}{1 + \dfrac{\epsilon_{ox}}{4\epsilon_{Si}}\dfrac{T_{Si}}{T_{ox}}} \left[\psi_c(y) + \frac{\epsilon_{ox}}{4\epsilon_{Si}}\frac{T_{Si}}{T_{ox}}(V_{gs} - V_{fb}) \right] \tag{5.7}$$

Now, $\psi(x,y)$ can be expressed as

$$\psi(x,y) = \left[1 + \frac{\epsilon_{ox}}{\epsilon_{Si}}\frac{x}{T_{ox}} - \frac{\epsilon_{ox}}{\epsilon_{Si}}\frac{x^2}{T_{ox}T_{Si}} \right] \cdot \frac{\psi_c(y) + \dfrac{\epsilon_{ox}}{4\epsilon_{Si}}\dfrac{T_{Si}}{T_{ox}}(V_{gs} - V_{fb})}{1 + \dfrac{\epsilon_{ox}}{4\epsilon_{Si}}\dfrac{T_{Si}}{T_{ox}}}$$
$$- \frac{\epsilon_{ox}}{\epsilon_{Si}}\frac{x}{T_{ox}}(V_{gs} - V_{fb}) + \frac{\epsilon_{ox}}{\epsilon_{Si}}\frac{x^2}{T_{ox}T_{Si}}(V_{gs} - V_{fb}) \tag{5.8}$$

Using Eq. (5.8), the two-dimensional Poisson's equation given in (5.1) can be written as:

$$\frac{d^2\psi_c(y)}{dx^2} + \frac{V_{gs} - V_{fb} - \psi_c(y)}{\lambda^2} = \frac{qN_A}{\epsilon_{Si}} \tag{5.9}$$

$$\lambda = \sqrt{\frac{\epsilon_{ox}}{\epsilon_{Si}}\left(1 + \frac{\epsilon_{ox}T_{Si}}{4\epsilon_{Si}T_{ox}}\right)T_{Si}T_{ox}} \tag{5.10}$$

This expression for characteristic length (λ) is derived for a double-gate FET. Increasing the number of gates in a transistor improves the electrostatic control, thereby decreasing the penetration of the drain field. Thus, λ is different for different multigate architectures. To develop the λ of a triple-gate structure, we need to include the gate control on the channel from the top and the sidewalls. For this, a three-dimensional (3D) Poisson's equation with appropriate boundary conditions needs to be solved. Similarly, we need to consider the control of all four gates in quadruple-gate FETs. Solving Poisson's equation for triple- or quadruple-gate FETs is complex and cannot be used in a compact model. The alternative approach is to use the existing λ of a DG-FET and a single gate to develop the λ for other multigate FETs. The compact form of unified λ developed for all multigate devices and used in BSIM-CMG is as follows [3,4]:

$$\lambda_c = \frac{1}{\sqrt{\left(\frac{1}{\lambda}\right)^2 + \left(\frac{\Lambda}{\lambda_{Hfin}}\right)^2}} \tag{5.11}$$

where

$$\lambda_{Hfin} = \sqrt{\frac{\epsilon_{Si}}{4\epsilon_{ox}}\left(1 + \frac{\epsilon_{ox}T_{Si}}{2\epsilon_{Si}T_{ox}}\right)H_{fin}T_{ox}} \tag{5.12}$$

and

$$\Lambda = 0 \quad \text{for DG} - \text{FET (GEOMOD} = 0) \tag{5.13}$$

$$= \frac{1}{2} \quad \text{for triple} - \text{gate FET (GEOMOD} = 1) \tag{5.14}$$

$$= 1 \quad \text{for surround} - \text{gate FET (GEOMOD} = 2, 3) \tag{5.15}$$

5.2.2 Channel potential

The electrostatic potential profile along the channel is developed using the 2D Poisson equation (5.1) and taking a small rectangular box of width δy and height $\frac{T_{Si}}{2}$ in the channel. After the application of Gauss's law on the rectangular box at position y along the channel, we get the following:

$$\frac{T_{Si}}{2} \cdot [E_y(y) - E_y(y + \delta y)] + \delta y \left[E_x \left(\frac{-T_{Si}}{2} \right) - E_x(0) \right] = q \frac{T_{Si} \delta y N_A}{2\epsilon_{Si}} \tag{5.16}$$

To simplify Eq. (5.16), divide both sides by δy and the result can be written as follows:

$$\frac{T_{Si}}{2} \cdot \frac{[E_y(y) - E_y(y + \delta y)]}{\delta y} + \left[E_x \left(\frac{-T_{Si}}{2} \right) - E_x(0) \right] = q \frac{T_{Si} N_A}{2\epsilon_{Si}} \tag{5.17}$$

With the use of relationship $E_y = -\frac{\partial \psi_s(y)}{\partial y}$; and by replacing $E_x\left(\frac{-T_{Si}}{2}\right)$ and $E_x(0)$ with their values from Eqs. (5.4), (5.5), we can get the following relation from Eq. (5.17):

$$\lambda^2 \cdot \frac{\partial^2 \psi_s(y)}{\partial y^2} = \psi_s(y) - \left(V_{gs} - V_{fb} - \frac{q N_A T_{Si}}{2 C_{ox}} \right) \tag{5.18}$$

where λ is the characteristic length. The surface potentials at the source and drain ends are $\psi_{s0} = V_s + V_{bi}$ and $\psi_{s0} = V_{ds} + V_{bi}$, respectively. Solving Eq. (5.18), we obtain $\psi_s(y)$, and using Eq. (5.7), we can get $\psi_c(y)$ as follows [5]:

$$\psi_c(y) = V_{SL} + (V_{bi} - V_{SL}) \frac{\sinh\left(\frac{L-y}{\lambda}\right)}{\sinh\left(\frac{L}{\lambda}\right)} + (V_{bi} + V_{ds} - V_{SL}) \frac{\sinh\left(\frac{y}{\lambda}\right)}{\sinh\left(\frac{L}{\lambda}\right)} \tag{5.19}$$

where V_{bi} is the built-in potential at the source end and V_{SL} is equivalent to the center potential for a long-channel transistor.

$$V_{SL} = V_{gs} - V_{fb} - \frac{q N_A}{\epsilon_{Si}} \lambda^2 \tag{5.20}$$

As $\sinh(0) = 0$, it is apparent that ψ_c reduces to V_{bi} for $y = 0$ and to V_{DS} for $y = L$. Note the similarity between the center potential given by Eq. (5.19) and the surface potential obtained from similar 2D analysis in bulk MOSFETs [5–7]. For a long-channel transistor ($L \gg \lambda$) in the middle of the channel, the terms in the ratios in Eq. (5.19) are both $e^{-L/2\lambda}$, which is very small; hence, ψ_c approaches

V_{SL}, as expected. For short-channel transistors (L is not large compared to λ), ψ_c will be larger than V_{SL}.

The threshold voltage (V_{th}) is the value of the gate voltage earlier which the surface potential is approximately pinned in strong inversion [5,6]. In all BSIM models, this is denoted by ψ_{st} (see Eq. 5.22). The potential equation (5.19) is minimum at approximately $x = L/2$, and the inversion charge density is also minimum at this point [5,7].

The difference in threshold voltage (ΔV_T) between short- and long-channel transistors is equivalent to the difference in ψ_c in these transistors at the onset of strong inversion such that the value of $\psi_c(x = L/2)$ from Eq. (5.19) is equal to its long-channel value of ψ_{st}.

$$\Delta V_T = -\frac{2(V_{bi} - V_{SL}) + V_{ds}}{2\cosh\left(\dfrac{L}{2\lambda}\right) - 2} \tag{5.21}$$

This expression has two important effects modeled through a single equation: first, the length dependence of V_{th} and second, drain bias effect on V_{th}. To provide more flexibility in the fitting of experimental data, BSIM-CMG models these through separate ΔV_T terms along with different parameters.

5.2.3 Threshold voltage roll-off

The decrease of threshold voltage with decrease in channel length is called threshold voltage roll-off. For V_{th} roll-off modeling at small V_{ds}, we can put $V_{ds} = 0$ in Eq. (5.21). Also, center potential at the source end (V_{SL}) can be taken as surface potential at the source side in the subthreshold region (ψ_{st}). Thus,

$$\psi_{st} = 0.4 + PHIN + \Phi_B \tag{5.22}$$

The V_{th} roll-off in BSIM-CMG is modeled as

$$\Delta V_{th,SCE} = -\frac{\dfrac{1}{2} \cdot DVT0}{\cosh\left(DVT1 \cdot \dfrac{L_{eff}}{\lambda}\right) - 1} \cdot (V_{bi} - \psi_{st}) \tag{5.23}$$

where two parameters, DVT0 and DVT1, have been added for flexibility in parameter extraction. An important point to keep in mind during implementation is numerical robustness. The $\cosh(x)$ used earlier is a smooth function and symmetric around $x = 0$. For large x, it goes to 1 and can cause divide by zero problem. To avoid this issue, BSIM-CMG uses the following approximation for $x > 40$.

$$\frac{\dfrac{1}{2}}{\cosh(x) - 1} = \frac{\dfrac{1}{2}}{\dfrac{e^x + e^{-x}}{2} - 1} \approx -\frac{\dfrac{1}{2}}{\dfrac{e^x}{2}} = e^{-x} \tag{5.24}$$

5.2.4 DIBL effect on threshold voltage

In short-channel devices, the drain voltage starts affecting the peak of the barrier height seen by the carriers, as shown in Chapter 1. This is called drain-induced barrier lowering (DIBL). Decrease in barrier height results in a decrease in threshold voltage. As mentioned earlier, Eq. (5.21) also captures the drain bias effect on threshold voltage. Another impact of DIBL is seen on $I_{ds} - V_{ds}$ characteristics, where it causes a finite slope in $I_{ds} - V_{ds}$ and increases output conductance. This is discussed in Section 5.11.

To model DIBL, Eq. (5.21) is modified as follows with two extra parameters for more flexibility in parameter extraction:

$$\Delta V_{th,DIBL} = -\frac{0.5 \cdot ETA0_a}{\cosh\left(DSUB \cdot \dfrac{L_{eff}}{\lambda}\right) - 1} \cdot V_{dsx} + DVTP0 \cdot V_{dsx}^{DVTP1} \qquad (5.25)$$

The second term on the right-hand side of Eq. (5.25) is added to model long-channel DIBL, also called drain-induced threshold shift (DITS) [5,8].

5.2.5 Reverse short-channel effect

To reduce SCEs in a given technology, halo implants are used near the source and drain. Although FinFETs are considered low-doped devices, there could be intentional halo doping in the channel for threshold voltage control or unintentional doping caused by punch-through implants. Halo doping results in an increase in threshold voltage with decreasing channel length as average doping in the channel increases. This is called reverse SCE.

The reverse SCE model is taken from BSIM4 and is given by

$$\Delta V_{th,RSCE} = K1RSCE \cdot \left[\sqrt{1 + \frac{LPE0}{L_{eff}}} - 1\right] \cdot \sqrt{\psi_{st}} \qquad (5.26)$$

Finally, all effects are combined in a single threshold voltage shift as follows:

$$\Delta V_{th,all} = \Delta V_{th,SCE} + \Delta V_{th,DIBL} + \Delta V_{th,RSCE} + \Delta V_{th,temp} \qquad (5.27)$$

$$V_{gsfb} = V_{gs} - \Delta \Phi - \Delta V_{th,all} - DVTSHIFT \qquad (5.28)$$

Here, $\Delta V_{th,temp}$ is the effect of temperature on threshold voltage and DVTSHIFT is a parameter to model variability.

Fig. 5.2 shows the validation of SCEs for variation in different physical parameters. Fig. 5.3 shows the roll-off in V_{th} extracted from the model for different T_{ox} values and compares it against the V_{th} extracted from 2D TCAD simulations. The smaller the T_{ox}, the closer the gate electrode is to the inversion layer enforcing a stronger electrostatic gate control and hence smaller V_{th} roll-off.

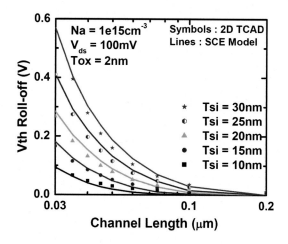

FIG. 5.2

Threshold voltage roll-off of DG-FETs for different TSi. Model demonstrates excellent scalability when compared against 2D TCAD simulations.

Reproduced from M.V. Dunga, Nanoscale CMOS Modeling (Ph.D. thesis), University of California, Berkeley, 2008.

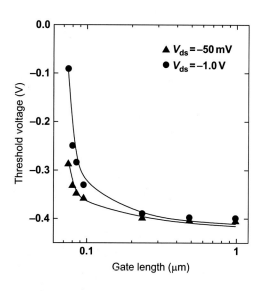

FIG. 5.3

Model validation with the experimental data of threshold voltage variation due to DIBL at $V_{ds} = -1.0\,V$ as compared to $V_{ds} = -1.0\,V$ of a lightly doped body ($N_{ch} = 2 \times 10^{15}\,cm^{-3}$) for a p-type SOI FinFET with $N_{fin} = 20$, $EOT = 2.0\,nm$, $H_{fin} = 60\,nm$, and $T_{fin} = 22\,nm$.

From Y.S. Chauhan, D.D. Lu, S. Venugopalan, M.A. Karim, A. Niknejad, C. Hu, Compact models for sub-22 nm MOSFETs, in: Workshop on Compact Modeling, Boston, USA, June 2011.

5.3 Subthreshold slope degradation

SCEs also degrade the subthreshold slope. This degradation in subthreshold slope is due to gate bias dependence of the V_{SL} term. This is modeled similar to BSIM4 as follows:

$$\psi_{st} = 0.4 + PHIN + \Phi_B \tag{5.29}$$

$$C_{dsc} = \frac{0.5}{\cosh\left(DVT1SS \cdot \dfrac{L_{eff}}{\lambda}\right) - 1} \cdot (CDSC + CDSCD \cdot V_{dsx}) \tag{5.30}$$

$$n = \begin{cases} \cdot \left(1 + \dfrac{CIT + C_{dsc}}{(2C_{si}) \parallel C_{ox}}\right) & \text{if } GEOMOD \neq 3 \\[3mm] \cdot \left(1 + \dfrac{CIT + C_{dsc}}{C_{ox}}\right) & \text{if } GEOMOD = 3 \text{ (nanowire)} \end{cases} \tag{5.31}$$

This n is multiplied by the thermal voltage V_{tm} within the surface potential Eq. (3.11). Although it is an empirical formulation, it gives the desired effect as observed experimentally. Fig. 5.4 shows the validation of the subthreshold slope model with experimental data for different lengths in a PMOS FinFET.

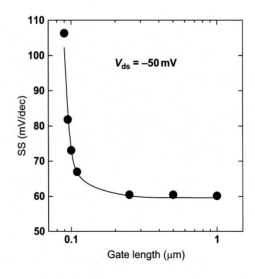

FIG. 5.4

Model comparison with experimental data of subthreshold slope (SS) for a p-type SOI FinFET with $N_{fin} = 20$, $EOT = 2.0$ nm, $H_{fin} = 60$ nm, and $T_{fin} = 22$ nm.

5.4 Quantum mechanical V_{th} correction

The channel thickness in the FinFETs plays an important role in the device characteristics. To control SCEs, the thickness of the channel is reduced. The reduction in film thickness induces a significant amount of geometrical confinement and converts these devices into a 2D system. This geometrical confinement becomes more and more significant with a reduction in film thickness. It is a well-known fact from quantum mechanics that confinement brings quantization in the energy level. The ground state subband energy level always lies above the conduction band edge. The energy difference between the conduction band edge and ground state subband energy levels increases as we decrease film thickness. This difference pushes the Fermi level up and brings it closer to the ground state energy. This indicates that for the same bias condition, the quantum mechanical inversion charge density Q_i^{qm} will be less than the classical inversion charge density Q_i^{cl}. As a result, the quantum mechanical threshold voltage of the device becomes higher as compared to the conventional threshold value. The change in the threshold voltage due to energy quantization in BSIM-CMG is introduced through $\Delta V_{t,QM}$. In the derivation of $\Delta V_{t,QM}$, the Schrodinger equation is solved assuming a linear potential profile of the solution of the Poisson's equation across the body/channel assuming Maxwell-Boltzmann statistics in the weak region. To get accurate V_{th} for thin film thicknesses, two energy levels E_0 and E_1 are used in the $\Delta V_{t,QM}$ derivation given here. $QMFACTOR_i$ also serves as a switch here.

If $GEOMOD \neq 3$ then

$$E_0 = \frac{\hbar^2 \pi^2}{2m_x \cdot TFIN^2} \tag{5.32}$$

$$E_0' = \frac{\hbar^2 \pi^2}{2m_x' \cdot TFIN^2} \tag{5.33}$$

$$E_1 = 4E_0 \tag{5.34}$$

$$E_1' = 4E_0' \tag{5.35}$$

$$\gamma = 1 + \exp\left(\frac{E_0 - E_1}{kT}\right) + \frac{g'm_d'}{gm_d} \cdot \left[\exp\left(\frac{E_0 - E_0'}{kT}\right) + \exp\left(\frac{E_0 - E_1'}{kT}\right)\right] \tag{5.36}$$

$$\Delta V_{t,QM} = QMFACTOR_i \cdot \left[\frac{E_0}{q} - \frac{kT}{q} \ln\left(\frac{g \cdot m_d}{\pi \hbar^2 N_c} \cdot \frac{kT}{TFIN} \cdot \gamma\right)\right] \tag{5.37}$$

If $GEOMOD = 3$ (nanowire) then

$$E_{0,QM} = \frac{\hbar^2 (2.4048)^2}{2m_x \cdot R^2} \tag{5.38}$$

$$\Delta V_{t,QM} = QMFACTOR_i \cdot \frac{E_{0,QM}}{q} \tag{5.39}$$

Fig. 5.5 shows the comparison of drain current with and without QME. Also, see the discussion on quantum confinement in Section 5.8.

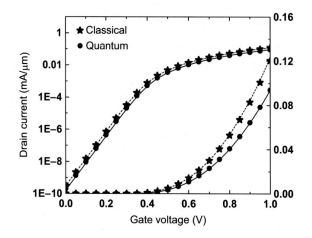

FIG. 5.5

I_{ds}-V_{gs} comparison with and without quantum mechanical effect.

5.5 Vertical field mobility degradation

The vertical field-dependent mobility model of BSIM-CMG includes the mobility degradation due to different scattering mechanisms, such as phonon scattering, surface roughness scattering, and coulomb scattering. The model has been taken from BSIM4 but uses charges instead of threshold voltage. The effect of mobility degradation is captured through the D_{mob} term in the drain current. The unified mobility model used in BSIM-CMG can be expressed as

$$\mu_{eff} = \frac{U0}{D_{mob}} \tag{5.40}$$

$$D_{mob} = \begin{cases} 1 + UA \cdot (E_{eff})^{EU} + \dfrac{UD}{\left[\frac{1}{2} \cdot \left(1 + \frac{q_{ia}}{q_b}\right)\right]^{UCS}} & \text{BULKMOD} = 0 \\[4ex] 1 + (UA + UC \cdot V_{eseff}) \cdot (E_{eff})^{EU} + \dfrac{UD}{\left[\frac{1}{2} \cdot \left(1 + \frac{q_{ia}}{q_b}\right)\right]^{UCS}} & \text{BULKMOD} = 1 \end{cases} \tag{5.41}$$

where

$$q_{ia} = \frac{q_{is} + q_{id}}{2} \tag{5.42}$$

$$E_{eff} = 10^{-8} \cdot \left(\frac{q_b + \eta \cdot q_{ia}}{\epsilon_{ratio} \cdot EOT}\right) \tag{5.43}$$

$$D_{mob} = \frac{D_{mob}}{U0MULT} \tag{5.44}$$

In Eq. (5.41), UA, UC, and UD are used for phonon/surface roughness scattering, body effect coefficient for mobility, and columbic scattering, respectively. U0MULT is a multiplier parameter used for variability modeling and its default value is 1.0.

5.6 Drain saturation voltage, V_{dsat}

Current saturation can happen due to pinch-off, velocity saturation, or even from source-side velocity limiting. Drain saturation voltage (V_{dsat}) is needed to get an effective drain voltage V_{dseff}, which in turn is used to calculate the surface potential at the drain side (ψ_d). Note that V_{dsat} is a function of applied V_{gs} only. In other words, V_{dsat} depends only on the surface potential at the source side (ψ_s) and not on V_{ds}. Also, the presence of source-drain resistance in the transistor affects V_{dsat}. The V_{dsat} expression in BSIM-CMG is adopted from the BSIM4 with the required modification. For the convenience of the reader, we are presenting the detailed derivation of V_{dsat} for $RDSMOD = 0$ and $RDSMOD = 1$.

5.6.1 Extrinsic case (RDSMOD = 1 and 2)

In the extrinsic case, source-drain resistances are modeled using external resistors, that is, RDSMOD = 1 and 2 (see Chapter 7).

When sufficiently high drain voltage is applied, the lateral electric field is large enough to cause carrier velocity to saturate near the drain end. In this case, the channel has different behavior of carrier velocities adjacent to the source and the drain ends. Adjacent to the source end, velocity depends on the lateral field, while adjacent to the drain end, it saturates and becomes field independent. At the boundary between the two portions, the channel voltage is equal to the drain saturation voltage V_{dsat} and the lateral electric field is equal to E_{sat} [9,10]. To use the drain current expression of BSIM4 in the derivation of V_{dsat} for BSIM-CMG, we replace $A_{bulk} = 1$ and $V_{gsteff} + 2\frac{kT}{q} = KSATIV \cdot$ $\left(V_{gsfbeff} - \psi_s + 2\frac{kT}{q}\right)$. $KSATIV$ is a model parameter and has a default value of 1.0. The modified drain expression in a strong inversion regime of operation at any point y in the channel along the channel length is expressed as follows:

$$I_{ds} = W_{eff} \cdot C_{ox}\left(KSATIV \cdot \left(V_{gsfbeff} - \psi_s + 2\frac{kT}{q}\right) - V(y)\right) \cdot v(y) \qquad (5.45)$$

where $v(y)$ is carrier velocity at any point y in the channel and it depends on the lateral electric field in the channel as follows:

$$v(y) = \begin{cases} \mu \cdot E_y & \text{if} \quad E_y < E_{sat} \\ v_{sat} & \text{if} \quad E_y > E_{sat} \end{cases} \qquad (5.46)$$

where μ is the mobility of the charge carriers in the channel and includes the effect of lateral electric field as follows:

$$\mu = \frac{\mu_{\text{eff}}}{1 + \dfrac{E(y)}{E_{\text{sat}}}} \tag{5.47}$$

The drain current before the saturation condition (i.e., in linear or triode regime of operation) in the channel can be obtained with the help of Eqs. (5.45)–(5.47);

$$E_y = \frac{dV(y)}{dy} = f(I_{\text{ds}}, \psi_{\text{s}}, V_{\text{gsfbeff}}, V_{\text{ds}}) \tag{5.48}$$

$$\int_0^{V_{\text{dsat}}} 1 \cdot dV(y) = \int_0^{L_{\text{eff}}} f(I_{\text{ds}}, \psi_{\text{s}}, V_{\text{gsfbeff}}, V_{\text{ds}}) \cdot dy \tag{5.49}$$

$$I_{\text{ds}} = \mu_{\text{eff}} \cdot C_{\text{ox}} \cdot \frac{W_{\text{eff}}}{L_{\text{eff}}} \cdot \frac{1}{1 + \dfrac{V_{\text{ds}}}{E_{\text{sat}} \cdot L_{\text{eff}}}} \cdot \left(KSATIV \cdot \left(V_{\text{gsfbeff}} - \psi_{\text{s}} + 2\frac{kT}{q} \right) - \frac{V_{\text{ds}}}{2} \right) \cdot V_{\text{ds}} \tag{5.50}$$

Now, using Eqs. (5.45), (5.50) and exploiting the drain current continuity at the boundary of linear and saturation region of operation (at $V_{\text{ds}} = V_{\text{dsat}}$), we can write

$$\mu_{\text{eff}} \cdot C_{\text{ox}} \cdot \frac{W_{\text{eff}}}{L_{\text{eff}}} \cdot \frac{1}{1 + \dfrac{V_{\text{dsat}}}{E_{\text{sat}} \cdot L_{\text{eff}}}} \cdot \left(KSATIV \cdot \left(V_{\text{gsfbeff}} - \psi_{\text{s}} + 2\frac{kT}{q} \right) - \frac{V_{\text{dsat}}}{2} \right) \cdot V_{\text{dsat}}$$
$$= W_{\text{eff}} \cdot C_{\text{ox}} \left(KSATIV \cdot \left(V_{\text{gsfbeff}} - \psi_{\text{s}} + 2\frac{kT}{q} \right) - V_{\text{dsat}} \right) \cdot v_{\text{sat}} \tag{5.51}$$

or

$$\frac{1}{L_{\text{eff}}} \cdot \frac{1}{1 + \dfrac{V_{\text{dsat}}}{E_{\text{sat}} \cdot L_{\text{eff}}}} \cdot \left(KSATIV \cdot \left(V_{\text{gsfbeff}} - \psi_{\text{s}} + 2\frac{kT}{q} \right) - \frac{V_{\text{dsat}}}{2} \right) \cdot V_{\text{dsat}}$$
$$= \left(KSATIV \cdot \left(V_{\text{gsfbeff}} - \psi_{\text{s}} + 2\frac{kT}{q} \right) - V_{\text{dsat}} \right) \cdot \frac{v_{\text{sat}}}{\mu_{\text{eff}}} \tag{5.52}$$

In Eq. (5.52), after replacing $\dfrac{v_{\text{sat}}}{\mu_{\text{eff}}}$ with $\dfrac{E_{\text{sat}}}{2}$, $E_{\text{sat}} \cdot L_{\text{eff}}$ with E_{satL} and rearranging, we get a solution for V_{dsat} in a compact form as follows [6]:

$$V_{\text{dsat}} = \frac{E_{\text{satL}} \cdot KSATIV \cdot \left(V_{\text{gsfbeff}} - \psi_{\text{s}} + \dfrac{2kT}{q} \right)}{E_{\text{satL}} + KSATIV \cdot \left(V_{\text{gsfbeff}} - \psi_{\text{s}} + \dfrac{2kT}{q} \right)} \tag{5.53}$$

5.6.2 Intrinsic case (RDSMOD = 0)

In the intrinsic case, part of source-drain resistances is modeled using an internal resistor, that is, RDSMOD = 0 (see Chapter 7).

As the resistance effect is included in the current equation in this case (see Dr term in Eq. 5.109), V_{dsat} also changes from the intrinsic case. The series resistances cause a fraction of the applied drain voltage to drop across them. As a result, V_{dsat}

increases as compared to Eq. (5.53). V_{dsat} in the case of RDSMOD = 0 in BSIM-CMG is implemented in a similar way to BSIM4 [9,10] to reduce the complexity. The modified intrinsic drain current (without considering source/drain resistances) expression of I_{ds0} after putting $\lambda = 1$, $A_{bulk} = 1$, and $V_{gsteff} + 2\frac{kT}{q} = KSATIV \cdot \left(V_{gsfbeff} - \psi_s + 2\frac{kT}{q}\right)$ in the BSIM4 model can be written as follows:

$$I_{ds0} = \frac{W_{eff} \cdot C_{ox} \cdot KSATIV \cdot \left(V_{gsfbeff} - \psi_s + 2\frac{kT}{q}\right)}{\left(1 + \frac{V_{ds}}{E_{sat} \cdot L_{eff}}\right)} \cdot \mu_{eff} \cdot V_{ds}$$

$$\times \left[1 - \frac{V_{ds}}{2KSATIV \cdot \left(V_{gsfbeff} - \psi_s + 2\frac{kT}{q}\right)}\right]$$

(5.54)

Here, KSATIV is a model parameter and it has a default value of 1.0. After considering the source/drain resistance $R_{ds}(V)$ and applying Ohm's law, the drain current can be written as follows:

$$I_{ds} = \frac{V_{ds}}{R_{ch} + R_{ds}}$$

(5.55)

or

$$I_{ds} = \frac{I_{ds0}}{1 + \frac{R_{ds} \cdot I_{ds0}}{V_{ds}}}$$

(5.56)

where $R_{ch} = \frac{V_{ds}}{I_{ds0}}$ when $R_{ds} = 0$. The drain current in the saturation region of operation can be expressed as follows:

$$I_{ds,sat} = W_{eff} \cdot q_{ch}(V(y) = V_{dsat}) \cdot VSAT$$

(5.57)

$$I_{ds,sat} = W_{eff} \cdot C_{ox} \cdot KSATIV \cdot \left(V_{gsfbeff} - \psi_s + 2\frac{kT}{q}\right) \cdot VSAT$$

$$\times \left[1 - \frac{V_{dsat}}{KSATIV \cdot \left(V_{gsfbeff} - \psi_s + 2\frac{kT}{q}\right)}\right]$$

(5.58)

At $V_{ds} = V_{dsat}$, the channel current expressed by Eq. (5.56) will be equal to the drain current of Eq. (5.58) (i.e., $I_{ds,sat}$).

$$\frac{I_{ds0}(V_{dsat})}{1 + \frac{R_{ds} \cdot I_{ds0}}{V_{dsat}}} = W_{eff} \cdot C_{ox} \cdot KSATIV \cdot \left(V_{gsfbeff} - \psi_s + 2\frac{kT}{q}\right) \cdot VSAT$$

$$\times \left[1 - \frac{V_{dsat}}{KSATIV \cdot \left(V_{gsfbeff} - \psi_s + 2\frac{kT}{q}\right)}\right]$$

(5.59)

After putting the value of I_{ds0} from Eq. (5.59), the previous equation can be written as follows:

$$\frac{\dfrac{W_{\text{eff}} \cdot C_{\text{ox}} \cdot KSATIV \cdot \left(V_{\text{gsfbeff}} - \psi_s + 2\dfrac{kT}{q}\right)}{\left(1 + \dfrac{V_{\text{dsat}}}{E_{\text{sat}} \cdot L_{\text{eff}}}\right)} \cdot \mu_{\text{eff}} \cdot V_{\text{dsat}} \times \left[1 - \dfrac{V_{\text{dsat}}}{2KSATIV \cdot \left(V_{\text{gsfbeff}} - \psi_s + 2\dfrac{kT}{q}\right)}\right]}{1 + \dfrac{R_{\text{ds}}}{V_{\text{dsat}}} \cdot \dfrac{W_{\text{eff}} \cdot C_{\text{ox}} \cdot KSATIV \cdot \left(V_{\text{gsfbeff}} - \psi_s + 2\dfrac{kT}{q}\right)}{\left(1 + \dfrac{V_{\text{dsat}}}{E_{\text{sat}} \cdot L_{\text{eff}}}\right)} \cdot \mu_{\text{eff}} \cdot V_{\text{dsat}} \times \left[1 - \dfrac{V_{\text{dsat}}}{2KSATIV \cdot \left(V_{\text{gsfbeff}} - \psi_s + 2\dfrac{kT}{q}\right)}\right]}$$

$$= W_{\text{eff}} \cdot C_{\text{ox}} \cdot KSATIV \cdot \left(V_{\text{gsfbeff}} - \psi_s + 2\frac{kT}{q}\right) \cdot VSAT$$

$$\times \left[1 - \frac{V_{\text{dsat}}}{KSATIV \cdot \left(V_{\text{gsfbeff}} - \psi_s + 2\dfrac{kT}{q}\right)}\right]$$

(5.60)

In Eq. (5.60), put $\mu_{eff} = \frac{2 \cdot VSAT}{E_{sat}}$, $E_{sat} \cdot L_{eff} = E_{satL}$ and after some manipulation and rearranging, we get the following form of Eq. (5.60)

$$\frac{V_{dsat} \cdot \left(2 \cdot KSATIV \cdot \left(V_{gsfbeff} - \psi_s + 2\frac{kT}{q}\right) - V_{dsat}\right)}{E_{satL} + V_{dsat} + W_{eff} \cdot C_{ox} \cdot VSAT \cdot \left(2 \cdot KSATIV \cdot \left(V_{gsfbeff} - \psi_s + 2\frac{kT}{q}\right) - V_{dsat}\right) \cdot R_{ds}}$$

$$= \left(\cdot KSATIV \cdot \left(V_{gsfbeff} - \psi_s + 2\frac{kT}{q}\right) - V_{dsat}\right) \quad (5.61)$$

$$W_{eff} \cdot C_{ox} \cdot VSAT \cdot R_{ds} \cdot V_{dsat}^2 - 3W_{eff} \cdot C_{ox} \cdot VSAT \cdot R_{ds} \cdot KSATIV \cdot \left(V_{gsfbeff} - \psi_s + 2\frac{kT}{q}\right)$$

$$\times V_{dsat} - \left(KSATIV \cdot \left(V_{gsfbeff} - \psi_s + 2\frac{kT}{q}\right) + E_{satL}\right) \cdot V_{dsat}$$

$$\times 2W_{eff} \cdot C_{ox} \cdot VSAT \cdot KSATIV \cdot \left(V_{gsfbeff} - \psi_s + 2\frac{kT}{q}\right)^2$$

$$+ E_{satL} \cdot KSATIV \cdot \left(V_{gsfbeff} - \psi_s + 2\frac{kT}{q}\right) = 0 \quad (5.62)$$

Eq. (5.62) is now converted into the standard form of $a \cdot x^2 - b \cdot x + c = 0$, and it can be written as follows:

$$\frac{a}{2} \cdot V_{dsat}^2 - b \cdot V_{dsat} + c = 0 \quad (5.63)$$

where

$$a = 2 \cdot W_{eff} VSAT C_{ox} \cdot R_{ds} \quad (5.64)$$

$$b = E_{satL} + KSATIV\left(V_{gsfbeff} - \psi_s + 2\frac{kT}{q}\right)\left(1 + \frac{3}{2}T_a\right) \quad (5.65)$$

$$c = KSATIV\left(V_{gsfbeff} - \psi_s + 2\frac{kT}{q}\right)\left[E_{satL} + T_a \cdot KSATIV \cdot \left(V_{gsfbeff} - \psi_s + 2\frac{kT}{q}\right)\right] \quad (5.66)$$

The solution of V_{dsat} in terms of T_a, T_b, and T_c can be written as follows:

$$V_{dsat} = \frac{T_b - \sqrt{(T_b)^2 - 2 \cdot T_a T_c}}{T_a} \quad (5.67)$$

Once V_{dsat} has been evaluated, V_{dseff} is calculated using the following formulation:

$$V_{dseff} = \frac{V_{ds}}{\left[1 + \left(\frac{V_{ds}}{V_{dsat}}\right)^{MEXP}\right]^{1/MEXP}} \quad (5.68)$$

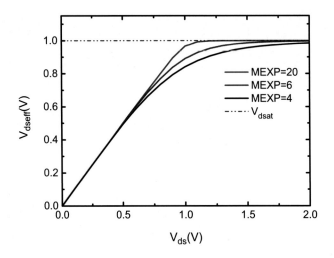

FIG. 5.6

The effective drain-source voltage V_{dseff} (*solid lines*) versus V_{ds} for various values of MEXP.

The interpolation function used in V_{dseff} for a smooth transition from V_{ds} to V_{dsat} ensures symmetry around $V_{dseff} = 0$. Another beauty of this function is that it forces V_{dseff} to be zero, when $V_{ds} = 0$ (see Fig. 5.6), without getting affected by compiler's numerical precision, etc.

5.7 Velocity saturation model

The effect of velocity saturation on the drain current in BSIM-CMG is included through D_{vsat}. The velocity saturation model in BSIM-CMG is similar to BSIM6 bulk MOSFETs [11,12].

$$E_{sat1} = \frac{2 \cdot VSAT1 \cdot D_{mob}}{\mu_0} \tag{5.69}$$

where VSAT1 (= VSAT by default) has been provided for extra flexibility for the separate tuning of V_{dsat} and current.

$$D_{vsat} = \frac{1 + \left[DELTAVSAT + \left(\frac{\Delta q_i}{E_{sat1} \cdot L_{eff}} \right)^{PSAT} \right]^{1/PSAT}}{1 + (DELTAVSAT)^{1/PSAT}} + \frac{1}{2} \cdot PTWG \cdot q_{ia} \cdot \Delta \psi^2 \tag{5.70}$$

where DELTAVSAT, PSAT, and PTWG are velocity saturation parameters. D_{mob} is the vertical field-induced mobility degradation given by Eq. (5.41). The default value of the PSAT parameter is 2. In Eq. (5.70), the second term containing $\Delta \psi^2$ is added empirically to have more flexibility in the model for better fitting of the g_{msat} behavior.

5.7.1 Nonsaturation effect

Some devices do not exhibit prominent or abrupt velocity saturation. The parameters A1 and A2 are used to tune this nonsaturation effect to fit $I_{d,sat}$ or $g_{m,sat}$.

$$T0 = \max\left[\left(A1 + \frac{A2}{q_{ia} + 2.0 \cdot \frac{nkT}{q}}\right) \cdot \Delta q_i^2, -1\right] \tag{5.71}$$

$$N_{sat} = \frac{1 + \sqrt{1 + T0}}{2} \tag{5.72}$$

$$D_{vsat} = D_{vsat} \cdot N_{sat} \tag{5.73}$$

5.8 Quantum mechanical effects

The QMEs in multigate field effect devices are more influential in the device characteristics than conventional bulk MOSFETs. Multigate devices have additional effects arising due to geometrical confinement along with electrical confinement. It is well known that QMEs cause the average charge location to shift away from the interface (called charge centroid shift) [13]. In BSIM-CMG, the threshold voltage shift due to bias-dependent ground state subband energy is considered in surface potential calculation as given by Eq. (5.32). The charge centroid shift in BSIM-CMG is modeled through variation in oxide thickness and width [14] and QME is activated by QMTCENIV and QMTCENCV parameters for I-V and C-V characteristics, respectively. The charge centroid in the inversion region can be expressed as [10,13,14]

$$T_{cen} = \frac{T_{cen0}}{1 + \left(\dfrac{q_{ia} + ETAQM \cdot q_{ba}}{QM0}\right)^{PQM}} \tag{5.74}$$

where ETAQM, QM0, and PQM are parameters that vary the location of the charge centroid with applied bias conditions. T_{cen0} used in equation earlier is calculated from geometry parameters, that is, fin height and fin width in case of DGFET and TGFET; and radius of nanowire in case of nanowire. Using coupled Poisson-Schrodinger equations, it has been shown that T_{cen0} varies from $0.25T_{FIN}$ for thick fin to $0.351T_{FIN}$ for thin fin in DGFET. To model this dependence, empirical functions are used in BSIM-CMG and are given by [14]

$$\frac{T_{cen0}}{T_{FIN}} = 0.25 + (0.351 - 0.25)\exp\left(\frac{T_{FIN}}{T_0}\right) \tag{5.75}$$

where T_0 is a fitting parameter. Similarly, the nanowire has the following relation between T_{cen0} and nanowire radius R

$$\frac{T_{cen0}}{R} = 0.334 + (0.576 - 0.334)\exp\left(\frac{R}{R_0}\right) \tag{5.76}$$

where R_0 is a fitting parameter. The functions used for T_{cen0} are validated for different thicknesses and radii with TCAD data and show an excellent match, as shown in Fig. 5.7. The variation of the charge centroid with gate voltage (or inversion charge) given by Eq. (5.74) is also validated with TCAD simulations, as shown in Fig. 5.8. After a certain critical charge in the channel, the centroid shifts toward the gate-oxide-channel interface with increasing gate bias.

The shifting of the centroid causes the width to vary with the applied bias, that is, width increases as the centroid shifts toward the interface with an increase in gate voltage, saturating in deep strong inversion. The effective width for I-V characteristics is calculated from QMTCENIV and T_{cen}. Similarly, the effective width and effective oxide thickness for C-V characteristics are calculated from QMTCENIV and T_{cen}.

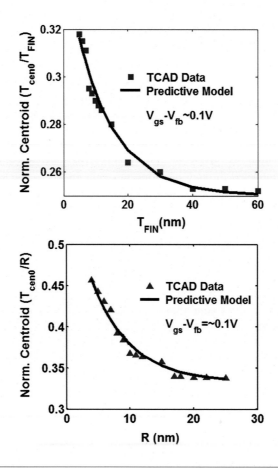

FIG. 5.7

Dependency of the charge centroid on channel thickness for DG-FET and nanowire-FET at low-gate voltages [14].

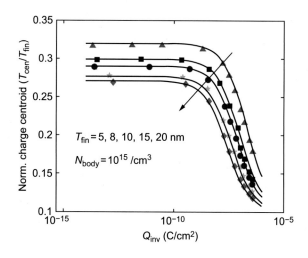

FIG. 5.8

Shifting of the charge centroid with inversion charge in the channel. The figure shows shifting of charge centroid toward the Si/SiO$_2$ interface after critical charge values depending on the channel thickness [14].

5.8.1 Effective width model

The effective width for different multigate structures is given here, which are used in current and capacitance modules. In these calculations, the effective width changes from W_{eff0} and $W_{\text{eff,CV0}}$, only when QMTCENIV and QMTCENCV are *nonzero*, respectively. By default, C-V calculations use the same parameter values as used in the I-V model, unless value of QMTCENCV is separately provided in the model card.

$$W_{\text{eff}} = W_{\text{eff0}} - \Lambda \cdot QMTCENIV \cdot T_{\text{cen}} \tag{5.77}$$

$$W_{\text{eff,CV}} = W_{\text{eff,CV0}} - \Lambda \cdot QMTCENCV \cdot T_{\text{cen}} \tag{5.78}$$

$$\Lambda = 0 \quad \text{for DG} - \text{FET (GEOMOD} = 0) \tag{5.79}$$

$$= 4 \quad \text{for triple} - \text{gate FET (GEOMOD} = 1) \tag{5.80}$$

$$= 8 \quad \text{for quadruple} - \text{gate FET (GEOMOD} = 2) \tag{5.81}$$

$$= 2\pi \quad \text{for surround} - \text{gate FET (GEOMOD} = 3) \tag{5.82}$$

5.8.2 Effective oxide thickness/effective capacitance

A QME on C-V characteristics is enabled by the nonzero value of QMTCENCV. When QMTCENCV$\neq$0, the effective insulator capacitance $C_{\text{ox,eff}}$ is used for C-V calculation but I-V characteristics always use C_{ox}. For $C_{\text{ox,eff}}$ evaluation, T_{cen} is added to the scaled physical oxide thickness (TOXP) to account for any dielectric material.

If $QMTCENCV \neq 0$ then

$$C_{\text{ox,eff}} = \begin{cases} \dfrac{3.9 \cdot \epsilon_0}{TOXP\dfrac{3.9}{EPSROX} + T_{\text{cen}} \cdot \dfrac{QMTCENCV}{\epsilon_{\text{ratio}}}} & GEOMOD \neq 3 \\[2em] \dfrac{3.9 \cdot \epsilon_0}{R \cdot \left[\dfrac{1}{\epsilon_{\text{ratio}}} \ln\left(\dfrac{R}{R - T_{\text{cen}}}\right) + \dfrac{3.9}{EPSROX} \ln\left(1 + \dfrac{TOXP}{R}\right) \right]} & GEOMOD = 3 \end{cases}$$

(5.83)

If $QMTCENCV = 0$ then

(5.84)

$$C_{\text{ox,eff}} = C_{\text{ox}}$$

5.8.3 Charge centroid calculation for accumulation

If $QMTCENCV \neq 0$ then

$$C_{\text{ox,acc}} = \begin{cases} \dfrac{3.9 \cdot \epsilon_0}{TOXP\dfrac{3.9}{EPSROX} + \dfrac{T_{\text{cen0}}}{1 + \left(\dfrac{q_{i,\text{acc}}}{QM0ACC}\right)^{PQMACC}} \cdot \dfrac{QMTCENCVA}{\epsilon_{\text{ratio}}}} & GEOMOD \neq 3 \\[3em] \dfrac{3.9 \cdot \epsilon_0}{R \cdot \left[\dfrac{1}{\epsilon_{\text{ratio}}} \ln\left(\dfrac{R}{R - \dfrac{T_{\text{cen0}}}{1 + \left(\dfrac{q_{i,\text{acc}}}{QM0ACC}\right)^{PQMACC}}} \right) + \dfrac{3.9}{EPSROX} \ln\left(1 + \dfrac{T_{\text{oxp}}}{R}\right) \right]} & GEOMOD = 3 \end{cases}$$

(5.85)

If $QMTCENCV = 0$ then

(5.86)

$$C_{\text{ox,acc}} = \dfrac{3.9 \cdot \epsilon_0}{EOTACC}$$

The small-signal (AC) gate capacitance of a double-gate (DG) structure with an intrinsic silicon channel and a silicon-oxide gate dielectric (T_{FIN} = 15 nm, T_{ox} = 1.5 nm, and L = 10 μm) was simulated using TCAD, as shown in Fig. 5.9. The device was biased in the linear region (drain voltage, V_{ds} = 50 mV). Two scenarios were simulated to evaluate the impact of the quantum mechanical (QM) charge centroid—one without QM effects and one with QM effects. As expected, the gate capacitance is reduced when QM confinement effects are considered.

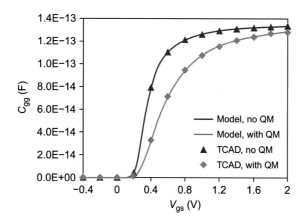

FIG. 5.9

Small-signal gate capacitance (C_{gg}) versus gate voltage for a long-channel DG-FET with TFIN = 15 nm, Tox = 1.5 nm [14].

5.9 Lateral nonuniform doping model

Lateral nonuniform doping along the length of the channel leads to I-V and C-V displaying different threshold voltages. However, the consistent surface potential-based I-V and C-V model does not allow for the usage of different V_{th} values. A straightforward method would be to recompute the surface potentials at the source and drain ends twice for I-V and C-V separately breaking the consistency but at the expense of computation time. The following model has been introduced as a multiplicative factor to the drain current (I-V) to allow for that V_{th} shift. This model should be exercised after the C-V extraction step to match the V_{th} for the subthreshold region in the $I_{d(\mathrm{lin})} - V_g$ curve. The parameter K0 is used to fit the subthreshold region, while the parameter K0SI helps reclaim the fit in the inversion region.

$$M_{\mathrm{nud}} = e^{\left(-\frac{K0(T)}{K0SI(T) \cdot q_{\mathrm{ia}} + 2.0 \cdot \frac{nkT}{q}} \right)} \tag{5.87}$$

5.10 Body effect model for bulk FinFET (BULKMOD=1)

A small amount of body bias effect can be captured from the following model. This model is applicable only for I-V and not for C-V.

$$V_{\mathrm{esx}} = V_{\mathrm{es}} - 0.5 \cdot (V_{\mathrm{ds}} - V_{\mathrm{dsx}}) \tag{5.88}$$

$$V_{\mathrm{eseff}} = \min(V_{\mathrm{esx}}, 0.95 \cdot PHIBE_i) \tag{5.89}$$

$$dVth_{\mathrm{BE}} = \sqrt{PHIBE_i - V_{\mathrm{eseff}}} - \sqrt{PHIBE_i} \tag{5.90}$$

$$M_{ob} = \exp\left(-dVth_{BE}\frac{K1(T) + K1SAT(T) \cdot V_{dsx}}{K1SI(T) \cdot q_{ia} + 2.0 \cdot \frac{nkT}{q}}\right) \tag{5.91}$$

Note: The lateral nonuniform doping model and the body effect model are empirical and have their limits as to how much V_{th} shift can be achieved without distorting the I-V curve. Overusage could lead to negative g_m or negative g_{ds}. The lateral nonuniform doping model could be used in combination with the mobility model to achieve high V_{th} shift between C-V and I-V curved to avoid any distortion of higher-order derivatives.

5.11 Output resistance model

The analog behavior of transistor, to a large extent, depends on the output resistance as gain is proportional to $\frac{g_m}{g_{ds}}$. Therefore, it is extremely important to correctly model it for different biases and dimensions. The continuous shrinking of geometrical dimensions has resulted in significant increase in the SCEs on the output resistance and has posed additional modeling challenges to the model engineers. Earlier compact models only accounted for channel length modulation (CLM) and therefore were inaccurate. The first comprehensive analytical scalable compact model of output resistance of bulk MOSFET was proposed by Huang et al. [15].

Fig. 5.10 shows the variation of typical on-resistance of a short-channel bulk MOSFET with drain bias. As indicated in the figure, different SCEs affect different

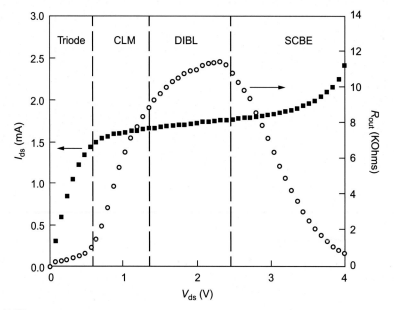

FIG. 5.10

Drain current and on-resistance variation with drain bias [15].

regions of characteristics. BSIM-CMG employs an early voltage (V_A)-based model for the output conductance similar to BSIM4 and BSIM6 [10,12,15].

Drain current is a weak function of drain voltage in saturation region. It can be approximated as

$$I_{ds}(V_{gs}, V_{ds}) = I_{ds}(V_{gs}, V_{dsat}) + \frac{dI_{ds}(V_{gs}, V_{ds})}{dV_{ds}}(V_{ds} - V_{dsat}) \tag{5.92}$$

$$= I_{dsat}\left(1 + \frac{V_{ds} - V_{dsat}}{V_A}\right) \tag{5.93}$$

where

$$I_{dsat} = I_{ds}(V_{gs}, V_{dsat}) \tag{5.94}$$

$$V_A = \frac{I_{dsat}}{\left(\dfrac{dI_{ds}}{dV_{ds}}\right)} \tag{5.95}$$

5.11.1 Channel length modulation

Fig. 5.11 shows that, in saturation, the MOSFET channel can be visualized as composed of two parts—one extending from the source to the saturation point and other from the saturation point to the drain. The saturation point of the channel is a function of drain bias and shifts toward the source as the drain bias is increased. As a result, the effective channel length reduces and drain current increases with drain voltage even in saturation. The reduction in channel length is a increasing function of V_{ds}-V_{dsat}. This effect is called CLM. The same arguments can safely be extended for multigate transistors as well.

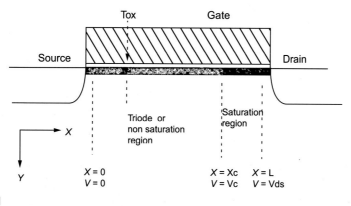

FIG. 5.11

Representation of bulk-MOSFET in saturation region.

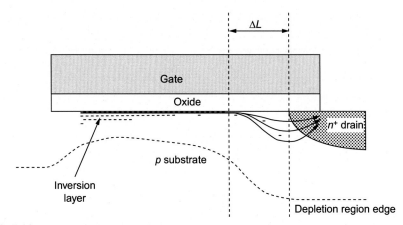

FIG. 5.12

Inversion layer in saturation in bulk-MOSFET.

Reproduced from Y. Tsividis, C. McAndrew, Operation and Modeling of the MOS Transistor, Oxford Series in Electrical and Computer Engineering, 2010.

As shown in Fig. 5.12, the inversion layer near the drain is located below the surface and accurate modeling of CLM requires 2D analysis in that region. The channel length reduction is given by [9,16,17]

$$\Delta L = Litl \cdot \sinh^{-1}\left(\frac{V_{ds} - V_{dsat}}{E_{sat} \cdot Litl}\right) \tag{5.96}$$

where the characteristic drain field length (*Litl*) is given by

$$Litl = \sqrt{\frac{\epsilon \cdot TOXE \cdot XJ}{EPSROX}} \tag{5.97}$$

Early voltage due to CLM is given by

$$V_{A,CLM} = \frac{I_{dsat}}{\left(\dfrac{dI_{ds}}{dL} \cdot \dfrac{dL}{dV_{ds}}\right)} \tag{5.98}$$

From quasi-2D analysis [16], it can be equivalently written as

$$V_{A,CLM} = C_{clm} \cdot (V_{ds} - V_{dsat}) \tag{5.99}$$

$$\frac{1}{C_{clm}} = \begin{cases} PCLM + PCLMG \cdot q_{ia} & \text{for } PCLMG_i > 0 \\[2mm] \dfrac{1}{\dfrac{1}{PCLM} - PCLMG \cdot q_{ia}} & \text{for } PCLMG_i < 0 \end{cases} \tag{5.100}$$

$$M_{clm} = 1 + \frac{1}{C_{clm}} \ln\left[1 + \frac{V_{ds} - V_{dseff}}{V_{dsat} + E_{satL}} \cdot C_{clm}\right] \tag{5.101}$$

M_{clm} is multiplied to I_{ds} in the final drain current expression.

5.11.2 Drain-induced barrier lowering

DIBL is another important SCE, which significantly affects output resistance. For a short-channel device, applied drain voltage reduces the barrier between the source and the channel, allowing more number of carriers to enter the channel compared to long-channel theory prediction. This manifests as reduction in threshold voltage and increase in drain current. The larger the drain bias, the more prominent the barrier lowering and thus current becomes a function of the drain voltage.

Although the DIBL effect is already captured through shift in the threshold voltage (see Eq. 5.25), it is not sufficient for accurate fitting of current and output resistance in saturation. Additional flexibility is added in the model using an early voltage concept for DIBL modeling [15].

$$
PVAG\,factor = \begin{cases} 1 + PVAG \cdot \dfrac{q_{\mathrm{ia}}}{E_{\mathrm{sat}}L_{\mathrm{eff}}} & \text{for } PVAG > 0 \\[2ex] \dfrac{1}{1 - PVAG \cdot \dfrac{q_{\mathrm{ia}}}{E_{\mathrm{sat}}L_{\mathrm{eff}}}} & \text{for } PVAG < 0 \end{cases} \tag{5.102}
$$

$$
\theta_{\mathrm{rout}} = \frac{0.5 \cdot PDIBL1}{\cosh\left(DROUT \cdot \dfrac{L_{\mathrm{eff}}}{\lambda}\right) - 1} + PDIBL2 \tag{5.103}
$$

$$
V_{\mathrm{ADIBL}} = \frac{q_{\mathrm{ia}} + 2kT/q}{\theta_{\mathrm{rout}}} \cdot \left(1 - \frac{V_{\mathrm{dsat}}}{V_{\mathrm{dsat}} + q_{\mathrm{ia}} + 2kT/q}\right) \cdot PVAG\,factor \tag{5.104}
$$

$$
M_{\mathrm{oc}} = \left(1 + \frac{V_{\mathrm{ds}} - V_{\mathrm{dseff}}}{V_{\mathrm{ADIBL}}}\right) \cdot M_{\mathrm{clm}} \tag{5.105}
$$

M_{oc} is multiplied to I_{ds} in the final drain current expression.

5.12 Channel current

Drain-to-source current including all effects is given by

$$
I_{\mathrm{ds}} = IDS0MULT \cdot \mu_0 \cdot C_{\mathrm{ox}} \cdot \frac{W_{\mathrm{eff}}}{L_{\mathrm{eff}}} \cdot i_{\mathrm{ds0}} \cdot \frac{M_{\mathrm{oc}}M_{\mathrm{ob}}M_{\mathrm{nud}}}{D_{\mathrm{mob}} \cdot D_{\mathrm{r}} \cdot D_{\mathrm{vsat}}} \times NFIN_{\mathrm{total}} \tag{5.106}
$$

where i_{ds0} is the core current model developed in Section 3.16. IDS0MULT is a multiplier to the drain current and is used for variability modeling. The "D_{r}" term is the effect of source-drain resistance on channel currently given as follows for different RDSMOD cases (see Chapter 8 for more details).

RDSMOD = 0 (**Internal bias dependent, external bias independent**)

$$
R_{\mathrm{source}} = R_{\mathrm{s,geo}} \tag{5.107}
$$

$$
R_{\mathrm{drain}} = R_{\mathrm{d,geo}} \tag{5.108}
$$

$$R_{ds} = \frac{1}{NFIN_{total} \times Weff0^{WR}} \cdot \left(RDSWMIN + \frac{RDSW}{1 + PRWGS \cdot q_{ia}} \right) \qquad (5.109)$$

$$D_r = 1.0 + NFIN_{total} \times \mu_0 \cdot C_{ox} \cdot \frac{W_{eff}}{L_{eff}} \cdot \frac{i_{ds0}}{\Delta q_i} \cdot \frac{R_{ds}}{D_{vsat} \cdot D_{mob}}$$

RDSMOD = 1 **(External)**

$$D_r = 1.0 \qquad (5.110)$$

RDSMOD = 2 **(Internal bias independent and bias dependent)**

$$R_{source} = 0.0 \qquad (5.111)$$

$$R_{drain} = 0.0 \qquad (5.112)$$

$$R_{ds} = \frac{1}{NFIN_{total} \times Weff0^{WR}} \cdot \left(R_{s,geo} + R_{d,geo} + RDSWMIN + \frac{RDSW}{1 + PRWGS \cdot q_{ia}} \right) \quad (5.113)$$

$$D_r = 1.0 + NFIN_{total} \times \mu_0 \cdot C_{ox} \cdot \frac{W_{eff}}{L_{eff}} \cdot \frac{i_{ds0}}{\Delta q_i} \cdot \frac{R_{ds}}{D_{vsat} \cdot D_{mob}}$$

$R_{s,geo}$ and $R_{d,geo}$ are the source and drain diffusion resistances.

References

[1] K. Suzuki, T. Tanaka, Y. Tosaka, H. Horie, Y. Arimoto, Scaling theory for double-gate SOI MOSFETs, IEEE Trans. Electron Devices 40 (12) (1993) 2326–2329.

[2] K. Suzuki, Y. Tosaka, T. Sugii, Analytical threshold voltage model for short channel n+/p+ double-gate SOI MOSFETs, IEEE Trans. Electron Devices 43 (5) (1996) 732–738.

[3] G. Pei, J. Kedzierski, P. Oldiges, M. Ieong, E. Kan, FinFET design considerations based on 3-D simulation and analytical modeling, IEEE Trans. Electron Devices 48 (8) (2002). 1441–1419.

[4] C.-H. Lin, Compact Modeling of Nanoscale CMOS (Ph.D. dissertation), University of California, Berkeley, 2007.

[5] Z.H. Liu, C. Hu, J.H. Huang, T.Y. Chan, M.C. Jeng, P.K. Ko, Y.C. Cheng, Threshold voltage model for deep-submicrometer MOSFETs, IEEE Trans. Electron Devices 40 (1) (1993) 86–95.

[6] C.C. Hu, Modern Semiconductor Devices for Integrated Circuits, Prentice Hall, 2010.

[7] Y. Tsividis, C. McAndrew, Operation and Modeling of the MOS Transistor, Oxford Series in Electrical and Computer Engineering, 2010.

[8] K. Cao, W. Liu, X. Jin, K. Vasanth, K. Green, J. Krick, T. Vrotsos, C. Hu, Modeling of pocket implanted MOSFETs for anomalous analog behavior, in: IEDM Tech. Dig., 1999, pp. 171–174.

[9] BSIM4 Technical Manual and Code [Online]. Available from: http://www-device.eecs. berkeley.edu/bsim/?page=BSIM4.

[10] W. Liu, C. Hu, BSIM4 and MOSFET Modeling for IC Simulation, World Scientific, 2011.

[11] Y.S. Chauhan, S. Venugopalan, M.-A. Chalkiadaki, M.A. Karim, H. Agarwal, S. Khandelwal, N. Paydavosi, J.P. Duarte, C.C. Enz, A.M. Niknejad, C. Hu, BSIM6: analog and RF compact model for bulk MOSFET, IEEE Trans. Electron Devices 61 (2) (2014) 234–244.

[12] BSIM6 Technical Manual and Code [Online]. Available from: http://www-device.eecs.berkeley.edu/bsim/?page=BSIM6.

[13] W. Liu, X. Jin, Y. King, C. Hu, An efficient and accurate compact model for thin-oxide-MOSFET intrinsic capacitance considering the finite charge layer thickness, IEEE Trans. Electron Devices 46 (5) (1999) 1070–1072.

[14] S. Venugopalan, M.A. Karim, S. Salahuddin, A.M. Niknejad, C.C. Hu, Phenomenological compact model for QM charge centroid in multigate FETs, IEEE Trans. Electron Devices 60 (4) (2013) 1480–1484.

[15] J.H. Huang, Z.H. Liu, M.C. Jeng, P.K. Ko, C. Hu, A physical model for MOSFET output resistance, in: IEEE International Electron Device Meeting, 1992, pp. 569–572.

[16] N.G. Einspruch, G. Gildenblat, Advanced MOS Device Physics, Academic Press, 1989.

[17] BSIM-CMG Technical Manual and Code [Online]. Available from: http://www-device.eecs.berkeley.edu/bsim/.

Further reading

Y.S. Chauhan, S. Venugopalan, M.-A. Chalkiadaki, M.A. Karim, H. Agarwal, S. Khandelwal, N. Paydavosi, J.P. Duarte, C.C. Enz, A.M. Niknejad, C. Hu, BSIM6: analog and RF compact model for bulk MOSFET, IEEE Trans. Electron Devices 61 (2) (2014) 234–244.

M.V. Dunga, Nanoscale CMOS Modeling (Ph.D. thesis), University of California, Berkeley, 2008.

Leakage currents

6.1 Introduction

During the last four decades, while IC makers reduced the physical size of planar Si MOSFETs to enhance their speed, power efficiency, and fabrication cost per transistor, an unfavorable consequence grew in parallel. Short-channel effects (SCEs) caused an increase in leakage current, which in turn led to an increase in leakage (static) power dissipation. Today, typical planar, bulk MOSFET scaling has reached a stage where static leakage power accounts for about half of the power dissipated in a device. The usual bulk MOSFET scaling is coming to an end, but not due to challenges with manufacture; rather, this is due to the fact that more scaling will not reduce power consumption and would rather increase it. As demonstrated in Chapters 1 and 4, the FinFET and GAA/nanosheet FET architectures, respectively, could significantly lower SCEs, persuading the industry to switch from planar to FinFET/GAA FET architectures.

For a nonzero drain voltage V_{ds}, the drain current I_d of a typical MOSFET is often plotted (in logarithmic scale) as a function of the gate voltage V_{gs}, and the value of the intersection with the drain current axis is what we call the off-state leakage current (see Fig. 5.1). Typically, the transistor's off-state leakage current for V_{ds} equal to the supply voltage V_{dd} is defined as the off-state current, I_{off}. At this bias point, the transistor should ideally be entirely off, but as will be shown later in this chapter, leakage currents always persist due to different physical leakage mechanisms. These leakage current sources include the gate-induced drain leakage (GIDL) current between the drain terminal and substrate terminal, the weak-inversion current between the drain terminal and source terminal, the substrate and drain junction leakage currents (both forward[a] and reverse diode currents), and a portion of the gate oxide tunneling current.

From those, the junction leakage and gate oxide tunneling currents extend to the transistor's on-state and add to impact ionization leakage, which becomes noticeable at on-state. Leakage currents between terminal pairs other than those involving the drain may also exist, such as gate-induced source leakage (GISL) between the source and substrate terminals. The models in BSIM-CMG are capable of simulating

---

[a] Forward leakage can occur under intentional forward well bias and under voltage spikes that may even lead to latch-up.

FinFET/GAA Modeling for IC Simulation and Design. https://doi.org/10.1016/B978-0-323-95729-8.00007-6

the leakage currents of all terminals[b] as well as various device architectures such as FinFET, GAA/nanosheet FETs, and multibridge FETs.

This chapter's first portion provides a review of the weak-inversion current with an emphasis on the terminology used in the field. A method for creating a GIDL/GISL model for BSIM-CMG that is comparable to the one originally used in BSIM4 is detailed in Section 5.2. Furthermore, Section 5.2 introduces the new model for trap-assisted tunneling (TAT)-based GIDL and parasitic FET GIDL/GISL leakage models for GAA/nanosheet FETs. The methods and formulas for gate oxide tunneling are covered in Section 5.3. The impact ionization concept is explained in Section 5.4. In Chapter 10, the junction leakage component will be thoroughly examined.

6.2 Weak-inversion current

Assuming a room-temperature N-channel MOSFET, there are always some electrons in the source diffusion area with sufficient energy to cross the source-channel barrier and reach the drain side (see Fig. 6.2). For $V_{ds} > 0$, these electrons will produce an I_d that is nonzero. This is the primary leakage mechanism in contemporary short-channel devices and is called the weak-inversion or subthreshold current. Since the number of these carriers is exponentially increased by an applied gate voltage when below the threshold voltage, the weak-inversion current is represented by a straight line with a finite slope in a semilog plot as shown in Fig. 6.1. The reverse of the slope of this line is known as subthreshold swing S and it has a unit of millivolts of the gate voltage per decade of the drain current. The ideal value of S is $\approx$60 mV/decade. This number serves as a fundamental limit and indicates that V_g must be raised by at least 60 mV in order to increase the current by a factor of 10 above the potential barrier. To beat this limit, the carriers must additionally tunnel through the barrier. Tunneling may be afforded by an MEMS switch [1] or a tunneling transistor (TFETs) based on gate-induced band-to-band tunneling (BTBT) [2].

Why is the subthreshold leakage a big concern in the recent CMOS technology nodes? As can be visualized in Fig. 6.1, reducing V_t, which is equivalent to shifting the entire curve to the left, or increasing S, which makes the slope shallower, will increase the off-state leakage current exponentially, and hence will increase the static power dissipation. Short-channel MOSFETs inherently have smaller threshold voltages due to 2D electrostatics, which originate from the proximity of the source and drain regions and their charge sharing with the gate (the effect known as V_t roll-off). The effect is enhanced at higher drain voltages in short-channel devices because the drain region is close enough that the drain voltage can affect and lower the source-channel barrier height at the channel-dielectric interface (the effect known

---

[b] Note that for FinFET on SOI (BULKMOD = 0), the substrate leakage current will flow out of the source, that is, the holes will be injected into the source and appear as an additional drain-source leakage.

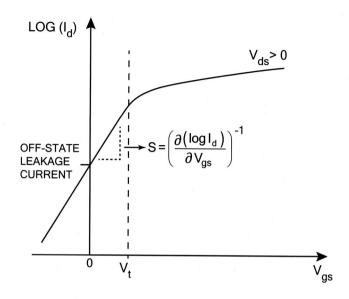

FIG. 6.1

The transistor's drain current (in logarithmic scale) as a function of its gate voltage. The off-state leakage current, threshold voltage, and subthreshold swing have been marked.

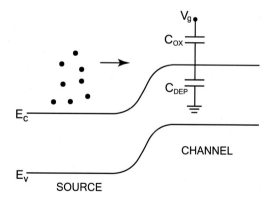

FIG. 6.2

The potential barrier at the source/channel determines the subthreshold current. The current can be increased by 10× for 60 mV decrease in the potential barrier or equivalently, for a $60 \times \left(1 + \frac{C_{DEP}}{C_{OX}}\right)$ mV increase in V_g. The thin body of the FinFET gets fully depleted for a small value of applied V_g; this makes $C_{DEP} = 0$ and S roughly 60 mV for FinFET/GAA FETs.

as drain-induced barrier lowering or DIBL). Typical values of DIBL for high performance, 32 nm node, planar bulk MOSFETs are around 100 mV/V; this means there is a 100 mV shift in V_t for 1 V of applied drain voltage. In very short-channel devices, S is also affected and becomes larger (the effect is known as subthreshold swing degradation). This is because the drain is so close that it can lower the source-channel barrier height for paths a few nanometers below the surface, resulting in subsurface leakage. Typical values of S for high performance, 32 nm node, planar bulk MOSFETs are in the range of 70–100 mV/decade.

The FinFET/GAA has proven to have excellent capabilities in managing SCEs and suppression of the off-state leakage current by offering a tighter electrostatic control around the channel. The values of DIBL and S for a high-performance, 5 nm-node FinFET/GAA are $\approx$32 mV/V and $\approx$65 mV/decade, respectively, leading to low values of I_{off} in the range of 20–100 nA/µm [3–5]. However, it is crucial for IC designs to carefully model phenomena that lower V_t or degrade S (and thereby worsen the off-state leakage current) in order to determine the subthreshold behavior in scaled FinFETs/GAA FETs. This is especially critical for low-power, processor/ mobile circuit applications. As discussed in Chapter 5, a drain current equation that is applicable for a FinFET/GAA FETs from weak inversion (subthreshold) to strong inversion is used by the BSIM-CMG model's core model. For the implementation of V_t roll-off, DIBL, and S degradation models, please refer to the real-device models described in Chapter 5.

6.3 Gate-induced source and drain leakages

Fig. 6.3 illustrates the cross-section of an N-channel, double-gate FinFET and its energy-band diagram for the gate-drain overlap region when a low gate voltage and a high drain voltage are applied. If the band bending at the oxide interface is

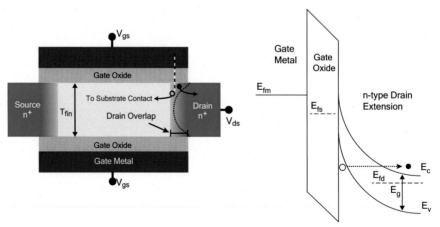

FIG. 6.3

(A) Schematic representation cross-section of the fin of a FinFET. (B) Schematic illustration of the energy-band diagram along the *dashed line* in (A). The substrate contact is normal to and below the page.

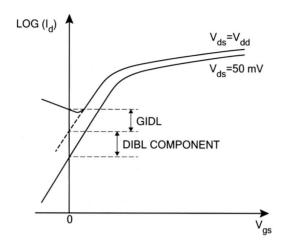

FIG. 6.4

Contributions of DIBL and GIDL to the transistor's off-state leakage current. The position of the dip caused by GIDL will vary around $V_{gs} = 0$ depending on V_{dd}, the channel material, doping, and trap density.

greater than or equal to the energy bandgap E_g of the drain material, a BTBT process will take place. The electrons in the valence band (VB) of the n-type drain will tunnel through the thinned band gap into the conduction band (CB) and they will be collected at the drain contact to be a part of the drain current, whereas the remaining holes will be collected at the substrate contact (the source contact in the case of Fin-FET/GAA FET on SOI) and will contribute to the substrate (source) leakage. This phenomenon, which was first elucidated and modeled by the researchers at UC Berkeley [6], discerns a potential major contributor to the off-state leakage current (see Fig. 6.4) and is called the GIDL current. Depending on the voltages applied, there might also exist a GISL current.

But, what are the prerequisites that must exist for the GIDL current to flow? First, there must be band bending greater than E_g so that the VB energy states overlap the CB energy states, as shown in Fig. 6.3. In that case the semiconductor surface in the gate-drain overlap region is in deep depletion with the band bending being much larger than $2\phi_B$.[c] The surface potential can exceed $2\phi_B$ because there is no inversion hole layer at the surface. There is no hole layer at the surface because any hole there would drift and diffuse to the body/substrate due to the built-in junction potential plus any substrate-drain reverse bias. However, with a forward-biased substrate-drain junction, the holes may remain at the interface and form an inversion layer and cause the band bending to be pinned at roughly $2\phi_B$, a value smaller than E_g, thereby suppressing the GIDL current. In the case of a FinFET/GAA on an SOI

[c] ϕ_B is the difference between the Fermi potential and the intrinsic potential in the drain.

substrate, holes build up in the floating body and raise the body potential until the body-source junction is slightly forward biased, enabling the GIDL-generated holes to be injected into the n^+ source. Second, the electric field needs to be large, that is, the tunneling barrier needs to be narrow. Compared to planar MOSFET, both of these two conditions are more difficult to meet in a FinFET/GAA FET because the potential at both sides of the thin fin is raised or lowered by the same V_g. Therefore, lightly doped and very thin fin FinFETs can have negligible GIDL leakage. Try to convince yourself of this by looking at Fig. 6.3 and remember the Poisson equation.

In addition, defects or traps in tunneling lead to trap-assisted BTBT by providing steppingstones along the tunneling path. Therefore, the GIDL current is larger in the presence of defects created by ion implantation. Using solid-source diffusion instead of implantation for drain creation or using a laser for the activation of dopants and annealing have been shown to reduce GIDL [7]. However, researchers at UC Berkeley have recently concluded that the TAT-based GIDL governs a large portion of total GIDL current in the state-of-the-art FinFET/nanosheet FET technologies [8,9]. The accurate simulation of leakage for low-power applications using FinFETs and nanosheet FETs require a revisit of GIDL physical mechanisms. Total GIDL is attributed to two physical mechanisms: (i) BTBT at a high vertical field in the gate-drain overlap region; (ii) TAT at a low vertical field in the gate-drain overlap region. For FinFET and nanosheet devices, however, a GIDL with only BTBT contribution is insufficient to capture the GIDL behavior because, for a high V_{dg}, band bending is significant enough that the tunneling width is small for the carriers to tunnel from the VB to the CB. However, for low V_{dg}, the tunneling width increases as a result of slight band bending, and as illustrated in Fig. 6.5, carriers are tunneled from VB to CB via mid-gap trap energy levels existing close to the oxide-semiconductor interface.

The total GIDL current considering band-to-band and TAT can be obtained by integrating the electron/hole generation rate over the device geometry,

$$
\begin{aligned}
I_{gidl} &= I_{\text{BTBT-GIDL}} + I_{\text{TAT-GIDL}} \\
&= W \int (G_{\text{BTBT}} + G_{\text{TAT}}) dx
\end{aligned}
\tag{6.1}
$$

where W is the width of the device and $G_{\text{BTBT}}(G_{\text{TAT}})$ is the generation rate for BTBT(TAT).

By integrating G_{BTBT} over the depletion region with WKB approximation, the BTBT current density is obtained as

$$
J = A \times E_s \times e^{\left(\frac{-B}{E_s}\right)}
\tag{6.2}
$$

where A is a preexponential constant related to the density of states of both the emitting and the receiving sides, B is a physical exponential parameter, which depends on E_g and the carrier's effective mass in the tunneling direction (≈ 20 MV/cm for silicon), and E_s is the surface electric field in the drain. Using Gauss's law at the onset of GIDL, when the band bending in the drain is equal to E_g, E_s is given by

$$
E_s = \left(\frac{V_{ds} - V_{gs} - E_g + V_{fbsd}}{\epsilon_{ratio} \cdot EOT} \right)
\tag{6.3}
$$

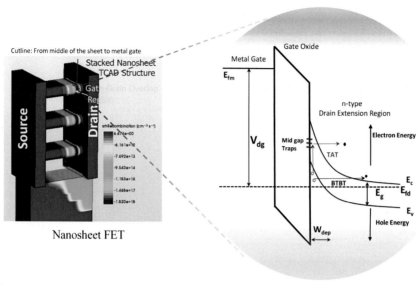

FIG. 6.5

Schematic diagram of the GIDL effect in FinFETs/nanosheet FETs. Due to the low vertical electric field, the generated carriers in the overlap region go to CB from VB through the mid-gap trap, resulting in TAT. BTBT tunneling is also represented.

In Eq. (6.3), V_{fbsd} is the flat-band voltage between the gate and the drain, ϵ_{ratio} is the ratio of the dielectric constant of the substrate material EPSRSUB over that of silicon dioxide, and EOT is the equivalent oxide thickness. The approximation of surface field E_s is very much sufficient for BTBT-based GIDL model. Eqs. (6.2), (6.3) lead to the following equation for the GIDL current in BSIM-CMG:

$$
\begin{aligned}
I_{\text{BTBT-GIDL}} = {} & AGIDL \cdot W_{\text{eff}} \cdot \left(\frac{V_{\text{ds}} - V_{\text{gs}} - EGIDL + V_{\text{fbsd}}}{\epsilon_{\text{ratio}} \cdot EOT} \right)^{PGIDL} \\
& \times \exp\left(-\frac{\epsilon_{\text{ratio}} \cdot EOT \cdot BGIDL(T)}{V_{\text{ds}} - V_{\text{gs}} - EGIDL + V_{\text{fbsd}}} \right) \times NFIN
\end{aligned}
\tag{6.4}
$$

where the constants A and B in Eq. (6.2) and E_g in Eq. (6.3) have been replaced by the model parameters AGIDL and BGIDL, and EGIDL, respectively, and PGIDL has been introduced for more flexibility in fitting the measured data. The GISL current is calculated in the same manner

$$
\begin{aligned}
I_{\text{BTBT-GISL}} = {} & AGISL \cdot W_{\text{eff}} \cdot \left(\frac{V_{\text{ds}} - V_{\text{gs}} - EGISL + V_{\text{fbsd}}}{\epsilon_{\text{ratio}} \cdot EOT} \right)^{PGISL} \\
& \times \exp\left(-\frac{\epsilon_{\text{ratio}} \cdot EOT \cdot BGISL(T)}{V_{\text{ds}} - V_{\text{gs}} - EGISL + V_{\text{fbsd}}} \right) \times NFIN
\end{aligned}
\tag{6.5}
$$

However, for TAT-based GIDL contribution the nonlinear dependence of peak maximum field on V_{dg} is given by

$$
E_{\text{smax}} = \frac{BTAT(V_{\text{ds}} - V_{\text{gs}})^2 - CTAT(V_{\text{ds}} - V_{\text{gs}}) - DTAT + V_{\text{fbsd}}}{\epsilon_{\text{ratio}} \times EOT}
\tag{6.6}
$$

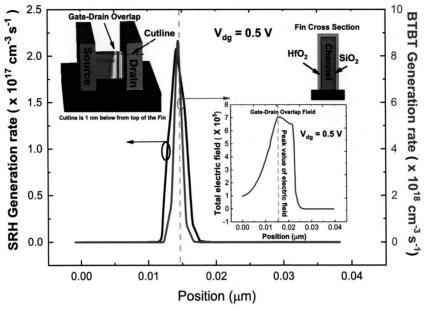

FIG. 6.6

TCAD simulation of FinFET. BTBT generation rate and SRH generation rates are plotted against gate-drain overlap region length, at $V_{dg} = 0.5$ V. Total electric field along the gate-drain overlap length is plotted in the inset graph, for $V_{dg} = 0.5$ V. The cutline in FinFET structure is taken at 1 nm below the Si-SiO$_2$ interface. Cutline starts at the middle of the Fin and ends inside the drain region.

The E_{smax} approximation is further proved using Sentaurus technology computer-aided design (TCAD) simulations, it is used to obtain the electrostatic field in gate-drain overlap region. TCAD simulation uses the FinFET with a single fin, an HfO$_2$ dielectric (effective oxide thickness (EOT) = 1.6 nm), a channel length of 25 nm, and a fin width of 15 nm. As shown in Fig. 6.6, both the BTBT generation rate and the TAT generation rate exhibit a maxima at $\approx$0.015 μm and these maximum extend slightly (<5 nm) on both sides of the maximum generation rate in gate-drain overlap region. This generation is due to peak electric field at the same position ($\approx$0.015 μm) (with 5 nm of excursion on both the sides of the peak electric field) in the gate-drain overlap region (refer inset graph in Fig. 6.6); for that reason, the analytical models for GIDL current in FinFET can be obtained by considering only the maximum electric field in the gate-drain overlap region. Validation of E_{smax} approximation with TCAD data of FinFET is shown in Fig. 6.7, the accurate matching of E_{smax} with TCAD maximum field for a different gate-drain voltage shows that the F_{max} approximation is also valid for FinFET/GAA devices.

In the strained technology and fully depleted Fin/sheet structure, the presence of traps and low gate-induced vertical field enhances the generation/recombination

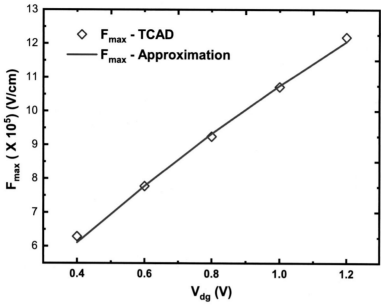

FIG. 6.7

Maximum electric field with V_{dg}. The accurate matching of F_{max} with TCAD maximum field for different gate-drain bias shows that the F_{max} approximation is also valid for FinFET/GAA devices.

rate ($G_{SRH} * \Gamma$) because of increased capture/emission probability (increased field-effect factor, Γ) [8]. Hence, the effective generation/recombination rate becomes $G_{SRH} * \Gamma$. By integrating effective field generation/recombination rate ($G_{SRH} * \Gamma$) over the gate-drain overlap depletion region and using maximum surface field approximation (E_{smax}) the total TAT current can be obtained as

$$I_{\text{TAT-GIDL}} = \int_0^{x_n} \Gamma(x) \cdot G_{SRH}(x) \cdot dx \qquad (6.7)$$

Using P-N junction diode theory, the rate of generation of electron-hole pair in the reverse bias condition is given by

$$G_{SRH} = \frac{\tau^{-1} \cdot n_i}{\left[2 \cdot \cosh\left(\dfrac{E_t - E_{fi}}{k_B T} \right) \right]} \qquad (6.8)$$

where τ is a carrier lifetime, n_i is the intrinsic carrier concentration, and G_{SRH} gives a maximum contribution of generation, when the fermi-level E_{fi} coincides with the mid-gap trap level E_t, and the generation rate decreases exponentially as E_t moves from the middle of the band gap in either direction; under the maximum electron-hole pair generation condition, generation rate in the gate-drain overlap region is given by

$$G_{SRH} = \tau^{-1} \cdot n_i \tag{6.9}$$

the resultant TAT current can be written as

$$I_{TAT} = \tau^{-1} \cdot n_i \int_0^{x_n} \Gamma(x)dx \tag{6.10}$$

where x is the vertical direction in the gate-drain overlap region, starting from Si-SiO$_2$ interface. Using all the mentioned approximations and integrating Eq. (6.10) with error function approximation, the TAT-GIDL current is given by [8]

$$I_{TAT\text{-}GIDL} = A_{TAT} \cdot W_\Gamma \cdot n_i(T) \cdot \Gamma_{max} \tag{6.11}$$

where ATAT is a model parameter, and theoretically it is equal to $\sigma \cdot V_{th} \cdot N_t$, σ is the capture cross-section for electrons and holes, and N_t is the trap density. W_Γ is the width over which TAT dominates and is given as

$$W_\Gamma = W_{dep} \cdot \left(D_{TAT} \cdot h_{max}^{3/2} \cdot + 1 \right)^{-1} \tag{6.12}$$

Here,

$$D_{TAT} = \frac{F_{nor}}{E_{smax}}$$
$$F_{nor} = \frac{4}{3} \cdot \frac{\sqrt{2 \cdot m^* \cdot \Delta E^3}}{q \cdot \hbar} \tag{6.13}$$
$$\Delta E = \frac{E_g}{2}$$

m^* is the effective mass of carriers. E_{smax} is the maximum field approximation, the transition from the low- to the high-field regime for TAT is taken care by h_{max} [8]. The integration of $\Gamma(x)$ has already been performed using the previous method and maximum field approximation; Γ_{max} is derived in Ref. [8] and also reproduced here:

$$\Gamma_{max} = \frac{E_{TAT} \cdot \exp(R_{TAT}) \cdot erfc[P_{TAT} \cdot (Q_{TAT} - 1)] \cdot \sqrt{\pi}}{2 \cdot P_{TAT}} \tag{6.14}$$

where

$$E_{TAT} = \frac{\Delta E}{K_B T}$$
$$P_{TAT} = \sqrt{\frac{3}{8} \cdot D_{TAT} \cdot h_{max}^{-1/2}}$$
$$Q_{TAT} = \frac{4}{3} \cdot \frac{E_{TAT}}{D_{TAT}} \cdot h_{max}^{1/2} - h_{max} \tag{6.15}$$
$$R_{TAT} = \frac{2}{3} \cdot \frac{E_{TAT}^2}{D_{TAT}} \cdot h_{max}^{1/2} - E_{TAT} \cdot h_{max} + \frac{D_{TAT}}{2} \cdot h_{max}^{3/2}$$

In the Γ expression, the low sensitive terms are neglected since the exponential is a rapidly varying function. The mathematical simplification with first-order approximation shows that Γ is exponentially dependent on the ratio of field in gate-drain overlap region to thermal voltage (V_t),

$$\Gamma \propto e^{\left(\frac{\text{Electric field in gate-drain overlap region}}{V_t}\right)} \tag{6.16}$$

Considering the approximation given in Eq. (6.16), Eq. (6.11) is approximated as the compact model suitable for TAT-GIDL expression as [10]

$$I_{\text{TAT-GIDL}} = ATATD \cdot W_{\text{eff}} \cdot n_i(T) \cdot$$
$$\times \exp\left(\frac{BTATD \cdot V_{\text{dg}}^2 - CTATD(T) \cdot V_{\text{dg}} - DTATD + V_{\text{fbsd}}}{V_t}\right) \tag{6.17}$$

$$CTATD(T) = CTATD \cdot (1 + TTATD \cdot (T - T_{\text{nom}})) \tag{6.18}$$

where W_{eff} is the parameter for the effective width of TAT. BTATD, CTATD, and DTATD are the model parameters and determine the effective field in the gate-drain overlap region (the effective electric field is modeled using a second-order polynomial expression [8]) and also absorb the effect of doping, different spacer design, gate-drain overlap length, and oxide thickness. V_{fbsd} is the flat band voltage between the gate and source/drain region. CTAT is a temperature-dependent parameter with TTAT as a temperature coefficient of a maximum field in the overlap region. The TAT-GISL current is calculated in the same manner

$$I_{\text{TAT-GISL}} = ATATS \cdot W_{\text{eff}} \cdot n_i(T) \cdot$$
$$\times \exp\left(\frac{BTATS \cdot V_{\text{dg}}^2 - CTATS(T) \cdot V_{\text{dg}} - DTATS + V_{\text{fbsd}}}{V_t}\right) \tag{6.19}$$

$$CTATS(T) = CTATS \cdot (1 + TTATS \cdot (T - T_{\text{nom}})) \tag{6.20}$$

The developed model is validated with the state-of-the-art pFinFET and nFinFET data, as shown in Fig. 6.9A and B, respectively. Nanosheet GIDL leakage current is also validated with TCAD simulation data in Fig. 6.10A. The developed model can also capture TCAD data for nFinFETs for a broad temperature range, as shown in Fig. 6.10B. The accurate matching of the model with data proves that TAT contributes at low field and BTBT contributes at high field. Figs. 6.9 and 6.10 show that the leakage currents for various temperatures begin to collapse at high fields due to the TAT being taken out by BTBT. However, because BTBT has a far weaker temperature dependence than TAT, very little temperature sensitivity is seen during high-field operations. Physically, the TAT expression in Eqs. (6.17), (6.19) also indicates that, as shown in Fig. 6.8, the tunneling probability decreases as temperature increases because the thermal energy of the carriers (thermally excited carriers can readily migrate from the trap to the CB by thermionic emission) masks the bias dependency (field). Tunneling from the trap to the CB is necessary because at the low temperature, the field in the gate-drain overlap area will predominate over the thermal excitation. Thus, it makes sense for exponentials to have temperature and field dependence.

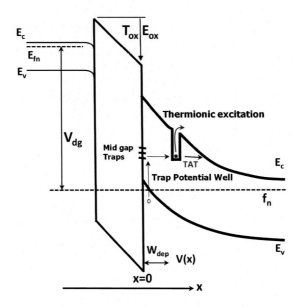

FIG. 6.8

Simplified energy-band diagram in a low-field gate-drain overlap region. The nonavailability of overlapping states between VB and CB pushes the electron to go to the CB via the mid-gap trap. However, at the higher temperature, the electron that is trapped in the trap potential is likely to escape to the CB rather than tunneling to the CB from the trap potential well and, hence, TAT is masked by thermionic excitation, and the bias dependence on anomalous body current is diminished in forward biased.

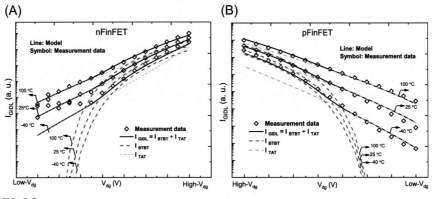

FIG. 6.9

Validation of the proposed GIDL model. (A) n-channel FinFET fabricated by industry laboratory, the *solid line* shows model fitting with measurement data. (B) p-channel FinFET fabricated by industry laboratory, *solid line* shows model fitting with measurement data.

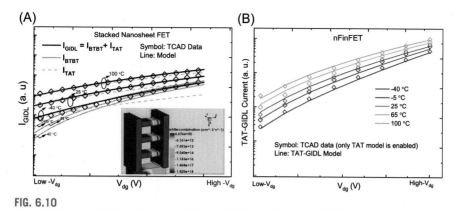

FIG. 6.10

Validation of the proposed GIDL model (BTBT + TAT) with measurement and TCAD data. (A) n-channel stacked nanosheet FET and (B) TAT-GIDL current validation for a broad range of temperatures.

In addition to the V_{dg} dependence present in Eqs. (6.4), (6.17), in bulk FinFETs/nanosheets (BULKMOD$\neq$0), the GIDL current is also affected by the substrate bias for small values of V_{de} (the drain to substrate voltage) as the deep depletion condition in the drain surface starts to fail. The total GIDL current is obtained by multiplying I_{gidl} from Eq. (6.1) with an empirical factor for modeling the low V_{de} effect as follows:

$$I_{gidl} = \begin{cases} (I_{\text{BTBT-GIDL}} + I_{\text{TAT-GIDL}}) \cdot \dfrac{V_{de}^3}{CGIDL_i + V_{de}^3} & \text{for } BULKMOD = 1 \\ (I_{\text{BTBT-GIDL}} + I_{\text{TAT-GIDL}}) \cdot V_{ds} & \text{for } BULKMOD = 0 \end{cases} \tag{6.21}$$

In Eq. (5.5), CGIDL is a nonnegative fitting parameter. A similar equation holds for GISL,

$$I_{gisl} = \begin{cases} (I_{\text{BTBT-GISL}} + I_{\text{TAT-GISL}}) \cdot \dfrac{V_{de}^3}{CGISL_i + V_{de}^3} & \text{for } BULKMOD = 1 \\ (I_{\text{BTBT-GISL}} + I_{\text{TAT-GISL}}) \cdot V_{ds} & \text{for } BULKMOD = 0 \end{cases} \tag{6.22}$$

Conventional GIDL/GISL current is calculated by GIDLMOD = 1 solely utilizing BTBT-based physics. The substrate parasitic GIDL/GISL components (Igidlb/Igislb) are calculated separately from the direct drain to source/source to drain GIDL/GISL components (Igidl/Igisl) for nanosheet devices with BULKMOD$\neq$0 and GIDLMOD = 2. To model TAT-based GIDL/GISL in addition to conventional BTBT for GAA/nanosheet devices, GIDLMOD = 3 is introduced in the BSIM-CMG model.

6.4 Gate oxide tunneling

For decades, to help the gate maintain its supremacy against the drain in controlling the source-to-channel barrier, the gate's silicon dioxide (silicon oxynitride) thickness was scaled down in proportion to L_g. In the early 2000s, the tunneling through the scaled silicon oxynitride started to dominate the transistor's off-state leakage current, making it intolerable. A thicker dielectric layer with a higher dielectric constant (K) was required. A thick, high-K gate oxide could retain the control of the gate over the channel with orders of magnitude reduction in dielectric leakage current compared with SiO_2 of the same EOT. Furthermore, a metal gate could eliminate the polysilicon gate depletion effect, which effectively increasing the gate dielectric thickness and thus reducing the gate control of the channel. High-K oxides, in general, were also found to form a better interface with metal gates than the traditional polysilicon gates. These led to the introduction of high-K metal-gate technology in the 45 nm node [11], which was extended to the successive nodes. However, the gate tunneling leakage through the gate oxide remains a significant and increasing concern as each new technology generation requires a smaller EOT.

6.4.1 Gate oxide tunneling formulation in BSIM-CMG

The gate oxide tunneling in the BSIM-CMG model inherits a similar formulation to that of BSIM4. Although the formulation has been derived for polysilicon-silicon oxide gate stack, it turns out to be accurate enough to be used for high-K metal-gate technology thanks to its flexibility. As illustrated in Fig. 6.11, the gate tunneling current is composed of several mechanisms: the gate-to-body leakage current I_{gb}, the leakage currents through gate-to-source and gate-to-drain overlaps I_{gs} and I_{gd}, and the gate-to-inverted channel tunneling current I_{gc}. Part of I_{gc} is collected by the source (I_{gcs}) while the rest goes to the drain (I_{gcd}). I_{gb}, I_{gs}, I_{gd}, and I_{gc} are determined from the MOS capacitor, dielectric leakage model described in Eq. (6.23). Then, I_{gc} is extended to nonzero V_{ds} and partitioned into I_{gcs} and I_{gcd}.

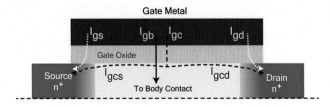

FIG. 6.11

Half cross-section of the fin of the FinFET/GAA FET shown in Fig. 6.3. The components of the tunneling current are shown. I_{gs} and I_{gd} are tunneling currents in the gate-to-source/drain overlap regions; I_{gb} flows between the gate and body; I_{gc} is the gate-to-channel tunneling current and it is partitioned into I_{gcs} and I_{gcd}, which flow out of the source and drain, respectively.

On the basis of the early work of Lee and Hu [12], the dielectric tunneling leakage current density of an MOS capacitor can be modeled as

$$J_g = A \times \left(\frac{TOXREF}{TOXG}\right)^{NTOX} \times \frac{(V_{ge} \times V_{aux})}{TOXG^2}$$

$$\times e^{(-B \times (\alpha - \beta \cdot |V_{ox}|) \times (1 + \gamma \cdot |V_{ox}|) \times TOXG)}$$

(6.23)

where $A = \frac{q^2}{(8\pi h \phi_b)}$, $B = \frac{(8\pi \sqrt{(2q \cdot m_{ox} \phi_b^{(3/2)})})}{3h}$, ϕ_b is the tunneling barrier height, m_{ox} is the effective carrier mass in the oxide, TOXG is the oxide thickness (different from TOXP to introduce more flexibility), TOXREF is the reference oxide thickness at which all the parameters are extracted, NTOX is a fitting parameter that defaults to 1, V_{aux} is an auxiliary function that represents the density of tunneling carriers as well as available energy states to tunnel into, and α, β, and γ are fitting parameters. Depending on the mode of operation (accumulation or depletion/inversion) and the gate tunneling component of interest, values of m_{ox}, ϕ_b, and V_{aux} will be different as explained in the following sections.

6.4.2 Gate-to-body tunneling current in depletion/inversion

Fig. 6.12 schematically demonstrates the dominant leakage mechanism between the gate and body in depletion/inversion, represented by I_{gbinv}. In both PMOS and NMOS, the electrons tunnel from the VB of the body into the gate material. For this case, the values of A, B, and V_{aux} for a Si-SiO$_2$ interface (silicon as the body and silicon oxide as the gate oxide) are calculated to be

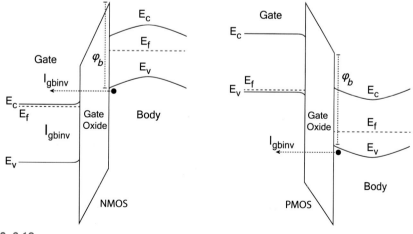

FIG. 6.12

In both NMOS and PMOS, tunneling of valence-band electrons from the body into the gate is the principal cause of the gate-to-body tunneling current in inversion, I_{gbinv}.

$$A = 3.75956 \times 10^{-7} \left(\frac{A}{V^2}\right) \tag{6.24}$$

$$B = 9.82222 \times 10^{11} \left(\frac{g}{Fs^2}\right)^{0.5} \tag{6.25}$$

and

$$V_{\text{aux,igbinv}} = NIGBINV \cdot \frac{kT}{q} \cdot \ln\left(1 + \exp\left(\frac{V_{\text{ox}} - EIGBINV}{NIGBINV \cdot \frac{kT}{q}}\right)\right) \tag{6.26}$$

In Eq. (6.26), NIGBINV and EIGBINV are model parameters. The total gate tunneling current I_{gbinv} is then given by

$$I_{\text{gbinv}} = W_{\text{eff0}} \cdot L_{\text{eff}} \cdot A \cdot \left(\frac{TOXREF}{TOXG}\right)^{NTOX} \cdot \frac{(V_{\text{ge}} \times V_{\text{aux,inv}})}{TOXG^2} \cdot NFIN_{\text{total}} \tag{6.27}$$
$$\times \exp\left(-B \cdot TOXG \cdot (AIGBINV(T) - BIGBINV \cdot q_{\text{ia}}) \cdot (1 + CIGBINV \cdot q_{\text{ia}})\right)$$

where α, β, γ, and V_{ox} in Eq. (6.23) have been replaced by the model parameters AIGBINV(T), BIGBINV, CIGBINV, and the average charge in the channel, q_{ia}, respectively.[d] The last approximation is valid since we assume that the fin is fully depleted and the body charge q_{ba} is a fixed value that can be incorporated into other model parameters.

6.4.3 Gate-to-body tunneling current in accumulation

In accumulation, the dominant leakage current between the gate and body I_{gbacc} is the tunneling of conduction-band electrons. In NMOS, the electrons tunnel from the CB of the gate material into the CB of the body and in PMOS in the reverse direction (see Fig. 6.13). For this case, the values of A, B, and V_{aux} for a polysilicon-silicon oxide-silicon structure are calculated to be

$$A = 4.97232 \times 10^{-7} \left(\frac{A}{V^2}\right) \tag{6.28}$$

$$B = 7.45669 \times 10^{11} \left(\frac{g}{Fs^2}\right)^{0.5} \tag{6.29}$$

and

$$V_{\text{aux,igbacc}} = NIGBACC \cdot \frac{kT}{q} \cdot \ln\left(1 + \exp\left(\frac{V_{\text{fb}} - V_{\text{ge}}}{NIGBACC \cdot \frac{kT}{q}}\right)\right) \tag{6.30}$$

---

[d] Note that all the charges in the BSIM-CMG model are normalized with respect to Cox. That is why one can substitute a voltage with a charge.

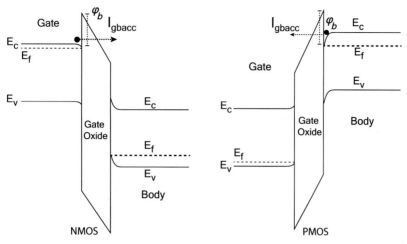

FIG. 6.13

In accumulation, conduction-band electrons tunnel from the gate into the body in NMOS and from the body into the gate in PMOS.

In Eq. (6.30), NIGBACC is a model parameter. The total gate tunneling current I_{gbacc} is then given by

$$I_{\text{gbacc}} = W_{\text{eff0}} \cdot L_{\text{eff}} \cdot A \cdot \left(\frac{TOXREF}{TOXG}\right)^{NTOX} \cdot \frac{(V_{\text{ge}} \times V_{\text{aux,igbacc}})}{TOXG^2} \cdot NFIN_{\text{total}}$$
$$\times \exp\left(-B \cdot TOXG \cdot (AIGBACC(T) - BIGBACC \cdot V_{\text{oxacc}}) \cdot (1 + CIGBACC \cdot V_{\text{oxacc}})\right)$$

(6.31)

where α, β, γ, and V_{ox} in Eq. (6.23) have been replaced by the model parameters **AIGBACC(T)**, **BIGBACC**, **CIGBACC**, and V_{oxacc}, respectively.

For BULKMOD$\neq$0, I_{gb} (i.e., $I_{\text{gbinv}} + I_{\text{gbacc}}$) simply flows from the gate into the substrate. For BULKMOD $= 0$, I_{gb} mostly flows into the source because the potential barrier for holes is typically lower at the source side. To ensure continuity when V_{ds} switches sign, I_{gb} is partitioned into a source component I_{gbs} and a drain component I_{gbd} using the following partitioning scheme:

$$I_{\text{gbs}} = (I_{\text{gbinv}} + I_{\text{gbacc}}) \cdot W_{\text{f}}$$

(6.32)

$$I_{\text{gbd}} = (I_{\text{gbinv}} + I_{\text{gbacc}}) \cdot W_{\text{r}}$$

(6.33)

$$W_{\text{f}} = \frac{1}{2} + \frac{1}{2} \times \tanh\left(\frac{(0.6 \times q \times V_{\text{ds}})}{kT}\right)$$

(6.34)

$$W_{\text{r}} = \frac{1}{2} - \frac{1}{2} \times \tanh\left(\frac{(0.6 \times q \times V_{\text{ds}})}{kT}\right)$$

(6.35)

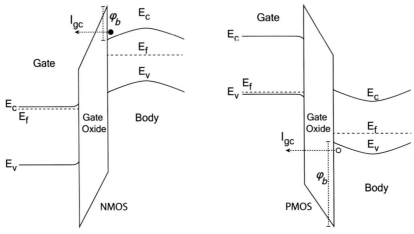

FIG. 6.14

In inversion for NMOS, conduction-band electrons tunnel from the channel into the gate, whereas in PMOS, the valence-band holes tunnel from the channel into the gate.

6.4.4 Gate-to-channel tunneling current in inversion

As shown in Fig. 6.14, in inversion, the electrons (holes in PMOS) tunnel from the inversion channel into the CB of the gate (VB for PMOS). This results in different values of A and B for NMOS and PMOS:

$$
A = \begin{cases} 4.97232 \times 10^{-7} \left(\dfrac{A}{V^2} \right) & \text{for} \quad \text{NMOS} \\[2mm] 3.42536 \times 10^{-7} \left(\dfrac{A}{V^2} \right) & \text{for} \quad \text{PMOS} \end{cases}
\tag{6.36}
$$

$$
B = \begin{cases} 7.45669 \times 10^{11} \left(\dfrac{g}{Fs^2} \right)^{0.5} & \text{for} \quad \text{NMOS} \\[2mm] 1.16645 \times 10^{12} \left(\dfrac{g}{Fs^2} \right)^{0.5} & \text{for} \quad \text{PMOS} \end{cases}
\tag{6.37}
$$

The auxiliary function $V_{\text{aux,igc}}$ can be shown to be

$$
V_{\text{aux,igc}} = V_{\text{ox}} \times (V_{\text{ge}} - 0.5 \cdot V_{\text{dsx}} + 0.5 \cdot V_{\text{es}} + 0.5 \cdot V_{\text{ed}})
\tag{6.38}
$$

The total gate-to-channel tunneling component at zero V_{ds} can be written as

$$
\begin{aligned}
I_{\text{gc0}} = {} & W_{\text{eff}} \cdot L_{\text{eff}} \cdot A \cdot \left(\frac{TOXREF}{TOXG} \right)^{NTOX} \cdot \frac{(V_{\text{ge}} \times V_{\text{aux,igc}})}{TOXG^2} \cdot NFIN_{\text{total}} \\
& \times \exp(-B \cdot TOXG \cdot (AIGC(T) - BIGC \cdot q_{\text{ia}}) \cdot (1 + CIGC \cdot q_{\text{ia}}))
\end{aligned}
\tag{6.39}
$$

To consider the drain bias effect, a current continuity equation is solved analytically along the channel, which extends I_{gc0} to nonzero V_{ds} and splits it into two components, I_{gcs} and I_{gcd}. For a detailed discussion on the derivation of this physical current partitioning factor, refer to Ref. [13]. The expressions for I_{gcs} and I_{gcd} are as follows:

$$I_{gcs} = I_{gc0} \cdot \frac{PIGCD \cdot |V_{dseff}| + e^{(PIGCD \cdot |V_{dseff}|)} - 1.0}{PIGCD^2 \cdot V_{dseff}^2} \tag{6.40}$$

$$I_{gcd} = I_{gc0} \cdot \frac{1.0 - (PIGCD \cdot |V_{dseff}| + 1.0) \times e^{(-PIGCD \cdot |V_{dseff}|)}}{PIGCD^2 \cdot V_{dseff}^2} \tag{6.41}$$

where PIGCD is a fitting parameter added for flexibility with a default value of unity.

6.4.5 Gate-to-source/drain current

The n^+ (p^+) gate to n^+ (p^+) source and drain currents I_{gs} and I_{gd} are principally caused by the tunneling of the conduction-band electrons in NMOS and VB holes in PMOS, as shown in Fig. 6.15. In NMOS, the electrons tunnel from the CB of the body into the gate. In PMOS, the holes tunnel from the VB of the body into the gate.

For this case, the parameters A and B are naturally equal to those given by Eqs. (6.36), (6.37), respectively. If the gate material is a metal, V_{aux} is also simplified to be equal to $|V_{gs}|$ and $|V_{gd}|$ for I_{gs} and I_{gd}, respectively.

The total gate-to-source extension tunneling component is

$$\begin{aligned}
I_{gs} = W_{eff} \cdot DLCIGS \cdot A \cdot & \left(\frac{TOXREF}{TOXG \times POXEDGE} \right)^{NTOX} \cdot NFIN_{total} \\
& \times \frac{(V_{gs} \times |V_{gs}|)}{(TOXG \times POXEDGE)^2} e^{(-B \cdot TOXG \times POXEDGE \cdot (AIGS(T) - BIGS \cdot |V_{gs}|) \cdot (1 + CIGC \cdot |V_{gs}|))}
\end{aligned} \tag{6.42}$$

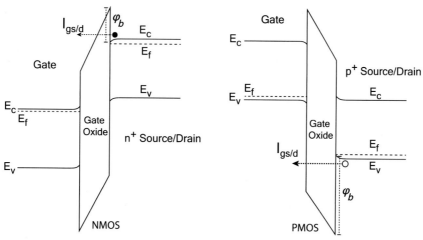

FIG. 6.15

In NMOS, the tunneling of the conduction-band electrons and in PMOS the tunneling of valence-band holes make the source/drain-gate overlap leakage. The figure shows the band diagrams for an inverted channel. In accumulation, the direction of the tunneling is reversed.

In Eq. (6.42), DLCIGS is the length of the gate-source overlap region and POX-EDGE is a factor for the gate oxide thickness in the source/drain extension regions.

Similarly, for the total gate-to-drain extension tunneling component we have

$$
\begin{aligned}
I_{gd} = W_{eff} \cdot DLCIGD \cdot A \cdot \left(\frac{TOXREF}{TOXG \times POXEDGE} \right)^{NTOX} \cdot NFIN_{total} \\
\times \frac{(V_{gd} \times |V_{gd}|)}{(TOXG \times POXEDGE)^2} e^{(-B \cdot TOXG \times POXEDGE \cdot (AIGD(T) - BIGD \cdot |V_{gd}|) \cdot (1 + CIGD \cdot |V_{gd}|))}
\end{aligned}
\tag{6.43}
$$

where DLCIGD is the length of the gate-drain overlap region.

6.5 Impact ionization

At a transistor on-state, because of the high-electric field near the drain end of the channel, carriers in this region can gain enough kinetic energy to ionize the lattice atoms when they collide. This collision frees an electron from the VB and leaves a hole behind. The generated hole will drift to the substrate and it will increase the substrate leakage. The released high-energy electron (hot carrier) is collected by the drain and it will be a part of the drain current. Also, there is a chance that the generated hot electron travels along the gate field and penetrates into the gate oxide. Hot carrier injection into the gate oxide over time can damage the oxide and cause reliability problems.

The local impact ionization current $I_{sub}(y)$ can be written as a function of the channel current (an increase in the number of carriers will increase the chance of collisions) and the strength of the local electric field (the stronger the electric filed, the higher the kinetic energy of carriers) as follows:

$$
I_{ii}(y) = I_{ds} \cdot A_i \cdot e^{\frac{-B_i}{E_l(y)}}
\tag{6.44}
$$

where A_i and B_i are two material constants and represent how often the impact ionization events take place and the critical field to trigger the events, respectively, and $E_l(y)$ is the longitudinal electric field along the transport direction. By integrating Eq. (6.44) along the length of the channel where velocity saturation happens, we can write the total impact ionization current as

$$
I_{ii} = I_{ds} \cdot A_i \int_{y=0}^{y='l} \cdot e^{\frac{-B_i}{E_l(y)}} dy
\tag{6.45}
$$

where $y = 0$ is the starting point of the velocity saturation region and $'l$ is the length of this region. The integration in Eq. (6.45) can be calculated (see Ref. [14], Chapter 5 for details) to give the following equation for the BSIM-CMG impact ionization model:

$$
I_{ii} = \frac{A_i}{B_i} \cdot I_{ds} \cdot (V_{ds} - V_{dsat}) \cdot e^{\frac{-B_i \cdot \lambda}{V_{ds} - V_{dsat}}}
\tag{6.46}
$$

where V_{dsat} is the saturation voltage (see Chapter 5) and λ is the characteristic length (see Chapter 5). The first impact ionization model (IIMOD = 1) has implemented Eq. (6.46) as

$$I_{ii} = \frac{ALPHA0(T) + ALPHA1(T) \cdot L_{eff}}{L_{eff}} (V_{ds} - V_{dseff}) \cdot e^{\frac{BETA0(T)}{V_{ds} - V_{dseff}}} \cdot I_{ds} \tag{6.47}$$

In Eq. (6.47), V_{dseff} is the effective drain voltage resulted from the smooth transition of V_{ds} to V_{dsat} (see Chapter 5), BETA0 is a fitting parameter, and the term $\frac{ALPHA0}{L_{eff}}$ has been introduced to improve the length dependence of I_{ii} over a wide range of channel lengths. There are approximations involved in deriving Eqs. (6.46), (6.47), including a linear dependence of $E_1(y)$ on $(V_{ds} - V_{dsat})$. The BSIM-CMG's second impact ionization model can be activated (IIMOD = 2) and used if a more flexible model is needed:

$$\begin{aligned} I_{ii} &= \frac{ALPHAII0(T) + ALPHAII1(T) \cdot L_{eff}}{L_{eff}} \cdot I_{ds} \\ &\cdot \exp\left(\frac{V_{diff}}{BETAII2 + BETAII1 \cdot V_{diff} + BETAII0 \cdot V_{diff}^2}\right) \end{aligned} \tag{6.48}$$

$$V_{diff} = V_{ds} - V_{dsatii} \tag{6.49}$$

$$V_{dsatii} = V_{gsStep} \cdot \left(1 - \frac{LII}{L_{eff}}\right) \tag{6.50}$$

$$V_{gsStep} = \left(\frac{ESATII \cdot L_{eff}}{1 + ESATII \cdot L_{eff}}\right)\left(\frac{1}{1 + SII1 \cdot V_{gsfbeff}} + SII2\right)\left(\frac{SII0(T) \cdot V_{gsfbeff}}{1 + SIID \cdot V_{ds}}\right) \tag{6.51}$$

Here, BETAII0, BETAII0, and BETAII0 are V_{ds}-dependent fitting parameters, LII is a channel length-dependent parameter, SII0, SII1, and SII2 are V_{gs}-dependent fitting parameters, and finally, ESATII is the channel saturation field with the default value of 1E7 V/m.

References

[1] H. Kam, V. Pott, R. Nathanael, J. Jeon, E. Alon, T.-J.K. Liu, Design and reliability of a micro-relay technology for zero-standby-power digital logic applications, in: 2009 IEEE International Electron Devices Meeting (IEDM), 2009, pp. 1–4.

[2] K. Jeon, W.Y. Loh, P. Patel, et al., Si tunnel transistors with a novel silicided source and 46 mV/dec swing, in: VLSI Symp. Tech. Dig., 2010, pp. 121–122.

[3] C.-H. Jan, U. Bhattacharya, R. Brain, S.-J. Choi, G. Curello, G. Gupta, W. Hafez, M. Jang, M. Kang, K. Komeyli, T. Leo, N. Nidhi, L. Pan, J. Park, K. Phoa, A. Rahman, C. Staus, H. Tashiro, C. Tsai, P. Vandervoorn, L. Yang, J.-Y. Yeh, P. Bai, A 22 nm SoC platform technology featuring 3-D tri-gate and high-k/metal gate, optimized for ultra low power, high performance and high density SoC applications, in: 2012 IEEE International Electron Devices Meeting (IEDM), 2012. pp. 3.1.1–3.1.4.

[4] D. Prasad, et al., Buried power rails and back-side power grids: Arm® CPU power delivery network design beyond 5 nm, in: 2019 IEEE International Electron Devices Meeting (IEDM), pp. 19.1.1–19.1.4.

[5] J. Ryckaert, et al., Enabling sub-5 nm CMOS technology scaling thinner and taller! in: 2019 IEEE International Electron Devices Meeting (IEDM), 2019. pp. 29.4.1–29.4.4.

[6] T.-Y. Chan, J. Chen, P.-K. Ko, C. Hu, The impact of gate-induced drain leakage current on MOSFET scaling, in: 1987 International Electron Devices Meeting, vol. 33, 1987, pp. 718–721.

[7] C.C. Hu, Modern Semiconductor Devices for Integrated Circuits, Prentice Hall, Upper Saddle River, NJ, 2009.

[8] C.K. Dabhi, A.S. Roy, Y.S. Chauhan, Compact modeling of temperature-dependent gate-induced drain leakage including low-field effects, IEEE Trans. Electron Devices 66 (7) (2019) 2892–2897, https:/doi.org/10.1109/TED.2019.2918332.

[9] C.K. Dabhi, A.S. Roy, L. Yang, Y.S. Chauhan, Anomalous GIDL effect with back bias in FinFET: physical insights and compact modeling, IEEE Trans. Electron Devices 68 (7) (2021) 3261–3267, https:/doi.org/10.1109/TED.2021.3083483.

[10] C.K. Dabhi, G. Pahwa, S. Salahuddin, C. Hu, Compact model for trap assisted tunneling based GIDL, in: 2022 Device Research Conference (DRC), Columbus, OH, 2022, pp. 1–2, https:/doi.org/10.1109/DRC55272.2022.9855798.

[11] C. Auth, M. Buehler, A. Cappellani, C.-H. Choi, G. Ding, W. Han, S. Joshi, B. McIntyre, M. Prince, P. Ranade, J. Sandford, C. Thomas, 45 nm high-k + metal-gate strain-enhanced transistors, Intel Technol. J. 12 (2) (2008) 77–86.

[12] W.-C. Lee, C. Hu, Modeling gate and substrate currents due to conduction- and valence-band electron and hole tunneling [CMOS technology], in: 2000 Symposium on VLSI Technology. Digest of Technical Papers, 2000, pp. 198–199.

[13] K.M. Cao, W.-C. Lee, W. Liu, X. Jin, P. Su, S.K.H. Fung, J.X. An, B. Yu, C. Hu, BSIM4 gate leakage model including source-drain partition, in: International Electron Devices Meeting 2000. Technical Digest. IEDM'00, 2000, pp. 815–818.

[14] W. Liu, C.C. Hu, BSIM4 and MOSFET Modeling for IC Simulation, World Scientific, Singapore, 2011.

Charge, capacitance, and nonquasi-static effects

7

Essential to IC design with FinFET is a compact model that can accurately predict the device's dynamic behavior in addition to DC/operating point analysis. This is because, in practical circuits, the transistor's terminal voltages are not static and vary with the applied, time-dependent, large or small signals. The functionality of such circuits is determined through AC and transient analyses. AC analysis needs the capacitances associated with the terminals of the transistor. To analyze the transient behavior, it is essential to model the charges stored in the transistor. The BSIM-CMG C-V model defines both the charges and the associated capacitances for the FinFET. This chapter accounts only for the intrinsic transistor charge and capacitances, and the details of source/drain and gate overlap regions as well as the real source and drain contacts will be discussed in Chapter 8. These will create parasitic overlap and fringe capacitances that should be added to the intrinsic model described here. This method is consistent with the historical approach to transistor modeling, such as with traditional MOSFETs [1] and bipolar transistors [2], where electrostatics and transport within an internal transistor are used to create a core model and external parasitics (which vary with overall device structure) are then added around this core.

Advanced RF electronics also requires that compact models accurately predict the device's high-frequency behavior at least up to the transistor's current-gain cutoff frequency (f_T), and preferably, somewhat beyond. This requires the incorporation of the nonquasi-static (NQS) effects in the compact model. We start this chapter by deriving the terminal charges and transcapacitances, and conclude by describing the BSIM-CMG's NQS models.

7.1 Terminal charges

7.1.1 Gate charge

The total gate charge Q_g is the integration of the local channel charge density $\acute{Q}_{ch}$ [a] over the effective channel length L

$$Q_g = -W \int_0^L \acute{Q}_{ch}(y) dy \tag{7.1}$$

---

[a]In this chapter, charge quantities with a prime denote charge densities per unit area.

FinFET/GAA Modeling for IC Simulation and Design. https://doi.org/10.1016/B978-0-323-95729-8.00008-8

where W is the total width of the FinFET and y is the transport direction along the channel. The value of $\acute{Q}_{ch}(y)$ from Gauss's law is given by

$$\acute{Q}_{ch}(y) = -C_{ox}\left[V_{gs} - V_{fb} - \psi(y)\right] \tag{7.2}$$

where C_{ox} is the oxide capacitance per unit area and ψ is the surface potential. To be able to evaluate the integration in Eq. (7.1) analytically, a closed-form expression for the variation of ψ along the channel is desired. We start with the current continuity

$$I_{ds}(L) = I_{ds}(y) \quad \forall \ \ 0 \le y \le L \tag{7.3}$$

and the following simplified—but adequately accurate—expression for the current

$$I_{ds}(y) = \frac{\mu W}{L}\left[h\left(\acute{Q}_{invs}\right) - h\left(\acute{Q}_{inv}(y)\right)\right] \tag{7.4}$$

In Eq. (7.4) the symbols are as follows: $h(Q) = \frac{Q^2}{2C_{ox}} + 2V_{tm}Q$; where V_{tm} is the thermal voltage given by $k_B T/q$, where k_B and T are the Boltzmann constant and the temperature, respectively; $\acute{Q}_{invs}$ is the inversion charge density at the source end; and $\acute{Q}_{inv}(y)$ is the inversion charge density at position y given by

$$\acute{Q}_{inv}(y) = -C_{ox}\left(V_{gs} - V_{fb} - \psi(y) - \frac{\acute{Q}_{bulk}}{C_{ox}}\right) \tag{7.5}$$

where $\acute{Q}_{bulk}$ is the fixed depletion charge density[b] and is given by $q \times$ NBODY $\times$ TFIN.

Using Eq. (7.4), the equality in Eq. (7.3) can be rewritten to read

$$\frac{h\left(\acute{Q}_{invs}\right) - h\left(\acute{Q}_{invd}\right)}{L} = \frac{h\left(\acute{Q}_{invs}\right) - h\left(\acute{Q}_{inv}(y)\right)}{y} \tag{7.6}$$

Substituting the values of $\acute{Q}_{invs}$ and $\acute{Q}_{invd}$ obtained from Eq. (7.5) into Eq. (7.6), we can write

$$\frac{(B - \psi_s - \psi_d)(\psi_d - \psi_s)}{L} = \frac{(B - \psi_s - \psi)(\psi - \psi_s)}{y} \tag{7.7}$$

where $B = 2\left(V_{gs} - V_{fb} - \acute{Q}_{bulk}C_{ox} + 2V_{tm}\right)$, and ψ_s and ψ_d are the source- and drain-end surface potentials, respectively. Eq. (7.7) can be solved to get $\psi(y)$; however, as it can be seen shortly, it would be more efficient to solve it in terms of y

$$y = \frac{L(B - 2\psi)(\psi - \psi_s)}{(B - \psi_s - \psi_d)(\psi_d - \psi_s)} \tag{7.8}$$

and differentiate it to get

$$dy = \frac{L(B - 2\psi)d\psi}{(B - \psi_s - \psi_d)(\psi_d - \psi_s)} \tag{7.9}$$

---

[b]The thin body of the FinFET gets fully depleted for a small value of applied V_{gs}.

Finally, substituting Eqs. (7.2) and (7.9) into Eq. (7.1) results in the following expression for the gate charge as a function of gate voltage, and source- and drain-end surface potentials:

$$\frac{Q_g}{WC_{ox}} = \int_0^L (V_{gs} - V_{fb})dy - \int_0^L \psi dy = (V_{gs} - V_{fb})L - \int_{\psi_s}^{\psi_d} \frac{\psi L(B - 2\psi)}{(B - \psi_s - \psi_d)(\psi_d - \psi_s)}d\psi$$

$$(7.10)$$

$$\frac{Q_g}{WLC_{ox}} = V_{gs} - V_{fb} - \int_{\psi_s}^{\psi_d} \frac{\psi(B - 2\psi)}{(B - \psi_s - \psi_d)\psi_{ds}}d\psi$$

$$= V_{gs} - V_{fb} - \frac{(\psi_s + \psi_d)}{2} + \frac{(\psi_d - \psi_s)^2}{6(B - \psi_s - \psi_d)}$$

$$(7.11)$$

7.1.2 Drain charge

The sum of the drain and the source charge should equal the total channel inversion charge. The task of dividing the channel inversion charge between the source and the drain is called charge partitioning. A simplistic partitioning scheme is the 50/50 partitioning, which divides the channel inversion charge equally between the source and the drain. This is only valid if the device is symmetric and the drain-source voltage V_{ds} is low. As V_{ds} increases above zero, this scheme becomes more and more erroneous. Another arbitrary assignment of charge is the 0/100 partitioning, which allots all the channel inversion charge to the source and the drain charge is set to zero. However, the commonly used and physically valid partitioning scheme is the 40/60 partitioning, also known as the Ward-Dutton partitioning [3]. According to this scheme, the drain charge is given as

$$Q_d = -WC_{ox} \int_0^L \frac{y}{L}\left(V_{gs} - V_{fb} - \psi - \frac{\acute{Q}_{bulk}}{C_{ox}}\right)dy$$

$$(7.12)$$

Substituting dy from Eq. (7.9) results in

$$Q_d = -WC_{ox}\left[\frac{V_{gs} - V_{fb} - \frac{\acute{Q}_{bulk}}{C_{ox}}}{2}L - \int_{\psi_s}^{\psi_d} \psi y \frac{(B - 2\psi)}{(B - \psi_s - \psi_d)(\psi_d - \psi_s)}d\psi\right]$$

$$(7.13)$$

Furthermore, substituting y from Eq. (7.8), we can write

$$\frac{-Q_d}{WLC_{ox}} = \frac{V_{gs} - V_{fb} - \acute{Q}_{bulk}/C_{ox}}{2} - \int_{\psi_s}^{\psi_d} \frac{\psi(B - 2\psi)(\psi - \psi_s)(B - \psi_s - \psi_d)}{(B - \psi_d - \psi_s)^2(\psi_d - \psi_s)^2}d\psi$$

$$(7.14)$$

The integration on the RHS of Eq. (7.14) can be evaluated to ultimately write

$$\frac{-Q_d}{WLC_{ox}} = \frac{V_{gs} - V_{fb} - \acute{Q}_{bulk}/C_{ox}}{2} - \frac{\psi_s + \psi_d}{4} + \frac{(\psi_d - \psi_s)^2}{60(B - \psi_s - \psi_d)}$$

$$+ \frac{(5B - 6\psi_s - 4\psi_d)(B - 2\psi_d)(\psi_s - \psi_d)}{60(B - \psi_s - \psi_d)^2}$$

$$(7.15)$$

7.1.3 Source charge

Based on the charge conservation, we can write

$$Q_s = -(Q_g + Q_d + Q_{bulk}) \tag{7.16}$$

where Q_{bulk} is the total bulk charge and it is given by

$$Q_{bulk} = -W \cdot L \cdot \acute{Q}_{bulk} \tag{7.17}$$

7.2 Transcapacitances

All transcapacitances are derived from the terminal charges to ensure charge conservation. The transcapacitances are calculated as

$$C_{ij} = \begin{cases} -\dfrac{\partial Q_i}{\partial V_j} & i \neq j \\[2mm] \dfrac{\partial Q_i}{\partial V_j} & i = j \end{cases} \tag{7.18}$$

where Q_i and V_j are the charge and voltage associated with the terminals i and j, respectively, while the voltages on the other terminals are held constant. In the BSIM-CMG model, the transcapacitances are calculated using the *ddx()* operator in Verilog-A and are reported as parts of the operating-point information. This operator eliminates the need to write analytical expressions for the derivatives that might contain errors introduced during the derivation or implementation.

Fig. 7.1 shows the transcapacitances C_{gg}, C_{sg}, and C_{dg} of an NMOS FinFET as a function of V_{gs} and for $V_{ds} = 0$. The transcapacitances start to increase as the device

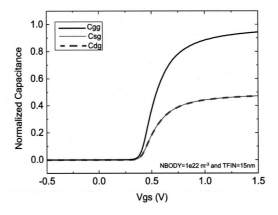

FIG. 7.1

The transcapacitances C_{gg}, C_{sg}, and C_{dg} of an NMOS FinFET as a function of the gate to source voltage V_{gs} and for $V_{ds} = 0$. The body doping is $1e22\,m^{-3}$ and the fin thickness is 15nm. The values have been normalized with respect to WLC_{ox}.

operating mode enters weak/moderate inversion. The C-V model partitions the channel charge equally between the source and drain, and at strong inversion the values of C_{sg} and C_{dg} reach $C_{gg}/2$. This is expected as for $V_{ds}=0$ the source and drain are electrically indistinguishable. However, for $V_{ds}=1.2\,\text{V}$, the Ward-Dutton partitioning scheme used in the model results in a 40/60 charge partitioning, and hence $C_{sg}\approx0.6\,C_{gg}$ and $C_{dg}\approx0.4\,C_{gg}$ (see Fig. 7.2).

Fig. 7.3 shows the transcapacitances C_{gg}, C_{sg}, C_{dg}, C_{gs}, and C_{gd} of an NMOS FinFET as a function of V_{ds} and for $V_{gs}=1.2\,\text{V}$. For $V_{ds}=0$, the source and drain terminals are electrically indistinguishable; therefore, $C_{sg}=C_{dg}$ and $C_{gs}=C_{gd}$. Furthermore, noting that $C_{sg}+C_{dg}=C_{gs}+C_{gd}=C_{gg}$, we have $C_{sg}=C_{dg}=C_{gs}=C_{gd}=C_{gg}/2$. As the drain voltage increases, $C_{sg}\approx0.6\,C_{gg}$ and $C_{dg}\approx0.4\,C_{gg}$. In saturation (i.e., $V_{ds}\geq V_{dsat}$), the value of $C_{gd}\to0$ ($C_{gs}\to C_{gg}$) because any extra drain voltage is dropped in the pinch-off region and has no effect on the channel inversion charge, or in other words, the drain is getting discounted from the channel.

7.3 NQS effects models

The operation of a MOSFET would be defined by a self-consistent solution of Poisson's equation which describes the electrostatics and the current-continuity equation which governs the dynamics:

$$\frac{\partial^2 \psi}{\partial y^2} = \frac{\rho}{\epsilon_{ch}} \tag{7.19}$$

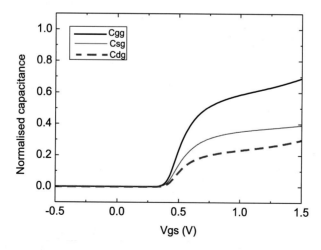

FIG. 7.2

The transcapacitances of Fig. 7.1 for $V_{ds}=1\,\text{V}$. The Ward-Dutton partitioning used in the model results in $C_{sg}\approx0.6\,C_{gg}$ and $C_{dg}\approx0.4\,C_{gg}$.

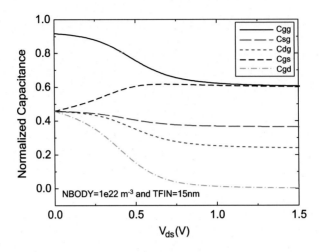

FIG. 7.3

The transcapacitances C_{gg}, C_{sg}, C_{dg}, C_{gs}, and C_{gd} of an NMOS FinFET as a function of the drain to source voltage V_{ds} and for $V_{gs} = 1.0$ V. The BSIM-CMG C-V model captures all the important physics as described in the text.

$$W \frac{\partial \acute{Q}(y,t)}{\partial t} = \frac{\partial I(y,t)}{\partial y} \tag{7.20}$$

Here, y is the direction along the channel, ψ is the surface potential, ρ is the volume charge density (both mobile and fixed charges, in cm^{-3}), $\acute{Q}$ is the channel mobile charge (in cm^{-2}), t is time, and $I(y,t)$ is the current in the channel. The gradual channel approximation is assumed in 1D Poisson's equation, i.e., the vertical field dominates over the lateral field along the channel. The continuity equation implies the fact that there would be no buildup of charge in any position along the channel.

In modern MOSFET compact models like BSIM-CMG, the quasi-static (QS) assumption is evoked (i.e., $\partial \acute{Q}/\partial t = 0$) and a steady-state drain current expression is derived to describe the DC operation. The AC or the small-signal operation is described by the terminal charges. The QS assumption implies that steady-state current and the channel charges establish instantaneously after terminal voltages are applied to the device. This is not true when the device is subject to high slew rate (high dV/dt) signals—either large voltage swings in a short time or very high-frequency signals with frequencies approaching or higher than the cut-off frequency f_T of the transistor. There is an inherent delay in the response of the transistor's currents and charges to applied voltages (often visualized as a distributed resistance-capacitance network). As research opens new avenues for circuit applications, today's circuit designs are either approaching f_t (such as Terahertz CMOS) or are being subject to high slew rate signals such as in CMOS RF power amplifiers. One needs to note that f_T is a function of the terminal voltages, especially gate voltage (see Fig. 2.12). Today, many circuit applications are subjected to near subthreshold gate voltages to lower operating power, and hence they will inherently have lower

f_T than that reported as a figure of merit for that device (which is usually at the highest operating voltage). For these cases, compact models are required to support the NQS mode of operation to accurately predict the circuit's behavior.

The BSIM-CMG model offers three different NQS models. Each of these can be turned on/off by NQSMOD switch. Setting NQSMOD=0 turns off all the NQS models and switches to QS calculations.

7.3.1 Relaxation time approximation model

A simple and elegant way to capture the NQS behavior of the channel is using the relaxation time approach and tracking the deficient or surplus charge in the channel, which is given by [4]

$$\frac{dQ_{def}}{dt} = \frac{dQ_{ch,eq}}{dt} - \frac{dQ_{def}}{\tau} \tag{7.21}$$

where the symbols are defined as follows: $Q_{def} = Q_{ch,nqs} - Q_{ch,eq}$ is the deficient/surplus channel charge; $Q_{ch,nqs}$ is the channel charge considering NQS effects; $Q_{ch,eq}$ is the channel charge at steady-state equilibrium or the QS charge; and τ is the relaxation time constant given as follows:

$$\tau = \left[\text{XRCRG1} \cdot \frac{\text{NF}}{\text{NFIN}} \cdot \frac{\mu_{eff} \cdot W}{L} C_{ox} \left(q_{ia} + \text{XRCRG2} \frac{kT}{q} \right) \right]^{-1} \tag{7.22}$$

In Eq. (7.22), q_{ia} is the average charge in the channel and model parameters XRCRG1 and XRCRG2 are used to improve the model's flexibility. Within a SPICE environment, Eq. (7.21) is implemented as a subcircuit whose node voltage tracks the deficient/surplus charge, Q_{def} (see Fig. 7.4). The source and drain terminal charges are then given by

$$Q_{d,nqs} = X_{part} \frac{Q_{def}}{\tau} \tag{7.23}$$

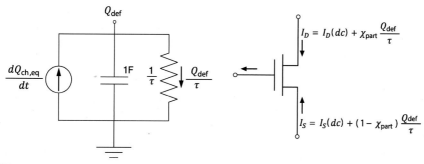

FIG. 7.4

A relaxation time approach for capturing nonquasi-static effects. The R-C subcircuit represents Eq. (7.21) for SPICE implementation and it is solved self-consistently with the MOSFET.

and

$$Q_{s,nqs} = \left(1 - X_{part}\right)\frac{Q_{def}}{\tau} \tag{7.24}$$

respectively. In Eq. (7.24), X_{part} is a bias-dependent partitioning fraction and it is approximated to its QS equivalent given by

$$X_{part} = \frac{Q_d}{Q_d + Q_s} \tag{7.25}$$

The calculated terminal charges $Q_{d,nqs}$ and $Q_{s,nqs}$ replace their QS equivalents Q_d and Q_s, respectively. The NQS body charge $Q_{bulk,nqs}$ is the same as Q_{bulk} since the body terminal charge does not experience any NQS effect (e.g., the holes in an NMOS traverse through a p-doped bulk region quickly). A more rigorous evaluation of this approach found it to be accurate up to $2 \times f_T$ when compared with TCAD-based simulations [5]. This NQS model can be activated by setting NQSMOD $= 2$.

Before using the model, the parameters XRCRG1 and XRCRG1 have to be extracted from the high-frequency measured data. The default values of 12 and 1 for XRCRG1 and XRCRG2, respectively, are obtained from the equivalent resistance of a distributed transmission line model with double-side contact (here source and drain).

7.3.2 Channel-induced gate resistance model

Another first-order model that helps to capture the NQS effects in the channel is the channel-induced gate resistance method [6]. This model is activated by setting NQSMOD $= 1$. In this method, the bias-dependent gate resistance R_{ii} is added at the gate terminal[c] whose value is proportional to the channel resistance (see Fig. 7.5). This resistance should not to be confused with the physical gate electrode resistance. The induced gate resistance is obtained from the relaxation time constant in Eq. (7.22) as follows:

$$R_{ii} = \frac{\tau}{W \cdot L \cdot C_{ox}} \tag{7.26}$$

While this method performs similar to the relaxation time approach in terms of accuracy up to f_T, it is restricted in terms of applicability. This model should be used in cases where the gate terminal is only excited by an input signal. NQS effects for applications such as passive mixers or common-gate LNAs where the source terminal is excited cannot be captured by this model. However, the relaxation time constant approach is valid for all terminal excitation as we recalculate all the terminal charges.

---

[c]A gate node is introduced between the intrinsic gate and the physical gate electrode for this purpose. This node collapses to the intrinsic gate if the user turns off this model.

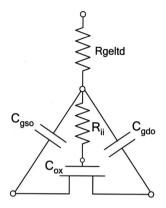

FIG. 7.5

A channel-induced gate resistance R_{ii} is added in series with the gate physical electrode resistance R_{geltd} to capture the nonquasi-static effects for a gate terminal excited FinFET. C_{gso} and C_{gdo} are the parasitic capacitances.

7.3.3 Charge segmentation model

If the channel length approaches infinitesimally small values, the carrier transit times through the channel tend to become small. For this device, the QS approximation is still valid. Using this concept, we can then visualize a transistor as a series of connected shorter channel length transistors or as we shall refer to them here as "charge segments" (owing to each smaller transistor carrying a part of the whole channel charge). Fig. 7.6 schematically shows such a representation. A charge-based version of the continuity equation (7.20) can be written as follows:

$$\frac{\partial q}{\partial t} + \mu_{\text{eff}} V_t \frac{\partial}{\partial y} \left[\frac{(2q+1)}{0.5 \left[1 + \sqrt{1 + 2\left(\frac{\mu_{\text{eff}}}{v_{\text{sat}}} \frac{\partial q}{\partial y}\right)^2} \right]} \frac{\partial q}{\partial y} \right] = 0 \qquad (7.27)$$

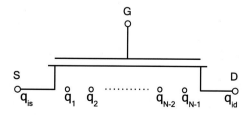

FIG. 7.6

$N_{\text{SEG}} = N$ charge segments in a MOSFET channel for NQS effects simulation. q_i represents the channel charge at the ith intermediate node.

This equation is valid at any point along the channel and includes velocity saturation effects. The solution to this continuity equation along the channel captures the NQS effects. One way to accomplish this in a SPICE simulator environment is through a simple series connection of QS transistors, commonly known as the channel segmentation approach [7,8]. If implemented within a compact model skillfully with care by not including short-channel effects for each segment (whose channel length is L/N_{SEG}, where N_{SEG} is the number of segments) or by not adding series resistance and parasitic capacitance multiple times for each segment, this method captures all the important features of NQS effects. However, often times this approach is ill-constrained leading to longer simulation times or nonconvergence.

A solution to the above continuity equation (7.27) where the continuity constraint is imposed not just on the first but also on the second and third derivatives is required. One such solution has been developed using spline-collocation method for a surface potential-based MOSFET model in Ref. [9], which we will adopt for a charge-based model here.

Starting from Eq. (7.27), we can further simplify it into the following form:

$$\frac{\partial q}{\partial t} + f\left(q, \frac{\partial q}{\partial y}, \frac{\partial^2 q}{\partial y^2}\right) = 0$$

$$f\left(q, \frac{\partial q}{\partial y}, \frac{\partial^2 q}{\partial y^2}\right) = \frac{\mu_{eff}V_t}{D_v}\left[2\left(\frac{\partial q}{\partial y}\right)^2 - \frac{2q+1}{D_v}\left(\frac{\mu_{eff}}{v_{sat}}\right)^2\left(\frac{\partial q}{\partial y}\right)^2\frac{\partial^2 q}{\partial y^2}\right] \quad (7.28)$$

$$D_v = \sqrt{1 + 2\left(\frac{\mu_{eff}}{v_{sat}}\right)^2\left(\frac{\partial q}{\partial y}\right)^2}$$

In order to solve this complex partial differential equation in a SPICE environment we need to convert it to a set of ordinary differential equations (ODEs). In Ref. [9], the first-order weighted residuals method was extended to assure current continuity up to the third order. Following a similar approach and assuming the channel is broken down into N_{SEG} segments, the charge in each segment can be expressed by a cubic equation. For example, the inversion charge in the nth segment is given by

$$q(y) = a_n y^3 + b_n y^2 + c_n y + d_n \qquad \frac{n-1}{N_{SEG}}L < y < \frac{n}{N_{SEG}}L \quad (7.29)$$

The boundary conditions for the charge at the source and drain ends are their respective QS solutions, i.e., $q(0) = q_s$ and $q(L) = q_d$. Applying the continuity conditions for q, $\partial q/\partial y$, and $\partial^2 q/\partial y^2$ at the node points in between any two charge segments and using the boundary conditions, one can derive a relation between the cubic equation coefficients a_n, b_n, c_n, and d_n for $n = 1$ to $N_{SEG} - 1$ and the charges at nodes $q_s, \ldots, q(y = nL/N_{SEG}), \ldots, q_d$. For example, for $N_{SEG} = 3$, the continuity conditions to be imposed are as follows:

$$q\left(\frac{L}{3}\right)^- = q\left(\frac{L}{3}\right)^+$$

$$q\left(\frac{2L}{3}\right)^- = q\left(\frac{2L}{3}\right)^+$$

$$\left.\frac{\partial q}{\partial y}\right|_{y=\frac{L^-}{3}} = \left.\frac{\partial q}{\partial y}\right|_{y=\frac{L^+}{3}}$$

$$\left.\frac{\partial q}{\partial y}\right|_{y=\frac{2L^-}{3}} = \left.\frac{\partial q}{\partial y}\right|_{y=\frac{2L^+}{3}} \qquad (7.30)$$

$$\left.\frac{\partial^2 q}{\partial y^2}\right|_{y=\frac{L^-}{3}} = \left.\frac{\partial^2 q}{\partial y^2}\right|_{y=\frac{L^+}{3}}$$

$$\left.\frac{\partial^2 q}{\partial y^2}\right|_{y=\frac{2L^-}{3}} = \left.\frac{\partial^2 q}{\partial y^2}\right|_{y=\frac{2L^+}{3}}$$

Additionally, at the boundaries

$$\left.\frac{\partial^2 q}{\partial y^2}\right|_{y=0} = \left.\frac{\partial^2 q}{\partial y^2}\right|_{y=L} - 0 \qquad (7.31)$$

Solution for the $N_{\text{SEG}} = 3$ case can be found in Ref. [9], which has been derived in the context of a double gate FET charge model. The values for the so derived coefficients remain the same. Only the function $f\left(q, \frac{\partial q}{\partial y}, \frac{\partial^2 q}{\partial y^2}\right)$ and the assumptions that go into a short-channel drain current equation change with the FET architecture.

The continuity equation at the nth node can be written as

$$\frac{\partial q_n}{\partial t} + f_n\left(q_n, \frac{\partial q_n}{\partial y}, \frac{\partial^2 q_n}{\partial y^2}\right) = 0 \qquad (7.32)$$

where q_n denotes the instantaneous channel charge at the nth node. It is noted that the derivatives of the charge, $\partial q/\partial y$ and $\partial^2 q/\partial y^2$, at the intermediate nodes can be expressed as functions of the node charges. This is because the coefficients a_n, etc., become linear functions of the node charges after applying Eqs. (7.30) and (7.31) [10].[d] Thus, the continuity equation at the nth node changes to

$$\frac{\partial q_n}{\partial t} + f_n(q_s, q_1, q_2, \ldots, q_n, \ldots, q_d) = 0 \qquad (7.33)$$

Continuity equations for the remaining $N_{\text{SEG}} - 1$ nodes can be written in a similar manner. Therefore, the partial differential equation (7.28) has been converted to $N_{\text{SEG}} - 1$ ODEs. These $N_{\text{SEG}} - 1$ ODEs are coupled, as the channel charge at the nth node depends on the charge at all the other nodes through the source function,

---

[d]A matrix-based evaluation of the coefficients of the node charges to express the charge derivatives for any N_{SEG} can be found in Ref. [11].

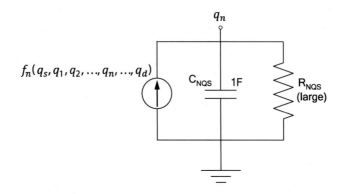

FIG. 7.7

R-C subcircuit representation of Eq. (7.33). The node voltage here represents the channel charge at the nth intermediate node.

$f_n()$. These $N_{\text{SEG}} - 1$ ODEs can be represented as $N_{\text{SEG}} - 1$ R-C subcircuits within a SPICE simulator (see Fig. 7.7). For SPICE implementation purposes, the resistor R_{NQS} is chosen to be a large value (1 K) to facilitate convergence, and the capacitance C_{NQS} is unity.

The terminal charges $Q_{\text{d,nqs}}$ and $Q_{\text{s,nqs}}$ at the drain and source ends, respectively, can be obtained by integrating the cubic segments using the Ward-Dutton partition scheme as follows:

$$Q_{\text{d,nqs}} = -WC_{\text{ox}} \int_0^L \frac{y}{L} q(y) dy$$

$$Q_{\text{d,nqs}} = -WC_{\text{ox}} \left[\frac{1}{90} q_s + \frac{1}{10} q\left(\frac{L}{3}\right) + \frac{4}{15} q\left(\frac{2L}{3}\right) + \frac{11}{90} q_d \right]$$

(7.34)

$$Q_{\text{s,nqs}} = -WC_{\text{ox}} \int_0^L \left(1 - \frac{y}{L}\right) q(y) dy$$

$$Q_{\text{s,nqs}} = -WC_{\text{ox}} \left[\frac{11}{90} q_s + \frac{4}{15} q\left(\frac{L}{3}\right) + \frac{1}{10} q\left(\frac{2L}{3}\right) + \frac{1}{90} q_d \right]$$

(7.35)

The overall gate terminal charge $Q_{\text{g,nqs}}$ can then be obtained as follows:

$$Q_{\text{g,nqs}} = -\left(Q_{\text{s,nqs}} + Q_{\text{d,nqs}} + Q_{\text{bulk}}\right)$$

(7.36)

The above expressions for $Q_{\text{d,nqs}}$, $Q_{\text{s,nqs}}$, and $Q_{\text{g,nqs}}$ replace their QS equivalent charges derived in the first section. This model can be selected by setting NQSMOD $= 3$.

References

[1] Y. Tsividis, Operation and Modeling of the MOS Transistor, second ed., McGraw-Hill, New York, 1999.

[2] R.L. Pritchard, Electrical Characteristics of Transistors, McGraw-Hill, New York, 1967.

[3] S.-Y. Oh, D. Ward, R. Dutton, Transient analysis of MOS transistors, IEEE J. Solid State Circuits 15 (4) (1980) 636–643.

[4] M. Chan, K. Hui, C. Hu, P.-K. Ko, A robust and physical BSIM3 non-quasistatic transient and AC small-signal model for circuit simulation, IEEE Trans. Electron Devices 45 (4) (1998) 834–841.

[5] Z. Zhu, G. Gildenblat, C. Mcandrew, I.-S. Lim, Accurate RTA-based nonquasistatic MOSFET model for RF and mixed-signal simulations, IEEE Trans. Electron Devices 59 (5) (2012) 1236–1244.

[6] X. Jin, J.-J. Ou, C.-H. Chen, W. Liu, M. Deen, P. Gray, C. Hu, An effective gate resistance model for CMOS RF and noise modeling, in: International Electron Devices Meeting Technical Digest (IEDM), 1998, pp. 961–964.

[7] A. Scholten, L. Tiemeijer, P. De Vreede, D.B.M. Klaassen, A large signal non-quasistatic MOS model for RF circuit simulation, in: International Electron Devices Meeting (IEDM) Technical Digest, 1999, pp. 163–166.

[8] M. Bucher, A. Bazigos, An efficient channel segmentation approach for a large-signal NQS MOSFET model, Solid State Electron. 52 (2) (2008) 275–281.

[9] H. Wang, X. Li, W. Wu, G. Gildenblat, R. van Langevelde, G.D.J. Smit, A. Scholten, D.B.M. Klaassen, A unified nonquasi-static MOSFET model for large-signal and small-signal simulations, IEEE Trans. Electron Devices 53 (9) (2006) 2035–2043.

[10] S. Sarkar, A.S. Roy, S. Mahapatra, Unified large and small signal non-quasi-static model for long channel symmetric DG MOSFET, Solid State Electron. 54 (11) (2010) 1421–1429.

[11] PSP 103.1 Documentation [Online], 2009. Available from: http:/pspmodel.asu.edu/psp_documentation.htm.

Parasitic resistances and capacitances

8

So far we have discussed the modeling of intrinsic device behavior of FinFET/GAA in BSIM-CMG. In this chapter we focus on the modeling of parasitic resistance and capacitances. Parasitic resistance is an important component, since its magnitude is comparable to channel resistance in scaled devices, and can seriously degrade transistor current drive. Parasitic capacitance is one of the key factors for circuit speed as well. In the FinFET/GAA devices, parasitic resistances and capacitances are difficult to model due to the complex three-dimensional (3D) geometry. We will discuss the modeling approach of BSIM-CMG to take care of this.

Parasitic resistances in the FinFET/GAA devices include the source/drain resistance and the gate resistance. The source/drain resistance is usually more significant compared to the gate resistance. This is especially true with the introduction of metal gate technology [1], in which highly conductive metal is used as the gate material. According to the 2022 International Roadmap for Devices and Systems (IRDS), series resistance degrades the saturation current by 40% or more [2]. An accurate model for source/drain resistance in FinFETs is needed. The gate resistance is present due to the finite conductivity of the gate material. At DC, the gate current is very small, so the gate resistance does not alter the transistor's DC behavior. However, it does impact the AC behavior, such as CMOS switching delay. A compact analytical model for gate resistance in FinFETs is available [3].

Several models for source/drain resistances in FinFETs have been proposed. Dixit [4] proposed a comprehensive source/drain resistance model for double-gate MOSFET with lithography-defined source and drain, and verified it with TCAD and measured data. Tekleab [5] extended it to consider more than one contact surface. However, these models are limited to rectangular source/drain contacts. FinFETs are expected to have several fins in parallel to have sufficient current driving capability. Research has shown that high layout density can be achieved if the multifins are enlarged and eventually merged using a selective epitaxial growth (SEG) process, forming a connected 3D raised source/drain contact [6, 7]. The cross-section of such a raised source/drain is likely nonrectangular. This needs to be considered. Moreover, it has been shown that the performance of fully depleted FinFET is the best with an underlapped source/drain [8, 9]. Devices with underlapped source and drain have large bias-dependent source/drain resistance. However, both Refs. [4, 5] consider bias-independent resistances only.

In BSIM-CMG, both bias- and geometry-dependent parasitic resistance for FinFETs are modeled. We will show that the model is applicable to raised source/drain FinFETs with nonrectangular source/drain cross-section. To verify the model

FinFET/GAA Modeling for IC Simulation and Design. https://doi.org/10.1016/B978-0-323-95729-8.00014-3

and demonstrate its predictivity, we will compare it with 3D TCAD simulations. In addition, we will also discuss the modeling of bias-dependent source and drain resistances in the FinFET with source/drain underlap.

Beyond intrinsic capacitances, BSIM-CMG also models both the bias-dependent overlap capacitance and the bias-independent fringe capacitance. The fringe capacitance is modeled starting from a 2D scenario, focusing on the FinFET structure. A 2D model consisting of a fin-to-gate capacitance and three source/drain contact to gate capacitance components are derived and validated against 2D TCAD simulations. A 3D extension is derived and shown to agree with 3D TCAD simulation of FinFET parasitic capacitances. A special mode for GAA devices is further developed to ensure GAA devices can be modeled as well.

8.1 FinFET device structure and symbol definitions

The parasitic resistance/capacitance models are developed starting from the FinFET structure. We consider FinFETs in which the SEG process is applied to merge individual fins. Single-fin FinFETs and multiple-fin (multifin) FinFETs will likely coexist on the same wafer. Therefore, we assume that both single-fin and multifin FinFETs will be subject to source/drain SEG, even though fin-merging is not necessary for single-fin FinFETs.

A 3D drawing of the raised source/drain FinFET is shown in Fig. 8.1. For simplicity, we are showing only one fin of a multifin FinFET. The channel portion of the fin is wrapped on three sides by the gate stack. The source and drain silicon are enlarged by SEG to reduce resistances. For multifin FinFETs, a larger source and drain also makes contacting easier. The thin region not covered by the gate is the extension region. It is not subject to SEG because it is protected by the spacer (not shown in the figure) during epitaxial growth. The metallic region on top of the raised source and drain is the silicide. For single-fin FinFETs, depending on the process, the silicide may wrap around the RSD on three sides.

Fig. 8.2 is the cross-sectional view of a FinFET along the source-to-drain direction. The insulating material on top of the fin and beneath the gate with height *TMASK* is the hard mask. In some FinFET processes, a hard mask is used for fin

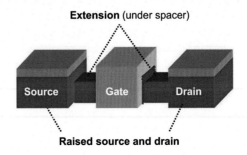

FIG. 8.1

Bird's-eye view of a raised source/drain FinFET.

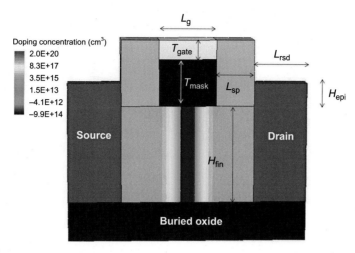

FIG. 8.2

Cross-section of a raised source/drain double-gate FinFET and symbol definition. This figure is a simulation structure in Sentaurus TCAD [10].

etching and is left on top of the fin and is never removed. Since the top surface of the fin channel does not conduct current, the device becomes a *double-gate FinFET*. In other processes, no hard mask is used. Instead, *TMASK* is set to T_{ox} and the top surface has current conduction. Such a FinFET is a *triple-gate FinFET*. FinFETs may also be classified by substrate type into SOI FinFETs and bulk FinFETs. For example, Fig. 8.2 shows an SOI FinFET, for which the fin is situated on top of the buried oxide. Table 8.1 lists the definitions of symbols.

Table 8.1 Symbol definition.

Parameter name	Definition
L_g	Gate length
LSP	Spacer thickness
LRSD	Raised source/drain length
HFIN	Fin height
TGATE	Gate height
HEPI	Height of epitaxial silicon above fin
FPITCH	Fin pitch of the multifin FinFET
TFIN	Fin thickness
CRATIO	Ratio of the corner area filled with silicon to the total corner area
NFIN	Total number of fins in the FinFET
A_{rsd}	Per-fin component of the raised source/drain area
ARSDEND	End component of raised source/drain area
DELTAPRSD	Correction term for silicide/epitaxial silicon interfacial length per fin
PRSDEND	End component of silicide/epitaxial silicon interfacial length

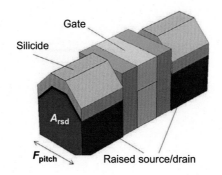

FIG. 8.3

Bird's-eye view of a FinFET with nonrectangular epi and top silicide. This figure is a simulation structure in Sentaurus TCAD [10].

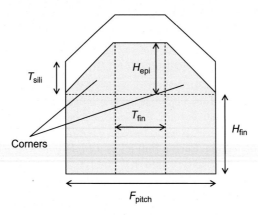

FIG. 8.4

Two-dimensional cross-section of a FinFET with nonrectangular epi and top silicide.

SEG results in faceted RSD [6]. The final RSD may look something like the drawing in Fig. 8.3. The corresponding cross-sectional diagram, cut in the direction parallel to the gate, is shown in Fig. 8.4.

The cross-sectional area of the raised source/drain is A_{rsd}. For generality, the source/drain resistance is modeled as a function of A_{rsd} regardless of its shape. A_{rsd} in a structure like Fig. 8.4 is given by

$$A_{rsd} = FPITCH \cdot HFIN + [TFIN + (FPITCH - TFIN) \cdot CRATIO] \cdot HEPI \tag{8.1}$$

where *CRATIO* is the ratio of the corner area filled with silicon to the total corner area. In the example given in Fig. 8.4, *CRATIO* is 0.5.

Most FinFET devices in a digital circuit will have multiple fins. For multifin devices, the source/drain resistance is modeled as a function of the total area and perimeter, which is given by

$$A_{rsd,total} = A_{rsd} \times NFIN + ARSDEND \tag{8.2}$$

$$P_{rsd,total} = (FPITCH + DELTAPRSD) \times NFIN + PRSDEND \qquad (8.3)$$

ARSDEND and *PRSDEND* are the end components associated with the first and last fins.

8.2 Modeling of geometry-dependent source/drain resistances in FinFETs

The FinFET source/drain resistance can be separated into three components, as illustrated in Fig. 8.5:

1. **Contact resistance** (R_{con}): The combined resistance due to the raised source/drain region bulk resistivity and the silicon/silicide interface resistance.
2. **Spreading resistance** (R_{sp}): The resistance due to current spreading from the source/drain extension into the raised source/drain.
3. **Extension resistance** (R_{ext}): The bias-dependent resistance in the thin source/drain extension region under the spacer.

We will discuss each of these components in Sections 8.2.1–8.2.3.

8.2.1 Contact resistance

The contact resistance model takes into account both the bulk resistivity in the raised source/drain region and the contact resistance at the silicon/silicide interface. Since the resistance is distributed, it is difficult to separate the two into individual resistors.

To consider the distributed effect, we partition the raised source/drain region into infinitesimally thin vertical slices (Fig. 8.6A). The slices are connected in a resistance network, as shown in Fig. 8.6B. For each slice, there is a bulk resistance component, ΔR_s, between adjacent slices, and a contact resistance component, ΔR_c, from each slice to the contact. The bulk resistance component is given by

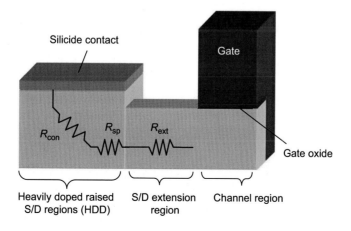

FIG. 8.5

Illustration of the three components of the FinFET source/drain resistance.

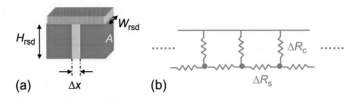

FIG. 8.6

Contact resistance derivation assumptions. (A) Illustration of source-drain contact structure and the infinitesimal slice of material with thickness of Δx considered. (B) Schematic of circuit that models the source-drain contact structure.

$$\Delta R_s = \rho \cdot \frac{\Delta x}{H_{rsd} \cdot W_{rsd}} \tag{8.4}$$

where ρ is the bulk resistivity given by

$$\rho = \frac{1}{q \cdot N_{rsd} \cdot \mu_{rsd}} \tag{8.5}$$

Here, N_{rsd} is the raised source/drain region doping concentration. We assume the raised source/drain region is in situ doped during SEG and is uniformly doped. The mobility μ_{rsd} is calculated using Masetti's model [11] as a function of N_{rsd}. The contact resistance component is given by

$$\Delta R_c = \frac{\rho_c}{\Delta x \cdot W_{rsd}} \tag{8.6}$$

where ρ_c is the specific contact resistivity in units of Ωcm^2.

Eqs. (8.4), (8.6) are valid only for rectangular contacts. To generalize it to any contact shape and multifin devices, we express them in terms of the raised source/drain cross-sectional area and the interface peripheral length:

$$\begin{cases} \Delta R_s = \rho \cdot \dfrac{\Delta x}{A_{rsd,total}} \\[2mm] \Delta R_c = \dfrac{\rho_c}{\Delta x \cdot P_{rsd,total}} \end{cases} \tag{8.7}$$

The transmission line model (TLM) [12] is applied to solve this problem. By solving a differential equation, we obtain the total contact resistance:

$$R_{con} = \rho \cdot \frac{L_T}{A_{rsd,total}} \cdot \frac{\eta \cdot \cosh(\alpha) + \sinh(\alpha)}{\eta \cdot \sinh(\alpha) + \cosh(\alpha)} \tag{8.8}$$

where

$$L_T = \sqrt{\frac{\rho_c \cdot A_{rsd,total}}{\rho \cdot P_{rsd,total}}} \tag{8.9}$$

$$\alpha = \frac{L_{rsd}}{L_T} \tag{8.10}$$

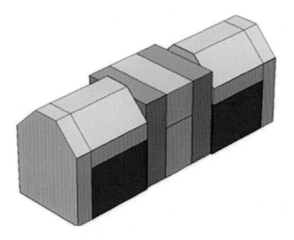

FIG. 8.7

FinFET with a nonrectangular epi and silicide on top and two ends. This figure is a simulation structure in Sentaurus TCAD [10].

$$\eta = \frac{\rho_c \cdot A_{\text{rsd,total}}}{\rho \cdot L_T \cdot A_{\text{term}}} \tag{8.11}$$

Here, A_{term} is the silicon/silicide area at the two ends of the FinFET for a structure like Fig. 8.7. A special case is for a FinFET without the end contacts, so that $A_{\text{term}} = 0$. Eq. (8.8) reduces to

$$R_{\text{con}} = \rho \cdot \frac{L_T}{A_{\text{rsd,total}}} \cdot \coth(\alpha) \tag{8.12}$$

Note that Eq. (8.12) is similar to, but more general than, the contact resistance formula in Refs. [4, 5], since it can model nonrectangular raised source/drain cross-sectional geometry.

8.2.2 Spreading resistance

When current flows from the source/drain extension region into the raised source/drain region, it spreads out gradually. We model the resistance increase due to such spreading as a new component, the spreading resistance. This spreading phenomenon is also known as current crowding.

The top view of the raised source/drain and extension regions is shown in Fig. 8.8, with the gray area representing the region through which the current flows. We assume a constant spreading angle θ in order to obtain a closed-form expression for spreading resistance.

We first consider a case where the cross-sections of the extension and the raised source/drain are both squares. In between, each current flow cross-section is also a square. We assume the side length of the squares increases linearly with position. Consequently, each slice in the spreading region with thickness Δx has resistance of

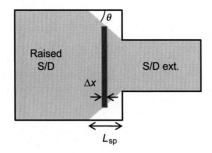

FIG. 8.8

Illustration of current spreading from the source/drain extension region into the raised source/drain region.

$$\Delta R = \frac{\rho \cdot \Delta x}{\left(\sqrt{A_{\text{fin}}} + 2 \cdot x \cdot \tan\theta\right)^2} \tag{8.13}$$

where A_{fin} is the cross-sectional area of the fin extension given by

$$A_{\text{fin}} = H_{\text{fin}} \cdot T_{\text{fin}} \tag{8.14}$$

The resistance in the spreading resistance is given by the integral from 0 to L_1:

$$R = \int_0^{L_1} \frac{\rho \cdot dx}{\left(\sqrt{A_{\text{fin}}} + 2 \cdot x \cdot \tan\theta\right)^2} \tag{8.15}$$

where L_1 satisfies the following relation:

$$2 \cdot L_1 \cdot \tan\theta = \sqrt{A_{\text{rsd}}} - \sqrt{A_{\text{fin}}} \tag{8.16}$$

After carrying out the integration, we obtain the total resistance in the spreading region

$$R = \frac{\rho \cdot \cot\theta}{2} \left(\frac{1}{\sqrt{A_{\text{fin}}}} - \frac{1}{\sqrt{A_{\text{rsd}}}}\right) \tag{8.17}$$

If we carry out the same analysis for a circular-shaped fin extension and a circular-shaped raised source/drain, we will obtain a similar result

$$R = \frac{\rho \cdot \cot\theta}{\sqrt{\pi}} \left(\frac{1}{\sqrt{A_{\text{fin}}}} - \frac{1}{\sqrt{A_{\text{rsd}}}}\right) \tag{8.18}$$

To be more general, we express the resistance in the spreading region as

$$R = \frac{\rho \cdot \cot\theta}{s} \left(\frac{1}{\sqrt{A_{\text{fin}}}} - \frac{1}{\sqrt{A_{\text{rsd}}}}\right) \tag{8.19}$$

where the shape parameter s depends on the shape of the fin extension and the raised source/drain.

We can also calculate R', the total resistance in the same region, if there were no spreading.

$$R' = \frac{\rho \cdot L_1}{A_{rsd}} = \frac{\rho \cdot \cot\theta}{s \cdot A_{rsd}} \left(\sqrt{A_{rsd}} - \sqrt{A_{fin}} \right) \tag{8.20}$$

Since the spreading resistance is defined as the increase in resistance, it is the difference between R and R', which is given by

$$R_{sp} = \frac{\rho \cdot \cot\theta}{s} \cdot \left(\frac{1}{\sqrt{A_{fin}}} - \frac{2}{\sqrt{A_{rsd}}} + \frac{\sqrt{A_{fin}}}{A_{rsd}} \right) \tag{8.21}$$

We define

$$R_0 = \rho \left[\frac{1}{\sqrt{A_{fin}}} - \frac{2}{\sqrt{A_{rsd}}} + \frac{\sqrt{A_{fin}}}{A_{rsd}} \right] \tag{8.22}$$

and let

$$R_{sp} = K \cdot R_0 \tag{8.23}$$

where the slope factor K is

$$K = \frac{\cot\theta}{s} \tag{8.24}$$

We hypothesize that K is insensitive to the device geometry in the range we are interested in.

To test the hypothesis, 3D TCAD simulations are performed to compute R_{sp}. We simulate test structures that consist of a uniformly doped silicon block with contacts on both sides, as illustrated in Fig. 8.9A. The specific contact resistivity is set to a very small value so that its effect is negligible.[a] The doping concentration is set to $N_{sd} = 2 \times 10^{20} \mathrm{cm}^{-3}$. The dimension of the left-side contact is fixed at $H_{rsd} = 60\mathrm{nm}$ and $W_{rsd} = 45\mathrm{nm}$. The height of the right-side contact varies from $H_{fin} = 30\mathrm{nm}$ to $H_{fin} = 60\mathrm{nm}$ in steps of 5nm, and the width varies simultaneously from $T_{fin} = 15\mathrm{nm}$ to $T_{fin} = 45\mathrm{nm}$ in steps of 5nm. For each case, five different raised source/drain lengths, L_{rsd}, are simulated. The spreading resistance is extracted by subtracting the resistance of the nonspreading case, where the nonspreading case is when $W_{rsd} = T_{fin}$ and $H_{rsd} = H_{fin}$.

In Fig. 8.9B, we plotted the spreading resistance versus R_0. We also fitted Eq. (8.23) to the data. The best fit is obtained at $K = 0.7$. The model and TCAD data agree reasonably well, suggesting that the constant-K is a reasonable assumption.

8.2.3 Extension resistance

Extension resistance (R_{ext}) refers to the resistance in the fin extension region under the spacer. The modeling of extension resistance requires knowledge of the doping profile in the extension region, which is often not accurately known in reality. The

---

[a] ρ_c is set to $10^{-12}\Omega\mathrm{cm}$ in TCAD simulation. In TCAD tools we are generally not allowed to set ρ_c to zero.

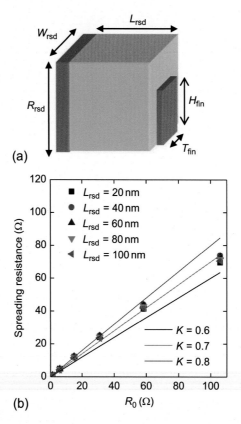

FIG. 8.9

Extraction of spreading resistance: (A) Illustration of the test structure and (B) extraction of the slope factor, K.

profile shape varies depending on process conditions. In addition, surface accumulation due to the fringe field originating from the gate creates bias dependency.

To simplify the problem, we make several assumptions about the spacer configuration and doping profile, as illustrated in Fig. 8.10. We assume the spacer (with length L_{sp}) consists of an offset spacer (with length L_{off}) and a main spacer. Implantation is performed after the offset spacer is deposited but before the main spacer forms. As a result, the doping is uniform under the main spacer, but decays following a Gaussian profile under the offset spacer. N_{ext} is the doping concentration at the offset spacer edge. Although an implanted extension is assumed, this model can also be applied to FinFETs with extension doping coming solely from in situ doped epitaxial source/drain.

The extension resistance is modeled in one bias-dependent accumulation resistance component and two bias-independent bulk resistance components, and combined into a resistance network as shown in Fig. 8.11.

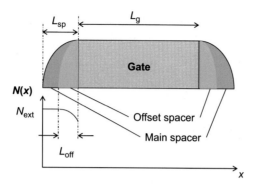

FIG. 8.10

Doping profile and spacer configurations considered for R_{ext} modeling.

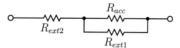

FIG. 8.11

Subcircuit for resistance modeling in the accumulation region.

The accumulation resistance (R_{acc}) represents the conductive path at the surface of the source/drain extension due to charge accumulation induced by the gate fringe fields [4]. The accumulation resistance is significant and needs to be properly considered, especially for devices with little or no source/drain to gate overlap and have a relatively small doping concentration at the gate edge. The accumulation resistance is modeled using the following expression:

$$R_{acc} = \frac{R_{acc0}}{H_{fin} \cdot (V_{gs(d)} - V_{fbsd})} \tag{8.25}$$

where V_{fbsd} is the flatband voltage at the source and drain. The accumulation resistance is inversely proportional to the conduction charge density, which is proportional to $V_{gs(d)} - V_{fbsd}$.

The bulk resistance of the fin extension is modeled in two separate components: R_{ext1} and R_{ext2}. R_{ext1} represents the bulk resistance of the fin beneath the accumulated part of the source/drain. We assume it is partially under the offset spacer and partially under the main spacer. R_{ext1} is given by

$$R_{ext1} = \frac{R_{ext1,0}}{H_{fin} T_{fin}} \tag{8.26}$$

We have lumped the complex doping profile and mobility change due to doping into the variable $R_{ext1,0}$.

Further away from the gate, there is no accumulation at the surface but only conductivity in the bulk of the fin. We model it with the following expression:

$$R_{\text{ext2}} = \frac{R_{\text{ext2,0}} \cdot (L_{\text{sp}} - \Delta L_{\text{ext}})}{H_{\text{fin}} T_{\text{fin}}} \tag{8.27}$$

where we have assumed R_{ext2} is located under the main spacer where the doping is horizontally uniform.

After combining the three resistance components into a network as shown in Fig. 8.11, we obtain the full expression for extension resistance:

$$R_{\text{ext}} = \left(\frac{R_{\text{acc0}}}{H_{\text{fin}}(V_{\text{gs(d)}} - V_{\text{fbsd}})} \right) \middle\| \left(\frac{R_{\text{ext1,0}}}{H_{\text{fin}} T_{\text{fin}}} \right) + \frac{R_{\text{ext2,0}} \cdot (L_{\text{sp}} - \Delta L_{\text{ext}})}{H_{\text{fin}} T_{\text{fin}}} \tag{8.28}$$

The earlier equation can be simplified to

$$R_{\text{ext}} = \frac{\dfrac{R_{\text{ext1,0}}}{H_{\text{fin}} T_{\text{fin}}}}{1 + \dfrac{R_{\text{ext1,0}}}{T_{\text{fin}} R_{\text{acc0}}} \cdot (V_{\text{gs(d)}} - V_{\text{fbsd}})} + \frac{R_{\text{ext2,0}} \cdot (L_{\text{sp}} - \Delta L_{\text{ext}})}{H_{\text{fin}} T_{\text{fin}}} \tag{8.29}$$

which, interestingly, turns out to have the same bias dependence as the BSIM4 model [13][b] :

$$Rs = \frac{1}{W_{\text{eff}}} \left[\frac{RSW}{1 + PRWG \cdot (V_{\text{gs}} - V_{\text{fbsd}})} + RSWMIN \right] \tag{8.30a}$$

$$Rd = \frac{1}{W_{\text{eff}}} \left[\frac{RDW}{1 + PRWG \cdot (V_{\text{gd}} - V_{\text{fbsd}})} + RDWMIN \right] \tag{8.30b}$$

8.3 Parasitic resistance model verification

In this section, the verification of the model with 3D TCAD simulation is presented. The details of the TCAD simulation setup, the separation of bias-dependent and bias-independent source/drain resistances, and the comparison of model versus TCAD simulation are discussed. In addition, we present our findings that the traditional transmission line method overestimates L_{eff}, and the physical L_{eff} must be extracted by other means.

8.3.1 TCAD simulation setup

Three-dimensional numerical simulations of the FinFET are carried out using TCAD. In this section, the details of the simulation setup are presented.

The simulation grid is created with Sentaurus Structure Editor [14] and mesh generation program Noffset3D [15]. A bird's-eye view of the FinFET simulation

---

[b] *PRWB* is set to 0. We consider the case when *RDSMOD* = 1, which activates the external resistance model.

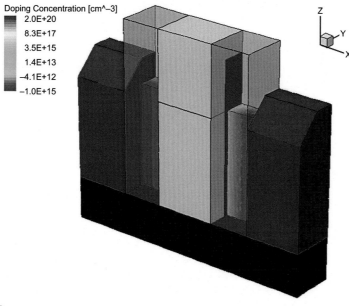

Doping Concentration [cm^–3]
2.0E+20
8.3E+17
3.5E+15
1.4E+13
−4.1E+12
−1.0E+15

FIG. 8.12

Bird's-eye view of the FinFET structure for TCAD simulation. The nitride spacer is intentionally made semitransparent to make the fin extension visible.

structure is shown in Fig. 8.12. Due to symmetry, only one-half of the FinFET needs to be simulated. This reduces the total number of grid points by half and speeds up the simulation with no change in accuracy. The spacer is intentionally made transparent to show the fin extension region. The raised source/drain has a wrapped-around contact. The top and sides of the raised source/drain as well as the source/drain end planes are in contact with silicide. Such contact is possible for FinFETs with unmerged epitaxial source/drain or SRAM FinFETs with a single fin per device. With a hard mask on top of the fin, the structure is a double-gate FinFET. Moreover, the device is situated on top of the buried oxide, therefore it is an SOI FinFET. The nominal geometry considered in this study is a linearly scaled version of a manufacturable FinFET technology [16]. Table 8.2 lists the nominal FinFET geometry and other simulation parameters for TCAD simulation.

The current-voltage characteristics are simulated using Sentaurus Device [10]. Doping-dependent mobility is modeled using the Masseti mode [11]. Mobility degradation at high vertical fields is accounted for with the Enhanced Lombardi Model [10, 17]. Velocity saturation is modeled using the Extended Canali Model [10, 18]. Quantum mechanical effects are taken into account with an equivalent oxide thickness approach. In other words, we lump the inversion layer thickness as part of the equivalent oxide thickness (EOT). The impact of quantum effects on source/drain Schottky barrier height lowering [19] is lumped into the specific contact resistivity

Table 8.2 Nominal FinFET geometry and other simulation parameters used for TCAD simulation.

Parameter name	Description	Nominal value
L_g	Physical gate length	15 nm
EOT	Equivalent oxide thickness	1.0 nm
T_{fin}	Fin thickness	10 nm
H_{fin}	Fin height	25 nm
H_{rsd}	Raised source/drain height	31 nm
T_{mask}	Oxide hard mask thickness	12 nm
L_{off}	Offset spacer width	6 nm
L_{rsd}	Raised source/drain length	14 nm
L_{sp}	Source/drain spacer width	10 nm
T_{epi}	Horizontal epi thickness	6 nm
N_{ext}	Source/drain extension doping	$2 \times 10^{20} cm^{-3}$
N_{rsd}	Raised source/drain epi doping	$2 \times 10^{20} cm^{-3}$
N_{body}	Fin body doping	$10^{15} cm^{-3}$
LDG	Lateral doping gradient in extension	2.5 nm/dec
ρ_c	Contact resistivity	$10^{-8} \Omega cm^2$
V_{dd}	Supply voltage	0.9 V

(ρ_c). For the nominal case, we set $\rho_c = 10^{-8} \Omega$ cm, as reported by ITRS [20]. Lower contact resistivity values may be achievable [19, 21]. Since the analytical model we developed is scalable, it can model those low ρ_c values as well.

8.3.2 Device optimization

We use a metal gate in the simulation and assume that threshold voltage (V_{th}) tuning through gate work function engineering is possible. We optimize the FinFET doping profile for maximum drive current (I_{on}) at a constant I_{off}/W_{eff} of 1nA/μm by altering the extension doping, N_{ext}, and the offset spacer width, L_{off} (Fig. 8.13). For all extension doping concentrations, I_{on} versus L_{off} are bell-shaped curves. When the offset spacer is too thin, the extension doping encroaches into the channel and degrades the subthreshold swing, forcing V_{th} to be very high for the same I_{off} and reducing I_{on}. On the other hand, when the offset spacer is too thick, the effective channel length becomes very large, the source/drain resistance becomes significant, and the on-current is degraded as well. Maximum I_{on} is achieved with $N_{ext} = 2 \times 10^{20} cm^{-3}$ and $L_{off} = 6$nm over the range of doping and offset spacer simulated. The corresponding metal gate work function for 1nA/μm off-current is 4.603 eV.

As we will show later, with $L_{off} = 6$nm, L_{eff} is a function of gate overdrive and ranges from about 16 to 20 nm for $V_{gs} < V_{dd}$. Therefore, the optimal device has an underlapped source/drain design. This conclusion is consistent with Refs. [22, 23], where the optimal design of FinFET source/drain is underlapped. We are assuming

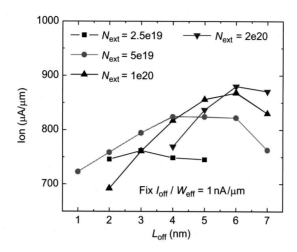

FIG. 8.13

Optimization of the offset spacer width (L_{off}) and fin extension doping (N_{ext}) for maximum on-current.

here that band-to-band tunneling leakage (or gate-induced drain leakage, GIDL) is small enough and its effect on I_{off} can be neglected.

8.3.3 Extraction of source and drain resistances

L_{eff} and the source/drain series resistance are often extracted by the transmission line method by plotting the channel resistance $R_{total} = V_{ds}/I_d$ versus the design gate length L_{des} at several gate overdrive values $V_{gs} - V_{th}$ at $V_{ds} = 50$mV and finding their interception point [24]. There are two potential issues:

1. The average channel mobility is a function of the gate length due to, for example, angled halo implant near the source/drain that suppresses subsurface leakage current.
2. The source/drain resistance is a function of gate bias.

The former is less of an issue for fully depleted FinFETs since a halo implant is usually unnecessary for FinFETs. On the other hand, the latter is more significant, as the source/drain resistance is highly bias-dependent. This is because the conductivity of the fin extension near the gate edge is modulated by the gate fringe field. The bias dependence is expected to be more significant for devices with little or no source/drain overlap, as is the nominal case in this study. As a consequence, L_{eff} and R_{ds} extracted using the transmission line method are likely to be quite different from their physical values. To illustrate this, we performed TCAD simulation for R_{total} versus L_g and fitted the results to linear curves (Fig. 8.14). The extrapolated curves intersect at approximately $L_g = -50$nm. If we had assumed there is no bias

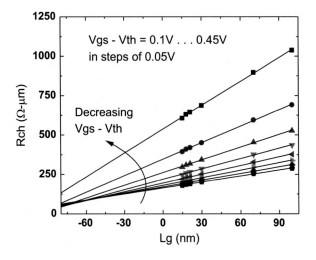

FIG. 8.14

The total channel resistance R_{total} versus physical gate length L_g at several gate overdrive, $V_{gs} - V_{th}$, for the optimized device described in Section 8.3.2. V_{th} is obtained by extrapolating $I_d - V_{gs}$ at the point with maximum g_m. *Symbols*: TCAD simulation; *Lines*: Linear fit.

dependence, we would have concluded that there is a 25 nm underlap on each side, which is unlikely given the device structure we have. Moreover, the intersection point itself is a function of bias.

In this study an alternative TCAD-based method for source/drain resistance extraction is used. The total channel resistance is given by

$$R_{total}(V_{gs}) = R_{ds}(V_{gs}) + \frac{L_g - \Delta L(V_{gs})}{\mu C_{ox} W_{eff} \left(V_{gs} - V_{th} - \frac{V_{ds}}{2} \right) V_{ds}} \tag{8.31}$$

While it is difficult to extract $R_{ds}(V_{gs})$ and $\Delta L(V_{gs})$ simultaneously, if $\Delta L(V_{gs})$ is known, we can find $R_{ds}(V_{gs})$ as the zero crossing of the R_{total} versus L_g curve. $\Delta L(V_{gs})$ is defined as the point at which the electron concentration at the inversion charge centroid is equal to the background doping, and can be extracted from TCAD simulation. This extraction method is carried out and the results are listed in Table 8.3.

L_{eff} is plotted versus gate overdrive, $V_{gs} - V_{th}$, in Fig. 8.15 for a 15 nm device. To smoothen out numerical error, we conduct a second-order polynomial fit, which gives

$$L_{eff}(nm) = 15.34 + 14.06 \cdot V_{gt} - 9.42 \cdot V_{gt}^2 \tag{8.32}$$

or

$$\Delta L(nm) = 0.34 + 14.06 \cdot V_{gt} - 9.42 \cdot V_{gt}^2 \tag{8.33}$$

Table 8.3 Extraction of ΔL from TCAD.

V_{gs}	$V_{gs} - V_{th}$	X_{DC}	$n_e(X_{DC})$	L_{eff}	ΔL
(V)	(V)	(nm)	(cm^{-3})	(nm)	(nm)
0.45	0.042	2.43	5.85×10^{17}	15.88	−0.88
0.495	0.087	2.22	1.09×10^{18}	16.51	−1.51
0.54	0.132	1.99	1.81×10^{18}	16.98	−1.98
0.585	0.177	1.76	2.85×10^{18}	17.58	−2.58
0.63	0.222	1.55	4.21×10^{18}	18.02	−3.02
0.675	0.267	1.37	5.87×10^{18}	18.49	−3.49
0.72	0.312	1.22	7.76×10^{18}	18.84	−3.84
0.765	0.357	1.09	9.83×10^{18}	19.11	−4.11
0.81	0.402	0.99	1.21×10^{19}	19.35	−4.35
0.855	0.447	0.91	1.55×10^{19}	19.69	−4.69
0.9	0.492	0.83	1.93×10^{19}	20.08	−5.08

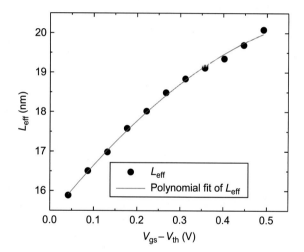

FIG. 8.15

Effective channel length as a function of gate overdrive.

With this model of $\Delta L(V_g)$, we can now calculate $R_{ds}(V_g)$. Fig. 8.16 shows R_{ds} versus V_g for two cases. For one case, we apply Eq. (8.33) and find the total source and drain resistance at $L_g = \Delta L$ (V_g-dependent L_{eff}). For another case, we assume $\Delta L = 0$ (constant L_{eff}). The results are similar, suggesting that the bias dependence of ΔL is not a significant factor for determining R_{ds}. From now on, we will extract the source/drain resistance using $\Delta L = 0$, which is perhaps more realistic since ΔL is not exactly known in experiments.

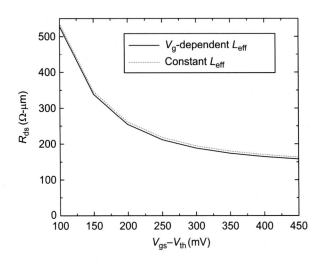

FIG. 8.16

Source/drain resistance as a function of gate overdrive.

In Section 8.2.3, we showed that the bias-dependent source/drain resistance is given as Eq. (8.30a). We further assume $RSW = RDW$ and $RSWMIN = RDWMIN$ for symmetry reasons and carry out fitting, which gives

$$R_{ds}(\Omega \, \mu m) = 107.5 + \frac{95.0}{1 + 7.54 \cdot (V_{gs} - V_{fbsd})} + \frac{95.0}{1 + 7.54 \cdot (V_{gd} - V_{fbsd})} \quad (8.34)$$

The previous expression fits TCAD data very well from $V_{gs} - V_{th} = 0.1$ V to $V_{gs} - V_{th} = 4.2$ V, suggesting the resistance network in Fig. 8.11 is a good description of the resistances in the extension region. The model fitting and TCAD are plotted together in Fig. 8.17, focusing on the low V_{gs} part.

Parameters $R_{ext1,0}$ and $R_{acc,0}$ can be extracted from the bias-dependent part of Eq. (8.34). On the other hand, the bias-independent part (with a value of 107.5) includes the extension resistance (R_{ext2}), the spreading resistance, and the contact resistance. Parameters $R_{ext2,0}$ and ΔL_{ext} can be extracted by the linear fitting of R_{ds} versus the spacer thickness (L_{sp}), as shown in Fig. 8.18. The model fits TCAD very well. The extracted model parameters are summarized in Table 8.4.

To verify Eq. (8.8), we also plot the total source/drain resistance versus raised source/drain length (L_{rsd}) in Fig. 8.19. The model agrees with TCAD well without the need to introduce fitting parameters for the contact resistance.

8.3.4 Discussion

Based on the fitted model, individual resistance components are plotted (Fig. 8.20). For the nominal case (with specific contact resistivity $\rho_c = 10^{-8} \Omega \, cm^2$), the

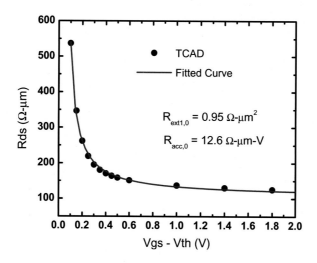

FIG. 8.17

Source/drain resistance as a function of gate overdrive ($V_{th} = 0.408$ V).

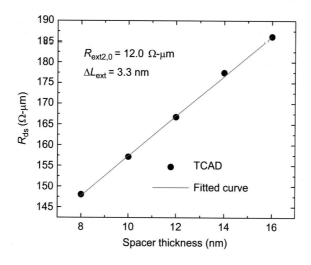

FIG. 8.18

Source/drain resistance as a function of spacer thickness at $V_{gs} = 0.9$ V, $V_{ds} = 50$mV. The channel resistance has been subtracted out.

extension resistance is larger than the contact resistance even at $V_{gs} = V_{dd}$. This is due to the underlapped source/drain junction design with large extension resistance considered in this study.

Fig. 8.20 also suggests that the spreading resistance is a relatively small component. Therefore, constant angle approximation is sufficient.

Table 8.4 Summary of extracted parameters.

Parameter name	Value	Units
$R_{ext1,0}$	0.95	$\Omega\ \mu m^2$
$R_{acc,0}$	12.6	$\Omega\ \mu m\ V$
$R_{ext2,0}$	12.0	$\Omega\ \mu m$
ΔL_{ext}	3.3	nm

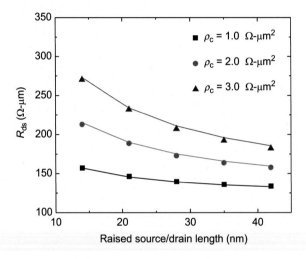

FIG. 8.19

Source/drain resistance as a function of raised source/drain length at $V_{gs} = 0.9V$, $V_{ds} = 50mV$. The channel resistance has been subtracted out. *Symbols*: TCAD simulation; *Lines*: model.

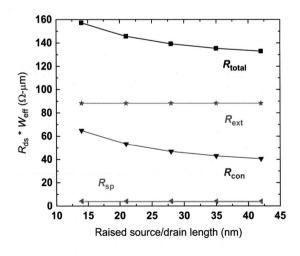

FIG. 8.20

Breakdown of source/drain resistance into individual components: contact resistance (R_{con}), extension resistance (R_{ext}), and spreading resistance (R_{sp}). *Lines* and *Symbols*: model.

The saturation of contact resistance at around 30 nm raised source/drain length (Figs. 8.19 and 8.20) suggests that further lengthening the source/drain has less benefits. In practice the gate pitch is fixed for a given technology, and the room for raised source/drain length optimization is limited.

8.4 Implementation considerations of parasitic resistance model

In this section we describe the practical implementation considerations of the contact, spreading, and extension resistance models in BSIM-CMG.

8.4.1 Physical parameters

The bulk resistivity parameter ρ affects both the contact and the spreading resistance models. The user can specify ρ via the parameter *RHORSD*. Otherwise, BSIM-CMG computes ρ using the Masetti's model [11] as described in Section 8.2.1.

The formulation of fin cross-sectional area A_{fin} and raised source and drain cross-sectional area A_{rsd} are implemented in BSIM-CMG as described in Section 8.1. Special consideration is given to the rare case where the top of the epitaxial source/drain is below the top of the fin (*HEPI* < 0). In such a case, A_{fin} is reduced to *TFIN* $\times$ (*HFIN* + *HEPI*).

8.4.2 Resistance components

Since the FinFET extension resistance model has the same mathematical form as the planar BSIM4 model, we adopt the same parameter naming conventions (rather than R_{ext1}, R_{ext2}, R_{acc}, etc. in Section 8.2.3).

Like BSIM4, two options for source/drain resistance implementation are provided in BSIM-CMG. If *RDSMOD* $= 0$, the extension resistance is implemented as a current degradation factor that is multiplied to drain current.[c] On the other hand, if *RDSMOD* $= 1$, the extension resistance is implemented as two resistors in series with the transistor.

To be consistent with prior BSIM models, we provide users the option to set *RGEOMOD* $= 0$ to specify geometry-dependent parasitic source/drain resistances via sheet resistance parameters *RSHS* and *RSHD*, where

$$R_{\text{s,geo}} = NRS \cdot RSHS \tag{8.35}$$

$$R_{\text{d,geo}} = NRD \cdot RSHD \tag{8.36}$$

---

[c] For *RDSMOD* $= 0$, the gate bias dependence term is modified from $V_{\text{gs(d)}} - V_{\text{fbsd}}$ to average channel inversion charge density q_{ia}. This is to avoid a hump in drain current behavior, which may appear because $V_{\text{gs(d)}} - V_{\text{fbsd}}$ in *RDSMOD* $= 0$ is forced to zero via a smoothing function when negative.

In such a case, the spreading and contact resistance models described in this chapter are not used for calculations.

If, on the other hand, the user sets *RGEOMOD* = 1, the spreading and contact resistance formulations described in this chapter will be used.

The spreading resistance model in BSIM-CMG is implemented as Eq. (8.23). The default *K* is set to about 0.4, which corresponds to $\theta = 55$ degrees. This value originated from an early study and is kept unchanged for backward compatibility reasons.

The contact resistance model in BSIM-CMG is implemented as Eq. (8.8). The raised source/drain area and the end contact are assumed to be the same ($A_{\text{rsd}} = A_{\text{term}}$) if the user sets parameter *SDTERM* = 1. Otherwise, if *SDTERM* = 0, no end contact is considered and the contact resistance formulation falls to Eq. (8.12).

Whether *RGEOMOD* = 0 or *RGEOMOD* = 1, $R_{\text{s, geo}}$ and $R_{\text{d, geo}}$ are implemented as resistors in series with the transistor. Note that if *RGEOMOD* = 1, five fitting parameters are introduced to capture extra geometry dependencies [25].

Comprehensive consideration of the complex geometry dependence of parasitic resistance in multiple-stacked GAA devices is yet to be developed.

Nevertheless, the general bias-dependence formula (Eqs. 8.30a, 8.30b), which models FinFET parasitic resistance as function of gate bias, is expected to work for GAA devices as well.

8.5 Gate electrode resistance model

The gate electrode resistance model can be switched on by setting *RGATEMOD* = 1. This introduces an internal node "ge." The gate electrode resistor (R_{geltd}) is placed between the external "g" node and the internal "ge" node.

The gate electrode resistance model takes into account the number of gate contacts, *NGCON*. *NGCON* = 1 indicates single-sided contact, while *NGCON* = 2 indicates double-sided contact. R_{geltd} is given by

$$R_{\text{geltd}} = \begin{cases} \dfrac{RGEXT + RGFIN \cdot NFIN/3}{NF} & \text{for } NGCON = 1 \\ \dfrac{RGEXT/2 + RGFIN \cdot NFIN/12}{NF} & \text{for } NGCON = 2 \end{cases} \quad (8.37)$$

8.6 FinFET/GAA parasitic capacitance models

In addition to the intrinsic capacitance as described in the previous chapter, BSIM-CMG models both the overlap capacitance (C_{ov}), which is bias-dependent, and the fringe capacitance (C_{fr}), which typically has weak or negligible bias dependence. We often use C_{ov} to capture the bias-dependent C-V characteristics of a given device over the bias range of interest. On the other hand, C_{fr} provides scaling of capacitance across device geometry. For C_{fr}, BSIM-CMG offers three models with different levels of complexity, which the users can select via model parameter *CGEOMOD*. In this chapter, we focus on *CGEOMOD* = 2/3, which are the most complete and physics-based model for FinFET/GAA parasitic capacitances, respectively. Its superior scalability with respect to geometrical parameters as compared to *CGEOMOD* = 0 and *CGEOMOD* = 1 makes it better suited for device variability modeling.

8.6.1 Connection of parasitic capacitance components

Both C_{fr} and C_{ov} have gate-to-source and gate-to-drain components. For *CGEOMOD* = 0 and *CGEOMOD* = 2/3, the respective fringe and overlap components are lumped together and connected between the gate and the inner source or drain nodes, as illustrated in Fig. 8.21.

To offer modeling flexibility, a special case is *CGEOMOD* = 1, where the fringe components are connected to the outer source/drain nodes, whereas the overlap components, being closer to the channel, are connected to the inner source/drain nodes (Fig. 8.22).

Details of *CGEOMOD* = 0 and *CGEOMOD* = 1 can be found in the BSIM-CMG technical manual [25].

8.6.2 Derivation of 2D fringe capacitance

We first discuss the case CGEOMOD = 2, focusing on the FinFET device structure. CGEOMOD = 3 for GAA devices is simply an extension to it, which we will discuss toward the end of this chapter.

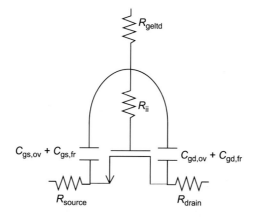

FIG. 8.21

R-C network for CGEOMOD = 0 and CGEOMOD = 2/3. In this illustration, the nonquasistatic resistance (R_{ii}), gate electrode resistance (R_{geltd}), and external source and drain resistance (R_{source} and R_{drain}) are all turned on by setting NQSMOD = 1, RGATEMOD = 1, and RDSMOD = 1.

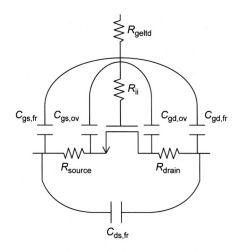

FIG. 8.22

R-C network for CGEOMOD = 1. In this illustration the nonquasistatic resistance (R_{ii}), gate electrode resistance (R_{geltd}), and external source and drain resistance (R_{source} and R_{drain}) are all turned on by setting NQSMOD = 1, RGATEMOD = 1, and RDSMOD = 1.

In planar MOSFETs perpendicular surfaces of the gate edge and the source/drain allow a simple conformal mapping equation to calculate the fringe capacitance (e.g., Ref. [26]). In a FinFET, however, to our best knowledge, no known exact 2D conformal mapping method exists, given a raised source and drain structure as illustrated in Fig. 8.23.

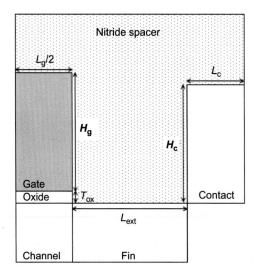

FIG. 8.23

Two-dimensional half-device cross-section for the FinFET, where the potential distribution in the nitride region must be solved to derive capacitance.

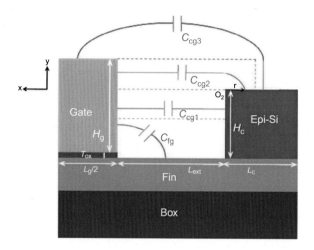

FIG. 8.24

Separation of fringe capacitance components into fin-to-gate (C_{fg}) and contact-to-gate (C_{cg1}, C_{cg2}, and C_{cg3}) components.

Considering Fig. 8.23, we adopt a solution where the total capacitance is decomposed into a fin-to-gate component, C_{fg}, and a fin-to-channel component C_{cg}. C_{cg} is further separated into C_{cg1}, C_{cg2}, and C_{cg3}, which have distinct electric field line trajectories (Fig. 8.24). Other parasitic components are expected to be relatively small.

Each component is calculated by summing infinitesimal capacitors, each having a capacitance of

$$\Delta C = \epsilon_{ox} \frac{\Delta A}{d} \tag{8.38}$$

where ΔA is the area of the infinitesimal capacitor and d is the length of the electric field line.

1. Fin-to-gate capacitance (C_{fg})

C_{fg} is associated with electric field lines originating from the top surface of the silicon fin and terminating on the gate edge. The length of each electric field line is the perimeter of a quarter ellipse, which is approximated with Euler approximation [27]:

$$\text{Perimeter of ellipse} = 2\pi \sqrt{\frac{a^2 + b^2}{2}} \tag{8.39}$$

where a is the length of the major axis and b is the length of the minor axis.

As shown in Fig. 8.25, each quarter ellipsoidal field line of C_{fg} has a major axis of l and a minor axis of $T_{ox} + h$, where h ranges from 0 to H_{max} and l ranges from 0 to L_{max}. We assume the ellipsoidal electric field lines do not go beyond L_{max} or H_{max}. Electric field lines that originate from the portion of the gate above H_{max} travel horizontally and land on the contact and

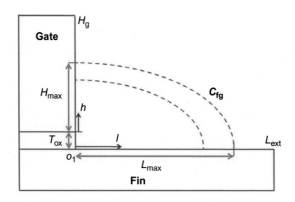

FIG. 8.25

Electric field lines in the fin-to-gate region.

become part of C_{cg1}, as we will show later. If H_g is less than H_{max} then C_{cg1} becomes zero. H_{max} is extracted from TCAD simulations by varying H_g and finding the point where C_{cg1} diminishes to zero.

Moreover, H_{max} is proportional to L_{ext} because increasing L_{ext} increases the distance between the gate and the contact and decreases the effect of the raised source/drain contact's inner surface on the fringe capacitance between the gate and the fin. Thus, we have

$$H_{max} = \frac{L_{ext}}{H_r} \tag{8.40}$$

$$L_{max} = \frac{L_{ext}}{L_r} \tag{8.41}$$

where $H_r = HR0 + HR1 \times \frac{H_g + T_{ox}}{H_c}$ and $L_r = LRA + LRB \times L_{ext} + LRC \times \frac{H_c}{H_g + T_{ox}}$.

By limiting h and l to H_{max} and L_{max}, the electric field lines in the gate-to-fin region have one axis of the ellipse varying from 0 to L_{max} and the other axis varying from T_{ox} to $H_{max} + T_{ox}$. To simplify the integration, the rates of change in the two axes are assumed to be proportional to each other. We may express h as a function of l:

$$h = \frac{H_{max}}{L_{max}} \cdot l = \frac{1}{H_r - L_r} \cdot l \tag{8.42}$$

The length of the electric field line is calculated from Eqs. (8.39), (8.42):

$$d(l) = \frac{1}{4}\left[2\pi \sqrt{\frac{\left(T_{ox} + \frac{1}{L_r H_r}\right)^2 + l^2}{2}} \right] \tag{8.43}$$

Using Eqs. (8.38), (8.43), the fin-to-gate capacitance per unit width becomes

$$C_{fg,sat} = \epsilon_{sp} \int_0^{L_{max}} \frac{1}{d(l)} dl = \epsilon_{sp} \frac{2\sqrt{2}}{\pi} \int_0^{L_{max}} \frac{1}{\sqrt{\left(T_{ox} + \frac{1}{H_r \cdot L_r}\right)^2 + l^2}} dl \qquad (8.44)$$

We integrate Eq. (8.44) and introduce a constant fitting parameter $CF1$. $C_{fg,sat}$ becomes

$$C_{fg,sat} = CF1 \cdot \frac{2\sqrt{2}\epsilon_{sp}r}{\pi\sqrt{r^2+1}}$$

$$\times \ln\left(\frac{\sqrt{(r^2+1)[(rT_{ox})^2 + 2rL_{max}T_{ox} + (r^2+1)L_{max}^2]} + rT_{ox} + (r^2+1)L_{max}}{rT_{ox}[1 + \sqrt{r^2+1}]}\right)$$

$$(8.45)$$

where $r = H_r \cdot L_r$.

Eq. (8.45) is calculated under the assumption that gate height (H_g) is larger than H_{max}, and, as a result C_{fg} does not change as we increase H_g. We denote this situation as the saturation condition. On the other hand, if H_g is less than H_{max}, the entire gate sidewall belongs to the fin-to-gate region. In this case C_{fg} is, to first order, a logarithmic function of H_g. We denote this as the log condition. $C_{fg,log}$ per unit length can be expressed as

$$C_{fg,log} - \epsilon_{sp}$$

$$\times \left[\frac{2\sqrt{2}}{\pi\sqrt{k+1}} \cdot \ln\left(\frac{\sqrt{k+1}\sqrt{(T_{ox} + H_g)^2 + (k \cdot H_g)^2} + T_{ox} + (k+1)H_g}{T_{ox}[(k+2) + \sqrt{(k+2)^2+1}]}\right) + WL\right]$$

$$(8.46)$$

where $k = \frac{L_{max}}{H_g + T_{ox}}$. WL is a fitting parameter.

A smoothing function with a fitting parameter $DELTA$ (δ) is used to describe the transition from $C_{fg,log}$ to $C_{fg,sat}$. The final C_{fg} is expressed as

$$C_{fg} = H_{fin}\left[C_{fg,sat} - \frac{(C_{fg,sat} - C_{fg,log} - \delta) + \sqrt{(C_{fg,sat} - C_{fg,log} - \delta)^2 + 4\delta C_{fg,sat}}}{2}\right] \qquad (8.47)$$

2. Source/drain contact to gate capacitance (C_{cg})

The contact-to-gate capacitance, C_{cg}, describes the capacitance between the gate and the epitaxially grown source/drain contact. The contact is assumed to be rectangular, as shown in Fig. 8.24. C_{cg} includes C_{cg1}, C_{cg2}, and C_{cg3}. Each component denotes the capacitance originating from different surfaces of the gate/contact. In addition, we introduce a fitting parameter $C0CG$ to provide a constant offset. The total C_{cg} can be expressed as

$$C_{cg} = H_{fin} \cdot (C_{cg1} + C_{cg2} + C_{cg3} + \epsilon_{sp} \cdot C0CG) \qquad (8.48)$$

In the following sections, expressions for each capacitance are developed under the assumption that the gate height is larger than the contact height ($H_g + T_{ox} > H_c$).

Note that out of the three C_{cg} components, only C_{cg1} and C_{cg2} are implemented in BSIM-CMG because, in practice, the region that C_{cg3} covers usually falls within the scope of adjacent structures and the corresponding capacitances need not be included.

(a) C_{cg1}:

C_{cg1} is a simple parallel plate capacitance between the gate and the contact, as illustrated in Fig. 8.24. Using Eq. (8.38) we derive the capacitance per unit height:

$$C_{cg1} = \epsilon_{sp} \frac{H_g - H_{max}}{L_{ext}} \quad \text{if } H_g \geq H_{max} \tag{8.49}$$

As mentioned earlier, if $H_g < H_{max}$, then the entire inner sidewall of the gate belongs to the fin-to-gate region. This also means

$$C_{cg1} = 0 \quad \text{if } H_g < H_{max} \tag{8.50}$$

A smoothing function combines Eqs. (8.49), (8.50) into a continuous equation:

$$C_{cg1} = \frac{1}{CNON} \cdot \ln\left[1 + \exp\left(CNON \cdot \epsilon_{sp} \frac{H_g - H_{max}}{L_{ext}}\right)\right] \tag{8.51}$$

where CNON ensures the exponential term is comparable to 1 when H_g is close to H_{max}.

However, Eq. (8.51) assumes C_{cg1} spans the entire height of the gate above H_{max}. If the contact is not tall enough, this will become physically incorrect. Therefore, final expression for C_{cg1} as implemented in BSIM-CMG is

$$C_{cg1} = \frac{1}{CNON} \cdot \ln\left[1 + \exp\left(CNON \cdot \epsilon_{sp} \frac{\min(H_c, H_g + T_{ox}) - H_{max}}{L_{ext}}\right)\right] \tag{8.52}$$

Note that H_g in Eq. (8.51) is replaced with $\min(H_c, H_g + T_{ox})$.

(b) C_{cg2}:

The electric field lines of C_{cg2} originate from the gate sidewall, travels a distance L_{ext} horizontally, and then follow a quarter circle until they terminates on top of the contact (Fig. 8.24). The quarter circle has a radius r centered at the corner of the contact (marked as O_2 in Fig. 8.24). Therefore, the expression for the length of the electric field lines is

$$d = L_{ext} + \frac{2\pi r}{4} \tag{8.53}$$

where the radius r varies from 0 to R.

$$R = \frac{1}{2}\Delta H \cdot \frac{H_c}{H_g + T_{ox}} = \frac{H_c}{2}\left|1 - \frac{H_c}{H_g + T_{ox}}\right| \tag{8.54}$$

$$\Delta H = |H_g + T_{ox} - H_c|$$

Here, ΔH is the difference of the gate height plus T_{ox} and the contact height. The capacitance in the region beyond R belongs to C_{cg3}. We perform integration from 0 to R to obtain C_{cg2} per unit height:

$$C_{cg2} = \epsilon_{sp} \int_0^R \frac{1}{L_{ext} + \frac{2\pi r}{4}} dr = \frac{2\epsilon_{sp}}{\pi} \ln\left(\frac{L_{ext} + 0.5\pi R}{L_{ext}}\right) \tag{8.55}$$

(c) C_{cg3}:

C_{cg3} describes the capacitance with electric field lines originating from the top of the gate and terminating on the top of the contact. This is modeled by two capacitances in series: C_{cg3a} and C_{cg3b}. C_{cg3b} is a simple parallel plate capacitance characterized by the area of the contact surface and the horizontal distance:

$$C_{cg3b} = \epsilon_{sp} \frac{L_c - R}{\Delta H} \tag{8.56}$$

C_{cg3a} is characterized by a similar equation to C_{fg} except with semicircular electric fields with diameters ranging from L_{ext} to $L_{ext} + L_c + L_g/2$. Only half of the gate length is considered because the other half would contribute to the same parasitic capacitance on the other side of the FinFET. Eq. (8.57) results from a similar derivation.

$$C_{cg3a} = \frac{2\epsilon_{sp}}{\pi[(L_c - R) + L_g/2]} \ln\left(\frac{L_{ext} + L_c + L_g/2}{L_{ext} + R}\right) \tag{8.57}$$

The equation for C_{cg3a} is derived assuming that the gate is thicker than the contact epitaxy. If the opposite is true, then the values for $L_g/2$ and L_c are swapped, resulting in the same equation. The total C_{cg3} is described by the series combination of the two capacitances.

$$C_{cg3} = \frac{1}{\frac{1}{C_{cg3a}} + \frac{1}{C_{cg3b}}} \tag{8.58}$$

As mentioned earlier in this section, C_{cg3} is not implemented in BSIM-CMG. Instead, capacitance in the region covered by C_{cg3} is modeled outside of BSIM-CMG.

(d) Generalization to contacts taller than the gate:

Special consideration for the corner case where $H_g + T_{ox} < H_c$ is required. In FinFETs, the gate is usually taller than the raised source/drain contact in the vertical direction, so we do not encounter this corner case. However, when we use the same model to describe parasitic capacitance on the side of the fin, H_g becomes the gate width on the side of the fin and H_c becomes the contact width over the original fin. In such a case, $H_g + T_{ox}$ can be less than H_c for a single-fin FinFET or for the last fin in a multifin FinFET. Several simple changes to the C_{cg} model are required to cover such cases. First, the width of the parallel plate capacitance in C_{cg1} becomes $(H_g + T_{ox}) - (H_{max} + T_{ox})$. A more general expression for C_{cg1} is

$$C_{\text{cg1}} = \frac{1}{CNON} \cdot \ln\left[1 + \exp\left(CNON \cdot \epsilon_{\text{sp}} \frac{\min(H_g + T_{\text{ox}}, H_c) - (H_{\text{max}} + T_{\text{ox}})}{L_{\text{ext}}}\right)\right] \quad (8.59)$$

Second, for C_{cg2}, a more general expression for R becomes

$$R = \frac{1}{2}\Delta H \cdot \min\left(\frac{H_g + T_{\text{ox}}}{H_c}, \frac{H_c}{H_g + T_{\text{ox}}}\right) \quad (8.60)$$

C_{cg3} does require minimal modification for the different geometry. All instances of L_c and $L_g/2$ are swapped with each other. C_{cg3a}, the semicircular portion, remains unchanged because the variables L_c and $L_g/2$ are commutative. The use of the absolute value in Eq. (8.55) also allows ΔH to be general. The term $L_c - R$ in the numerator of Eq. (8.56) is replaced by $\frac{L_g}{2} - R$. In general, L_c in Eq. (8.56) can be replaced by a variable L defined as

$$\begin{cases} L_c & \text{if } H_g + T_{\text{ox}} > H_c \\ L_g & \text{if } H_g + T_{\text{ox}} < H_c \end{cases} \quad (8.61)$$

Eq. (8.58) is still appropriate for C_{cg3}.

8.7 Modeling of FinFET/GAA fringe capacitance in 3D

In the previous section, we developed a model for the region between the gate and the raised source/drain contact. In 3D FinFETs (Fig. 8.26), the situation is more complex. To simplify the problem, we separate the 3D capacitances into a top component (C_{top}), two side components (C_{side}), and two corner components (C_{corner}), as illustrated in Fig. 8.27.

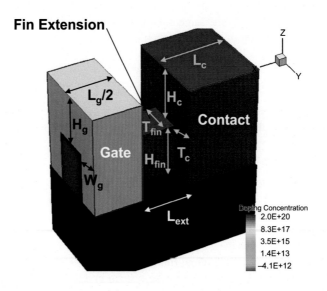

FIG. 8.26

Structure of a single fin in a 3D FinFET for TCAD simulations and symbol definitions.

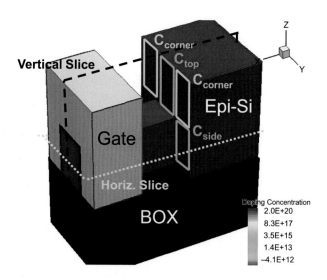

FIG. 8.27

Composition of simpler capacitances that sum up to the whole parasitic capacitance. Rectangle boxes C_{corner}, C_{top}, and C_{side} are starting surfaces of electric field lines for corresponding capacitance components.

The top component is exactly the 2D capacitance model multiplied by the fin width (T_{fin}). The side component also uses the 2D formulation, but with H_c replaced with T_c, and H_g replaced with W_g (see Fig. 8.26 for symbol definitions), and multiplied by the fin height (H_{fin}). C_{corner} is modeled as a simple parallel plate capacitor with the two plates separated by a distance of L_{ext}:

$$C_{corner} = \frac{\epsilon_{ox}}{L_{ext}} A_{corner} \tag{8.62}$$

where $A_{corner} = ((FPITCH - TFIN) \times H_c) \times NFIN + A_{rsd,end} + A_{sili,end}$. FPITCH is the fin pitch.

The total capacitance for a multifin FinFET with *NFIN* fins is modeled as

$$C_f = NFIN \times (2 \cdot CGEOE \cdot C_{side} + C_{top}) + C_{corner} \tag{8.63}$$

where CGEOE is a special fitting parameter for geometry-dependent parasitic capacitance.

Precisely, C_{side} and C_{top} are both calculated using the 2D capacitance model, expressed in capacitance per unit fin width or fin height.

$$C_{top} = C_{fringe,2D}(H_g, H_c) \times T_{fin} \qquad C_{side} = C_{fringe,2D}(W_g, T_c) \times H_{fin}$$

where three components are included in the 2D capacitance model:

$$C_{fringe,2D} = C_{cg1} + C_{cg2} + C_{fg}$$

Notice that the derivation in Section 8.6.2 focuses on fin-top capacitance (C_{top}). So $C_{fringe,2D}(H_g, H_c)$ uses the same notation as the derivation in Section 8.6.2. For the

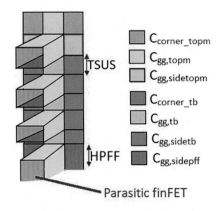

FIG. 8.28

Graphical representation of the cross-section of a stacked nanosheet FET showing the various parasitic capacitance components.

fin-side component C_{side}, we simply replace H_g with W_g and H_c with T_c in the calculation of individual components C_{cg1}, C_{cg2}, and C_{fg}.

For GAA devices (CGEOMOD = 3), parasitic capacitance components can be calculated using an extension of the 3D FinFET case (CGEOMOD = 2), as illustrated in Fig. 8.28. Top and side components are calculated as follows:

$$C_{gg,topm} = C_{fringe,2D}(H_g, H_c) \times W_{GAA} \qquad C_{gg,tb} = C_{fringe,2D}(TGATE, H_{c2})$$
$$\times W_{GAA} \qquad C_{sidetopm} = C_{sidetb} = C_{fringe,2D}(W_g, T_c) \times T_{GAA} \qquad C_{sidpff} = C_{fringe,2D}(W_g, T_c)$$
$$\times H_{PFF}$$

where W_{GAA} is the width of the GAA FET; T_{GAA} is the thickness of each GAA wire; H_{PFF} is the height of the bottom parasitic FinFET; H_{c2} is half of the spacing between wires, or $H_{c2} = \frac{1}{2}T_{SUS}$

The corner component is

$$C_{corner} = \frac{\epsilon_{ox}}{L_{ext}} A_{corner}$$

where A_{corner} is the total corner area:

$$A_{corner} = NFIN \times (A_{corner,topm} + 2N_{GAA}A_{corner,tb}) + A_{rsd,end} + A_{sili,end}$$
$$A_{corner,topm} = (FPITCH - W_{GAA}) \times H_c \qquad A_{corner,tb} = (FPITCH - W_{GAA}) \times H_{c2}$$

The total fringe capacitance is

$$C_f = C_{corner} + (C_{gg,topm} + 2N_{GAA}C_{gg,tb}) \times NFIN$$
$$+ CGEOE \times (C_{gg,sidetopm} + C_{gg,sidetb}(N_{GAA} - 1) + C_{gg,sidepff}) \times 2NFIN$$

Notice that in BSIM-CMG implementation, an additional factor is multiplied to the final C_f expression for better matching flexibility:

$$C_{f,final} = C_f \times (CGEOA + CGEOB \cdot T_{fin} + CGEOC \cdot FPITCH + CGEOD \cdot L_c) \times NF$$

where NF is the number of gate fingers.

8.8 Parasitic capacitance model verification

The analytical model for parasitic capacitance in FinFETs is developed with many degrees of freedom, allowing a variety of device architectures to be modeled. Inherent to the model assumptions, there are a few fitting parameters whose value needs to be determined. We determine these values by comparison with numerical simulations (TCAD). Figs. 8.29–8.31 verify the parasitic capacitance of each region by comparison with 2D TCAD. Figs. 8.32 and 8.33 verify the total gate parasitic capacitance by comparison with 3D TCAD. Fitting parameter values for the best match with TCAD are listed in Table 8.5. Here, only C0CG changes from capacitance to capacitance, while the other fitting parameters have the same values. Notice that the verification in this section is done with CGEOA = CGEOE = 1 and CGEOB = CGEOC = CGEOD = 0 so that these additional fitting parameters do not have an effect. In other words, we focus on the verification of the original model C_f without the additional terms for fitting.

In Fig. 8.29, the fin-to-gate capacitance (C_{fg}) per unit thickness is plotted versus H_{g} for different values of L_{ext}. As expected from the model, the saturation and log regions are shown: at first, C_{fg} grows logarithmically with H_{g} because $H_{\mathrm{g}} + T_{\mathrm{ox}}$ is still smaller than H_{max}. Then, C_{fg} becomes nearly constant because it goes into saturation region after $H_{\mathrm{g}} + T_{\mathrm{ox}} > H_{\mathrm{max}}$. Fig. 8.30 exhibits contact (S/D)-to-gate capacitance versus H_{c} at different L_{ext}, where we keep $H_{\mathrm{g}} + T_{\mathrm{ox}} = H_{\mathrm{c}}$. With $H_{\mathrm{g}} + T_{\mathrm{ox}} = H_{\mathrm{c}}$ we force C_{cg2} to zero according to Fig. 8.24. C_{cg3} is outside the simulation structure in this comparison. Fig. 8.31 shows the verification of total C_{cg}. In this case we simulated a structure as depicted in Fig. 8.24, where C_{cg2} and C_{cg3} are nonzero due to uneven $H_{\mathrm{g}} + T_{\mathrm{ox}}$ and H_{c}.

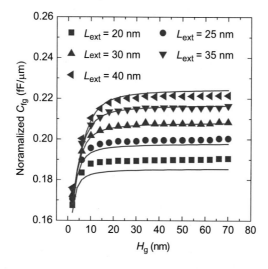

FIG. 8.29

Normalized C_{fg} (fin-to-gate capacitance per unit thickness) versus H_{g} at different L_{ext} (*Lines:* Model, *Symbols:* 2D TCAD).

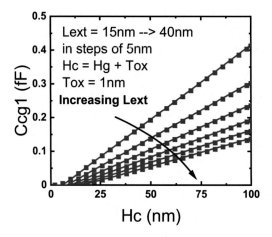

FIG. 8.30

Normalized C_{cg1} (C_{cg1} per unit thickness) versus H_c at different L_{ext} (*Lines*: Model, *Symbols*: 2D TCAD).

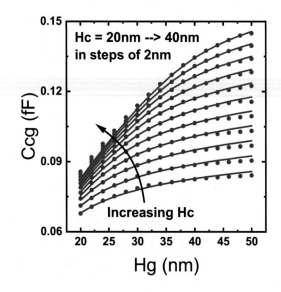

FIG. 8.31

Normalized C_{cg} (C_{cg} per unit thickness) versus H_g at different H_c. $C0CG = 0.4$ (*Lines*: Model, *Symbols*: 2D TCAD).

Figs. 8.32 and 8.33 verify the total gate parasitic capacitance in the middle-fin and end-fin cases from Fig. 8.34, respectively. The value of $C0CG$ is 0.47 in both situations. In the middle-fin case (Fig. 8.32), the total parasitic capacitance is plotted versus T_c (equals $W_g + T_{ox}$) at $H_{fin} = 20$, 30, and 40 nm. In the end-fin case (Fig. 8.33), the total parasitic capacitance is plotted versus W_g at different T_c.

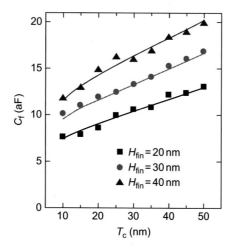

FIG. 8.32

Middle-fin case's total parasitic capacitance versus T_c at different H_{fin} (*Lines*: Model, *Symbols*: 3D TCAD).

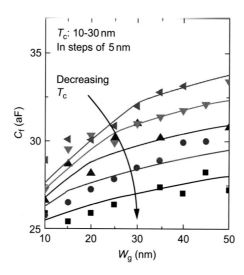

FIG. 8.33

End-fin case's total parasitic capacitance versus W_g at different T_c (*Lines*: Model, *Symbols*: 3D TCAD).

Table 8.5 Fitting parameters for the compact fringe capacitance model.

Parameter name	Value (TCAD calibration)	Value (BSIM-CMG)[a]
LRA	1.05	1.05
LRB	0.0025	0.0
LRC	−0.01	0.0
WR0	2.3	2.3
WR1	0.2	0.2
HL	2.6	12.27
CF1	0.935	0.70 for double-gate FinFET
		0.85 for tri-gate FinFET
DELTA(δ)	1.2×10^{-12}	1.2×10^{-12}
C0CG[b]	0.47	0.47
CNON	1.7×10^{12}	1.7×10^{12}

[a]Values implemented in BSIM-CMG are a result of additional calibration, which gives slightly different values from those used in the verification shown in this chapter.
[b]C0CG is a fitting parameter, the user is allowed to tune to match data.

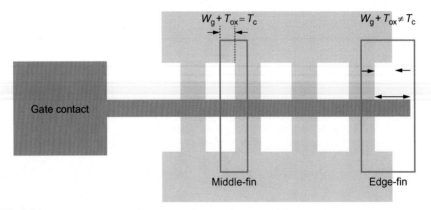

FIG. 8.34

Top view of a 4-fin FinFET with possible adaptations of the capacitance model to calculate the fringe capacitance in each section.

8.9 Summary

In BSIM-CMG, we model both parasitic resistances and parasitic capacitances in FinFETs and GAA devices.

As far as parasitic resistance is concerned, we have developed a simple model of gate electrode resistance, as well as a model for bias- and geometry-dependent parasitic source/drain resistance. The gate electrode resistance model is capable of modeling both single-sided contacts and double-sided contacts. As far as source and drain resistances are concerned, geometry dependence of contact, spreading, and extension resistances are well modeled. The transmission line-based contact resistance model predicts dependence on epi height, fin pitch, contact length, cross-sectional shape of the epi, etc., and models various contact silicidation schemes. Spreading resistance and extension resistance expressions are developed. The extension resistance exhibits bias dependence due to the fringe field from the gate. The model is validated against 3D TCAD simulations. The extension resistance model captures the gate bias dependence of source/drain resistance very well even for underlapped device where the resistance has strong bias dependency. Resistance breakdown analysis shows that the spreading resistance is negligibly small, whereas extension and contact resistances dominate.

A parasitic gate capacitance model was developed for a 2D slice of the FinFET structure and later extended to the complete 3D FinFET and GAA device structure. The 2D model is composed of a gate-to-fin region (C_{fg}) that is characterized by elliptical field lines from the gate to the fin, and a gate-to-contact region (C_{cg}), further separated into various surfaces on the contact or gate for the flux to pass through. C_{cg1} is essentially a parallel plate capacitance that levels off to zero at low gate or contact thicknesses. C_{cg2} has field lines emanating from the side of the gate to the top of the contact, which is characterized by a straight line plus a quarter-circle. C_{cg3} has field lines emanating from the top of the gate to the top of the contact. Its electric field lines can be modeled as a large half-circle plus a straight line. Although C_{cg3} was eventually not implemented in BSIM-CMG, as it was decided that this component should be handled by the layout parasitic extraction tool instead of the compact model itself. Finally, also, a parallel plate capacitance is used to describe C_{corner}. The total capacitance was verified with 2D and 3D TCAD simulations and was found to be accurate for a wide range of dimensions. The fitting parameters used were found to be suitable for the wide range of dimensions tested.

References

[1] K. Mistry, et al., A 45 nm logic technology with high-k+metal gate transistors, strained silicon, 9 Cu interconnect layers, 193 nm dry patterning, and 100% Pb-free packaging, in: International Electron Devices Meeting Technical Digest, 2007, pp. 247–250.

[2] More Moore, in: International Roadmap for Devices and Systems, 2022, Available: https://irds.ieee.org/.

[3] W. Wu, M. Chan, Gate resistance modeling of multifin MOS devices, IEEE Electron Device Lett. 27 (1) (2006) 68–70.

[4] A. Dixit, A. Kottantharayil, N. Collaert, M. Goodwin, M. Jurczak, K. De Meyer, Analysis of the parasitic S/D resistance in multiple-gate FETs, IEEE Trans. Electron Devices 52 (6) (2005) 1132–1140.

[5] D. Tekleab, P. Zeitzoff, Modeling and analysis of parasitic resistance in double gate Fin-FETs, in: Proceedings of the IEEE International SOI Conference, 2008, pp. 51–52.

[6] J. Kedzierski, M. Ieong, E. Nowak, T.S. Kanarsky, Y. Zhang, R. Roy, D. Boyd, D. Fried, H.-S.P. Wong, Extension and source/drain design for high-performance FinFET devices, IEEE Trans. Electron Devices 50 (4) (2003) 952–958.

[7] H. Shang, et al., Investigation of FinFET devices for 32 nm technologies and beyond, in: Symposium on VLSI Technology Digest of Papers, 2006, pp. 54–55.

[8] R.S. Sheony, K.C. Saraswat, Optimization of extrinsic source/drain resistance in ultra-thin body double-gate FETs, IEEE Trans. Nanotechnol. 2 (4) (2003) 265–270.

[9] V. Trivedi, J.G. Fossum, M.M. Chowdhury, Nanoscale FinFETs with gate-source/drain underlap, IEEE Trans. Electron Devices 52 (1) (2005) 56–62.

[10] Sentaurus Device, Synopsys Inc., 2008.

[11] G. Masetti, M. Severi, S. Solmi, Modeling of carrier mobility against carrier concentration in arsenic-, phosphorus-, and boron-doped silicon, IEEE Trans. Electron Devices 30 (7) (1983) 764–769.

[12] H.H. Berger, Model for contacts to planar devices, Solid-State Electron. 15 (1972) 145–158.

[13] BSIM4 Model, Department of Electrical Engineering and Computer Science, UC Berkeley, Available from: https://bsim.berkeley.edu/.

[14] Sentaurus Structure Editor, Synopsys Inc., 2008.

[15] Noffset 3D, Synopsys Inc., 2008.

[16] H. Kawasaki, et al., Demonstration of highly scaled FinFET SRAM cells with high-k metal gate and investigation of characteristic variability for the 32 nm node and beyond, in: International Electron Devices Meeting Technical Digest, 2008, pp. 237–240.

[17] C. Lombardi, S. Manzini, A. Saporito, M. Vanzi, A physically based mobility model for numerical simulation of nonplanar devices, IEEE Trans. Comput.-Aided Des. 7 (11) (1988) 1164–1171.

[18] C. Canali, G. Majni, R. Minder, G. Ottaviani, Electron and hole drift velocity measurements in silicon and their empirical relation to electric field and temperature, IEEE Trans. Electron Devices 22 (11) (1975) 1045–1047.

[19] R. Vega, T.-J. King, Three-dimensional FinFET source/drain and contact design optimization study, IEEE Trans. Electron Devices 56 (7) (2009) 1483–1492.

[20] Interconnect, in: International Technology Roadmap for Semiconductors 2010 Update, International Roadmap Committee, 2010. http://www.itrs.net/.

[21] N. Stavitski, M.J.H. van Dal, A. Lauwers, C. Vrancken, A.Y. Kovalgin, R.A.M. Wolters, Systematic TLM measurements of NiSi and PtSi specific contact resistance to n- and p-type Si in a broad doping range, IEEE Electron Device Lett. 29 (4) (2008) 378–381.

[22] R.S. Shenoy, K.C. Saraswat, Optimization of extrinsic source/drain resistance in ultra-thin body double-gate FETs, IEEE Trans. Nanotechnol. 2 (4) (2003) 265–270.

[23] V. Trivedi, J.G. Fossum, Source/drain-doping engineering for optimal nanoscale FinFET design, in: IEEE International SOI Conference, 2004, pp. 192–194.

[24] R.H. Dennard, F.H. Gaensslen, H.-N. Yu, V.L. Rideout, E. Bassous, A.R. Leblanc, Design of ion implanted MOSFET's with very small dimensions, J. Solid-State Circuits 9 (5) (1974) 256–268.

[25] S. Venugopalan, N. Paydavosi, J. Duarte, D. Lu, C.-H. Lin, M. Dunga, S. Yao, T. Morshed, A. Niknejad, C. Hu, BSIM-CMG 107.0.0 Multi-Gate MOSFET Compact Model Technical Manual, 2013.

[26] K. Suzuki, Parasitic capacitance of submicrometer MOSFET's, IEEE Trans. Electron Devices 46 (9) (1999) 1895–1900.

[27] S. Sykora, Approximations of Ellipse Perimeters, vol. I, Stan's Library, 2005. http://www.ebyte.it/library/docs/math05a/EllipsePerimeterApprox05.html.

Noise

9

9.1 Introduction

From a microscopic point of view, carrier transport in electronic devices is a stochastic process. Electrons and holes move randomly at their thermal velocity ($\sqrt{3kT/m} \approx 10^7$ cm/s). Transport models and BSIM-CMG I-V expressions described in previous chapters only model the average current. On the other hand, the stochastic nature of carrier movement gives rise to fluctuations in current. This phenomenon is described by noise models. In this chapter, we will focus on noise models in BSIM-CMG.

The physics of noise in FinFET/GAA devices is similar to that in planar MOS-FETs. For this reason, the noise model expressions in BSIM-CMG basically follow the well established formulations for planar MOSFETs, such as those in the BSIM4 model [1]. For FinFET/GAA devices, however, the impact of nonuniform oxide trap distribution on flicker noise in weak inversion is significant [2]. We discuss in this chapter how this is accounted for in BSIM-CMG. In addition, quantitative differences in mobility due to different channel surface orientations in FinFETs ($\langle 110 \rangle$), GAA ($\langle 110 \rangle$ and $\langle 100 \rangle$) versus planar devices ($\langle 100 \rangle$) are simply accounted for by adopting different mobility parameter values. Furthermore, unlike BSIM4, BSIM-CMG is a surface-potential-based model. Some quantities such as the threshold voltage are not available in BSIM-CMG. Therefore, BSIM-CMG offers a comprehensive version of thermal noise model ($TNOIMOD = 1$) that is re-derived for the new BSIM-CMG surface-potential-based I-V formulation. At the same time, the implementation of such new thermal noise model in Verilog-A is done in a way that the correlation coefficient between gate and drain thermal noise follows the correct theoretical value [3]. This is an improvement from the BSIM4 thermal noise model.

Noise components in BSIM-CMG include thermal noise associated with the channel, the source resistance, the drain resistance, and the gate electrode resistance, flicker or $1/f$ noise, and shot noise due to quantum mechanical tunneling current, as we will describe in the following sections.

FinFET/GAA Modeling for IC Simulation and Design. https://doi.org/10.1016/B978-0-323-95729-8.00013-1

9.2 Thermal noise

Thermal noise (*Nyquist noise* or *Johnson noise*) associated with a conductive (or resistive) element is given in terms of power spectral density of noise current:

$$\frac{S}{\Delta f} = 4kTG \tag{9.1}$$

where G is the conductance of the element. For a MOSFET operating in the linear region, the channel conductance can be expressed in the long-channel limit as

$$G = \frac{I_{ds}}{V_{ds}} = \mu_{eff} \frac{W}{L_{eff}} Q_{inv} = \frac{\mu_{eff} q_{inv}}{L_{eff}^2} \tag{9.2}$$

where Q_{inv} denotes the inversion charge density per unit area and q_{inv} is the total inversion charge in the device channel. The previous expression is well established for long-channel MOSFETs.

The total resistance is the sum of channel resistance and external source and drain resistances:

$$\frac{1}{G} = R_{dsi} + \frac{L_{eff}^2}{\mu_{eff} q_{inv}} \tag{9.3}$$

Substituting Eq. (9.3) into Eq. (9.1), we have

$$S_{id} = \frac{4kT \mu_{eff} q_{inv}}{\mu_{eff} q_{inv} R_{dsi} + L_{eff}^2} \tag{9.4}$$

The previous expression (9.4) is identical to the charge-based model in BSIM4 (selected by setting *tnoiMod* = 0 in BSIM4).

For short-channel devices, drain noise is usually larger than what the long-channel theory predicts. The noise increase for short-channel devices is typically accounted for with a multiplication factor γ. In BSIM-CMG γ is represented with the model parameter *NTNOI*. The final thermal noise expression including *NTNOI* becomes

$$S_{id} = NTNOI \cdot \frac{4kT \mu_{eff} q_{inv}}{\mu_{eff} q_{inv} R_{dsi} + L_{eff}^2} \tag{9.5}$$

Although Eq. (9.5) is derived for MOSFET operation in the linear region, the same expression is applicable to the saturation region as well. This is because the thermal noise in the saturation region coincides with that in the linear region near the onset of saturation (Fig. 9.1) [4].

Besides noise from drain to source (S_{id}), thermal noise due to carrier fluctuation in the channel can be coupled to the gate as well through the gate dielectric. This is known as *induced gate noise*. The theory for induced gate noise modeling is well established [5–9]. Induced gate noise is proportional to $\omega^2 L_{eff}^2$, where ω is 2π times the frequency of operation. Therefore, induced gate noise is important for longer channel devices or at very high frequencies (Fig. 9.2). A model for induced gate noise is available and has been implemented into BSIM-CMG [5].

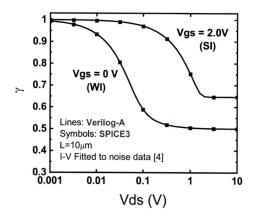

FIG. 9.1

γ versus drain voltage at weak-inversion region ($V_{gs} = 0$) and strong-inversion region ($V_{gs} = 2.0$) for a long-channel device. In weak-inversion drain, noise coincides with $2qI_d$. [5]

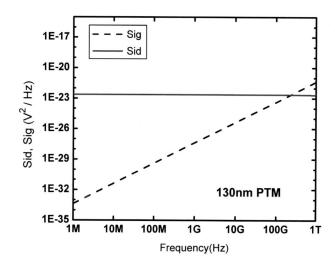

FIG. 9.2

Evaluation of drain current noise (S_{id}) and induced gate noise (S_{ig}) versus frequency for the 130nm BSIM4-based predictive technology model [10, 11] ($L = 130$nm, *tnoiMod* = 1, *fnoiMod* = 0). S_{ig} is much smaller than S_{id} until the frequency of operation is larger than 100 GHz.

As of BSIM-CMG version 111.2.1, both the simple charge-based thermal noise model (Eq. 9.5) and the more comprehensive induced gate noise model [5] are available for the user's selection. The former is selected by setting *tnoiMod* = 0; the latter is selected by setting *tnoiMod* = 1.

Unlike BSIM4, BSIM-CMG is surface-potential based. Its drain current formulation is re-derived. Likewise, it is better to re-derive thermal noise expressions. That was the basic thinking behind the new noise model (*tnoiMod* = 1) The principles for developing analytical expressions are well-established [7,8]. Notice that the results are three components: drain current noise Sid, induced gate noise Sig, and their complex correlation coefficient. It has been shown that the correlation coefficient has little practical impact for real applications [5,9]. In BSIM4, for *tnoiMod* = 1 a special noise component was added in parallel to the source resistor to generate the correct induced gate noise magnitude [1]. Although this way of implementation has resulted in an inaccurate correlation coefficient, no practical impact has ever been reported [5,9]. In BSIM-CMG, for *tnoiMod* = 1, an alternative way of implementing induced gate noise has been adopted [3], always requiring a new node, but generates induced gate noise correlating to drain current noise with tunable correlation coefficient. This way we can directly include the derived correlation coefficient expression in the model.

9.3 Flicker noise

Besides thermal noise, *flicker noise*, or 1/f noise, is another major source of MOS noise, especially at low frequencies. Its origin is found to be related to traps in the gate dielectric. There are two theories that explain the physical mechanism of 1/f noise: *carrier number fluctuation theory* and *mobility fluctuation theory*. In the carrier number fluctuation theory, the flicker noise is attributed to the random trapping and detrapping processes of charges in the oxide traps near the Si-SiO_2 interface. The charge fluctuation results in fluctuations in the channel carrier density, which in turn modulates the drain current [12]. In the mobility fluctuation theory [13], on the other hand, the current level increases/decreases as a result of the fluctuation in bulk mobility.

The *unified model* [14, 15], which accounts for both mechanisms, is adopted in BSIM-CMG. It was the same model used in BSIM4 [1]. In this section, we describe the development of the unified noise model.

The drain current in a MOSFET can be expressed as

$$I_d = W\mu q N \varepsilon_x \tag{9.6}$$

where ε_x is the lateral electric field. By taking the derivative and rearranging terms we have

$$\frac{1}{I_d}\frac{\delta I_d}{\delta N_t} = \underbrace{\frac{1}{N}\frac{\delta N}{\delta N_t}}_{\text{Carrier number fluctuation}} \pm \underbrace{\frac{1}{\mu}\frac{\delta \mu}{\delta N_t}}_{\text{Mobility fluctuation}} \tag{9.7}$$

where we mathematically separate the drain current variation in terms of the two underlying physical mechanisms. In Eq. (9.7), the channel carrier fluctuation, δN, and the trap number fluctuation, δN_t, are closely related via capacitive coupling:

$$R = \frac{\delta N}{\delta N_t} = -\frac{C_{inv}}{C_{ox} + C_{inv} + C_{it}} = -\frac{N}{N + N^*} \tag{9.8}$$

where

$$N^* = \frac{kT}{q^2}(C_{oxe} + CIT) \tag{9.9}$$

The mobility term in Eq. (9.7) can be expressed using the Matthiessen's rule:

$$\frac{1}{\mu} = \frac{1}{\mu_n} + \frac{1}{\mu_{ox}} = \frac{1}{\mu_n} + \alpha N_t \tag{9.10}$$

Differentiating with respect to N_t we have

$$\frac{\delta \mu}{\delta N_t} = -\alpha \mu^2 \tag{9.11}$$

Substituting Eqs. (9.8), (9.11) into Eq. (9.7), we obtain

$$\frac{\delta I_d}{I_d} = \left[\frac{1}{N} \cdot R \pm \alpha \mu\right] \delta N_t \tag{9.12}$$

We may express the local current fluctuation in terms of trap density fluctuation

$$S_{\Delta Id} = \left(\frac{R}{N} \pm \alpha \mu\right)^2 I_d^2 \cdot S_{\Delta Nt} \tag{9.13}$$

where the trap density fluctuation is given by

$$\begin{aligned}
S_{\Delta Nt} &= \int_{E_v}^{E_c} \int_0^W \int_0^{T_{ox}} 4N_t f_t (1 - f_t) \frac{\tau}{1 + \omega^2 \tau^2} dz dy d\varepsilon \\
&= N_t(E_{fn}) \cdot \frac{kTW\Delta x}{\gamma f}
\end{aligned} \tag{9.14}$$

where the Lorentzian noise spectrum associated with each individual trap is integrated. With time constant being an exponential function of position in the gate oxide

$$\tau = \tau_0(E) e^{\gamma z} \tag{9.15}$$

The integration has produced a $1/f$ term in Eq. (9.14). Details of the integration can be found in Ref. [15]. The local current fluctuation needs to be integrated again to obtain the noise spectral density at the drain terminal

$$\begin{aligned}
S_{id} &= \frac{1}{L^2} \int_0^L \frac{kTI_d^2}{\gamma f} N_t(E_{fn}) \left(\frac{R}{N} \pm \alpha \mu\right)^2 dx \\
&= \frac{1}{L^2} \int_0^L \frac{kTq\mu}{\gamma f} N_t(E_{fn}) \left(1 + \frac{\alpha \mu N}{R}\right)^2 \frac{R^2}{N} dV
\end{aligned} \tag{9.16}$$

We express the trap density as a parabolic function of carrier density with three parameters A, B, and C:

$$\begin{aligned}
N_t^*(E_{fn}) &= N_t(E_{fn})\left(1 \pm \frac{\alpha \mu N}{R}\right)^2 \\
&= A + BN + CN^2
\end{aligned} \tag{9.17}$$

We then carry out the integration in Eq. (9.16) and obtain the drain noise spectral density expression for the linear (triode) region:

$$S_{id} = \frac{kTq^2 I_d \mu}{a\gamma f L^2 C_{ox}} \left[A\ln\left(\frac{N_0 + N^*}{N_L + N^*}\right) + B(N_0 - N_L) + \frac{1}{2}C(N_0^2 - N_L^2) \right] \qquad (9.18)$$

where N_0 and N_l are the carrier densities at the source and the drain, respectively:

$$N_0 = \frac{C_{oxe} \cdot q_{is}}{q} \qquad (9.19)$$

$$N_L = \frac{C_{oxe} \cdot q_{id}}{q} \qquad (9.20)$$

The earlier expression for S_{id} is for noise in the triode region. For the pinch-off region, we simply substitute Eq. (9.17) into Eq. (9.16) and evaluate the integrand only at $x = L$ and substitute dx with ΔL_{clm}. The expression becomes

$$S_{id} = \Delta L_{clm} \cdot \frac{kTI_d^2}{\gamma f W L^2} \frac{A + BN_L + CN_L^2}{(N_L + N^*)^2} \qquad (9.21)$$

where ΔL_{clm} is the channel length modulation distance given by

$$\Delta L_{clm} = l \cdot \ln\left[\frac{1}{E_{sat,noi}} \cdot \left(\frac{V_{ds} - V_{dseff}}{l} + EM\right)\right] \qquad (9.22)$$

The final unified flicker noise expression for the strong inversion region is implemented in BSIM-CMG by adding Eqs. (9.18), (9.21):

$$S_{si} = \frac{kTq^2 \mu_{eff} I_{ds}}{C_{oxe} L_{eff,noi}^2 f^{EF} \cdot 10^{10}} \cdot FN1 + \frac{kTI_{ds}^2 \Delta L_{clm}}{W_{eff} \cdot NFIN_{total} \cdot L_{eff,noi}^2 f^{EF} \cdot 10^{10}} \cdot FN2 \qquad (9.23)$$

where $\gamma = 10^{10}$ is found via experiments. Although flicker noise is proportional to $1/f$ in theory, the adjustable parameter EF allows the model slope to slightly deviate from 1.0. $FN1$ and $FN2$ are given by

$$FN1 = NOIA \cdot \ln\left(\frac{N_0 + N^*}{N_1 + N^*}\right) + NOIB \cdot (N_0 - N_1) + \frac{NOIC}{2}(N_0^2 - N_1^2) \qquad (9.24)$$

$$FN2 = \frac{NOIA + NOIB \cdot N_1 + NOIC \cdot N_1^2}{(N_1 + N^*)^2} \qquad (9.25)$$

where $NOIA$, $NOIB$, and $NOIC$ are nothing but A, B, and C. For the weak-inversion region, the same S_{id} expression can be used, except that the charge-voltage relationship is different. The subthreshold noise expression can be found by evaluating Eq. (9.16) with

$$\frac{dN}{dV} = -\beta \qquad (9.26)$$

The weak-inversion S_{id} is

$$S_{wi} = \frac{NOIA \cdot kT \cdot I_{ds}^2 f_{nutd}}{W_{eff} \cdot NFIN_{total} \cdot L_{eff,noi} f^{EF} \cdot 10^{10} \cdot N^{*2}} \tag{9.27}$$

f_{nutd} is the factor associated with nonuniform trap distribution in energy and space [12]. Such nonuniformity is found to be prominent for FinFET/GAA technologies, causing the gate bias dependence of flicker noise even in weak inversion [12].

$$f_{nutd} = \frac{NOIA2/NOIA}{1 + \left(\frac{Q_{inv}}{QSREF}\right)^{MPOWER}}$$

The weak-inversion region (Eq. 9.27) and strong-inversion region (Eq. 9.23) formula are combined with the formula of harmonic means:

$$S_{id,flicker} = \frac{S_{wi}S_{si}}{S_{wi} + S_{si}} \tag{9.28}$$

The earlier equation (9.28) is implemented in BSIM-CMG and other BSIM models for flicker noise calculations.

9.4 Other noise components

Parasitic resistors in a MOS device also contribute to noise. For $RDSMOD = 0$, the thermal noise of source and drain resistance is considered in the channel thermal noise expression (9.5). On the other hand, for $RDSMOD = 1$, the source and drain resistances are modeled as resistor elements. In this case we model thermal noise associated with R_{source} and R_{drain} using Eq. (9.1):

$$\frac{\overline{i_{RS}^2}}{\Delta f} = 4kT \cdot \frac{1}{R_{source}} \tag{9.29}$$

$$\frac{\overline{i_{RD}^2}}{\Delta f} = 4kT \cdot \frac{1}{R_{drain}} \tag{9.30}$$

where $\overline{i_{RS}}$ and $\overline{i_{RD}}$ are noise current sources in parallel to R_{source} and R_{drain}.

Similarly, if the gate electrode resistance is modeled as a resistor element $RGATEMOD = 1$, we have

$$\frac{\overline{i_{RG}^2}}{\Delta f} = 4kT \cdot \frac{1}{R_{geltd}} \tag{9.31}$$

The noise associated with quantum mechanical tunneling is shot noise. Therefore, each gate current component has a corresponding shot noise component given by $2qI$, where I is the tunneling current:

$$\overline{i_{\text{gs}}^2} = 2q(I_{\text{gcs}} + I_{\text{gs}}) \tag{9.32}$$

$$\overline{i_{\text{gd}}^2} = 2q(I_{\text{gcd}} + I_{\text{gd}}) \tag{9.33}$$

If bulkMod = 1, gate tunneling current directly enters the substrate, in which case

$$\overline{i_{\text{gb}}^2} = 2q(I_{\text{gbinv}} + I_{\text{gbacc}}) \tag{9.34}$$

Otherwise (FinFET/GAA without a substrate),

$$\overline{I_{\text{gbs}}^2} = 2qI_{\text{gbs}}$$

$$\overline{I_{\text{gbd}}^2} = 2qI_{\text{gbd}}$$

9.5 Summary

We derived the thermal noise expression for BSIM-CMG, which expresses drain noise as a function of total inversion charge. Flicker noise is described by the unified model, which accounts for both carrier number fluctuation and mobility fluctuation. Shot noise generated from gate tunneling current is modeled as well. Finally, the thermal noise associated with each resistor element is properly accounted for in BSIM-CMG.

References

[1] BSIM4 MOSFET Model User's Manual, Available from: http://www-device.eecs.berkeley.edu/bsim/.

[2] P. Kushwaha, et al., Characterization and modeling of flicker noise in FinFETs at advanced technology node, IEEE Electron Device Lett. 40 (6) (2019) 985–988, https://doi.org/10.1109/LED.2019.2911614.

[3] C. McAndrew, G. Coram, A. Blaum, O. Pilloud, Correlated noise modeling and simulation, Workshop on compact modeling (2005).

[4] C.-H. Chen, M.J. Deen, Y. Cheng, M. Matloubian, Extraction of the induced gate noise, channel noise, and their correlation in submicron MOSFETs from RF noise measurements, IEEE Trans. Electron Devices 48 (12) (2001) 2884–2892, https://doi.org/10.1109/16.974722.

[5] D. Lu, Compact Models for Future Generation CMOS (Ph.D. thesis), UC Berkeley, 2011.

[6] A. Van der Ziel, Gate noise in field effect transistors at moderate high frequencies, Proc. IEEE 51 (3) (1963) 461–467.

[7] J.C.J. Paasschens, A.J. Scholten, R. van Langevelde, Generalizations of the Klaassen-Prins equation for calculating the noise of semiconductor devices, IEEE Trans. Electron Devices 52 (11) (2005) 2463–2472.

[8] A.J. Scholten, L.F. Tiemeijer, R. van Langevelde, R.J. Havens, A.T.A.Z. van Duijnhoven, V.C. Venezia, Noise modeling for RF CMOS circuit simulation, IEEE Trans. Electron Devices 50 (3) (2003) 618–632.

[9] J.-S. Goo, W. Liu, C. Chang-Hoon, K.R. Green, Z. Yu, T.H. Lee, R.W. Dutton, The equivalence of van der Ziel and BSIM4 models in modeling the induced gate noise of MOSFETs, in: International Electron Devices Meeting Technical Digest, 2000, pp. 811–814.

[10] Predictive Technology Model (PTM), Department of Electrical and Computer Engineering, University of Minnesota, Available from: https://mec.umn.edu/ptm [Online].

[11] Y. Cao, Predictive Technology Model for Robust Nanoelectronic Design, Springer, 2011, https://doi.org/10.1007/978-1-4614-0445-3.

[12] S. Christensson, I. Lundstrom, C. Svensson, Low frequency noise in MOS transistors - I. Theory, Solid State Electron. 11 (1968) 797.

[13] L.K.J. Vandamme, Model for 1/f noise in MOS transistors biased in the linear region, Solid State Electron. 23 (1980) 317.

[14] K.K. Hung, P. Ko, C. Hu, Y. Cheng, A unified model for the flicker noise in metal-oxide-semiconductor field-effect transistors, IEEE Trans. Electron Devices 37 (3) (1990) 654–665.

[15] K.K. Hung, P. Ko, C. Hu, Y. Cheng, A physics-based MOSFET noise model for circuit simulations, IEEE Trans. Electron Devices 37 (5) (1990) 1323–1333.

Junction diode I-V and C-V models

As the channel length scaled down, the traditional planar bulk-MOSFET incurred an increased amount of undesired subsurface leakage current from the source to drain. This region below the interface is less controlled by the gate and more by the source-drain electric fields. In planar transistors halo implants below the source and drain regions were used to mitigate this current. With decreasing channel length, we saw the need for an increasing number of halo implants that almost merged together from either side. However, with further scaling, ever higher halo implant dosage could not be used. Consequently, the semiconductor industry moved away from planar transistors, requiring the need for multigate transistors, where the 3D nature of the channel provided better gate control over the channel. Today, FinFETs on Bulk and SOI substrates are into mass production, and GAAFET is expected to follow suit in the next few years.

FinFETs, due to their 3D nature, are manufactured through a set of process steps that slightly differ from the traditional bulk planar MOSFETs. FinFETs on a bulk substrate have a region just below the fin that is less controlled by the gate. Electric field from the source/drain regions would extend into the region below the fin, continuing to create subsurface leakage (although lesser in magnitude compared to planar transistors at the same channel length). In order to weaken these fields some amount of punch-through stop (PTS) implant (a.k.a. ground plane doping) is employed for bulk FinFETs in production [1]. Fig. 10.1 illustrates a cross-section of a p-FinFET showing the presence of this implant just below the actual fin region. This implant present below the intrinsic fin region will laterally diffuse under the source/drain junction region. The use of high dose well implant will lead to an increase in junction tunneling current leakage component. Thus the magnitude of doping in this region must be less than or equal to the well doping. In Fig. 10.2 the cross-section of a p-FinFET under the source/drain region is illustrated leading to the creation of a double junction with two different n-type dopings—$p^+|n_{pts}|n_{well}$ as a general case. When the reverse bias applied to this junction is increased (e.g., through increased positive drain voltage), the depletion region edge will traverse through the n_{pts} region and could enter the n_{well} region too. This leads to a deviation in the behavior of source/drain junction diodes from that of an ideal uniformly doped $p^+|n$ kind of step junction diode observed in planar bulk MOSFETs. In terms of just the junction diode currents, the difference is imperceptible as the reverse-bias currents are rather very small in magnitude. For this reason, the junction diode current

FinFET/GAA Modeling for IC Simulation and Design. https://doi.org/10.1016/B978-0-323-95729-8.00012-X

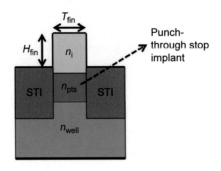

FIG. 10.1

Cross-section of a bulk p-FinFET showing punch-through stop implant below the fin. Gate and oxide regions not shown.

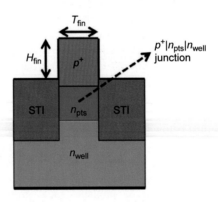

FIG. 10.2

Punch-through stop implant below the fin laterally diffuses below the S/D junction region leading to $p^+|n_{pts}|n_{well}$ type of junction. Gate and oxide regions now shown.

model in BSIM-CMG follows a modeling philosophy similar to that for planar bulk MOSFETs (such as in the BSIM4 model). A comprehensive junction diode current model that captures reverse breakdown as well as the effect of finite series parasitic resistance has been adopted in BSIM-CMG. However, reverse-bias junction diode capacitance has been observed to be markedly different in behavior compared to planar bulk MOSFETs. A new junction capacitance model has been developed to capture this process induced subtlety in the bulk FinFET junction region. Compared to BSIM4, the forward-bias junction diode capacitance model has been improved as well taking into consideration the accuracy of higher-order harmonic currents for RF CMOS IC designs.

10.1 Junction diode current model

From the Boltzmann transport model one can derive an analytic expression for the ideal junction current as follows:

$$I_{jn} = I_{jn0} \left[e^{\frac{qV_{jn}}{kT}} - 1 \right] \tag{10.1}$$

where V_{jn} is the voltage bias across the diode and I_{jn0} is the reverse-bias saturation current of the junction. In BSIM-CMG, similar to compact models for planar bulk MOS-FET, the source/drain-substrate junction diode regions are broken down into three distinct regions for the purpose of modeling: the junction bottom or the area component, the shallow trench isolation (STI) side junction or the perimeter component, and the MOSFET gate-edge component. These are captured through distinct set of parameters for both the source side and the drain side. In what follows we will restrict our discussion to source-side junction diode current. The drain-side junction diode current equations resemble the source-side junction although with a different set of parameters. Different set of parameters for the source and drain sides allow to capture any asymmetry in the source-side and drain-side junctions such as in a multifinger FinFET.

The reverse-bias saturation current is given by

$$I_{jns0} = ASEJ \cdot JSS + PSEJ \cdot JSWS + W_{eff} \cdot NF \cdot NFIN \cdot JSWGS \tag{10.2}$$

where ASEJ is the bottom area and PSEJ is the STI side perimeter of the junction. JSS, JSWS, and JSWGS are the bottom, STI side, and gate-edge reverse-bias saturation current densities. Under reverse-bias conditions, the ideal diode current equation (10.1) predicts a constant reverse saturation current. However, real junction diodes exhibit junction current that tend to increase with the magnitude of reverse bias (due to various leakage currents, discussed later), and after a certain reverse-bias voltage, there is an exponential increase in current due to a phenomenon known as breakdown. For large reverse-bias voltages, the depletion region of the junction experiences high electric fields that tend to accelerate electrons and holes, which, in turn, knock-off more electrons from the lattice of the material, thus contributing to an avalanche of current. Under similar conditions of high field, in some highly doped junctions, electrons could jump from one side of the junction to the other through a quantum mechanical effect known as tunneling. This reverse breakdown current is captured in the BSIM-CMG model as follows:

$$I_{jns} = I_{jns0} \left[e^{\frac{qV_{es}}{NJS \cdot kT}} - 1 \right] \cdot F_{breakdown} \tag{10.3}$$

where the breakdown factor $F_{breakdown}$ is empirically modeled as

$$F_{breakdown} = 1 + XJBVS \cdot e^{-\frac{q(BVS+V_{es})}{NJS \cdot kT}} \tag{10.4}$$

In the previous equation, V_{es} is the substrate to source terminal bias voltage. XJBVS, the breakdown coefficient, has a default value of 1. It can be conveniently set to 0 if one does not want to capture breakdown through the model. The parameter BVS represents the source to substrate breakdown voltage. The nonideality factor NJS is used to capture any inadequacies in the junction diode current model w.r.t. a real diode, such as additional leakage currents due to traps in the junction region (discussed later).

Under forward-bias conditions, the junction diode current equation (10.3) is purely exponential function, which could potentially cause convergence issues in a SPICE simulator environment. Realistic diodes are far from ideal and the quasineutral regions of the diode present a finite series resistance restricting the diode current behavior to a linear dependence on junction bias voltage. Similar is the situation under reverse breakdown conditions as well. In BSIM-CMG the effect of the series resistances are captured by making the junction diode currents smoothly transition to a linear behavior w.r.t. voltage bias after the current exceeds a certain absolute value. For the forward-bias region, this value is set by a parameter IJTHSFWD and for the reverse-bias region, it is determined through the parameter IJTHSREV in the model. Hence, the junction diode current formulation can be divided into three regions— less than IJTHSREV, in between IJTHSREV and IJTHSFWD, and greater than IJTHSFWD (Fig. 10.3). For the region in between IJTHSREV and IJTHSFWD, the diode current is described by Eq. (10.3). Let us now consider the case where

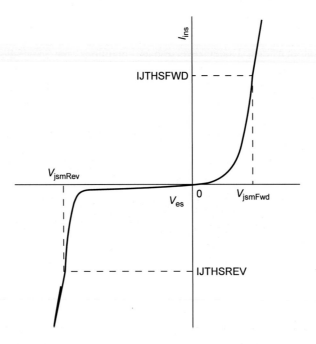

FIG. 10.3

Representative figure of junction diode current showing the three regions as described in the BSIM-CMG model.

the diode current is greater than IJTHSFWD. The bias-voltage, V_{jsmFWD}, at which this occurs can be found by substituting IJTHSFWD and V_{jsmFWD} in Eq. (10.3). Regrouping similar terms one obtains a quadratic equation as follows:

$$X^2 - BX - C = 0 \tag{10.5}$$

where

$$X = e^{\frac{q \cdot V_{\text{jsmFWD}}}{NJS \cdot kT}}$$

$$B = 1 + \frac{IJTHSFWD}{I_{\text{jns0}}} - XJBVS \cdot e^{-\frac{q \cdot BVS}{NJS \cdot kT}}$$

$$C = XJBVS \cdot e^{-\frac{q \cdot BVS}{NJS \cdot kT}}$$

One can then solve for the value of V_{jsmFWD} as

$$V_{\text{jsmFWD}} = \frac{NJS \cdot kT}{q} \log\left(\frac{B + \sqrt{B^2 + 4C}}{2}\right) \tag{10.6}$$

A first-order Taylor expansion of Eq. (10.3) at $V_{\text{es}} = V_{\text{jsmFWD}}$ is then used to obtain the junction diode current expression for junction diode current greater than IJTHSFWD, which is given as follows:

$$I_{\text{jns}} = IJTHSFWD + k_{\text{slopeFwd}} \cdot (V_{\text{es}} - V_{\text{jsmFwd}}) \tag{10.7}$$

where the slope factor k_{slopeFwd} can be obtained from the first-order derivative of Eq. (10.3) w.r.t. V_{es} at $V_{\text{es}} = V_{\text{jsmFWD}}$.

$$k_{\text{slopeFwd}} = \frac{qI_{\text{jns0}}}{NJS \cdot kT}\left[e^{\frac{qV_{\text{jsmFwd}}}{NJS \cdot kT}} + XJBVS \cdot e^{-\frac{BVS + V_{\text{es}}}{NJS \cdot kT}}\right] \tag{10.8}$$

For reverse bias, the diode current equation in Eq. (10.3) captures reverse breakdown. However, for large reverse bias (negative V_{es}), the term $F_{\text{breakdown}}$ dominates and increases exponentially. The effect of series resistance is captured by making the diode junction current linear in a fashion similar to the way we dealt with forward-bias behavior. Under reverse breakdown, the diode junction current can be approximated as follows:

$$I_{\text{jns,rev}} = I_{\text{jns0}}F_{\text{breakdown}} \tag{10.9}$$

The bias voltage at which we transition to a linear behavior V_{jsmRev} can be found by substituting $I_{\text{jns,rev}} = IJTHSREV$ and $V_{\text{es}} = V_{\text{jsmRev}}$ in the previous equation. We then arrive at

$$V_{\text{jsmRev}} = -BVS - \frac{NJS \cdot kT}{q} \log\left(\frac{IJTHSREV - I_{\text{jns0}}}{XJBVS \cdot I_{\text{jns0}}}\right) \tag{10.10}$$

With a first-order Taylor expansion of Eq. (10.9) at $V_{\text{es}} = V_{\text{jsmRev}}$, the reverse-bias junction diode current for values less than IJTHSREV is given by

$$I_{\text{jns}} = \left[e^{\frac{qV_{\text{es}}}{NJS \cdot kT}} - 1\right][IJTHSREV + k_{\text{slopeRev}}(V_{\text{es}} - V_{\text{jsmRev}})] \tag{10.11}$$

Finally, the overall source-side diode junction current in the BSIM-CMG model is given as

$$
I_{jns} = \begin{cases}
\left[e^{\frac{qV_{es}}{NJS \cdot kT}} - 1\right]\left[IJTHSREV + k_{slopeRev}(V_{es} - V_{jsmRev})\right] & V_{es} < V_{jsmRev} \\[2ex]
I_{jns0}\left[e^{\frac{qV_{es}}{NJS \cdot kT}} - 1\right] \cdot F_{breakdown} & V_{jsmRev} < V_{es} < V_{jsmFwd} \\[2ex]
IJTHSFWD + k_{slopeFwd} \cdot (V_{es} - V_{jsmFwd}) & V_{es} > V_{jsmFwd}
\end{cases}
$$

(10.12)

10.1.1 Reverse-bias additional leakage model

Junction diode in MOSFETs is known to exhibit three different leakage current phenomena.

First, the electron and hole generation-recombination trap centers are known to lead to Shockley–Read–Hall (SRH) current. These trap centers are present due to impurities and physical lattice defects in the silicon that are formed during the junction formation. Semiconductor fabrication steps like annealing strive to keep the density of these trap centers below a certain limit. Of these traps, only the traps that fall into the depletion region of the junctions contribute to the leakage current. In the forward-bias mode of operation these traps facilitate in net electron-hole recombination, while under reverse bias, there is a net electron-hole generation. The built-in electric field in the depletion regions helps separate/recombine the electron-hole pair leading to the SRH leakage current.

Heavily doped ($10^{19} cm^{-3}$) junctions also exhibit leakage current due to band-to-band tunneling (BTBT). BTBT-based current is usually observed when the built-in electric field exceeds 10 MV/cm. This quantum mechanical process occurs without the aid of traps across a physically thin energy barrier region. This component of current is proportional to $E_{dep,max}^2 e^{\frac{1}{E_{dep,max}}}$ where $E_{dep,max}$ denotes the magnitude of the maximum built-in electric field in the depletion region. This electric field is higher under reverse-bias than forward-bias operation, and hence the BTBT current as well. However, such high amounts of doping are usually avoided in carefully engineered modern CMOS devices to keep this leakage component well under control. For this purpose, we will not aim to capture this component in our model.

Finally, the trap centers discussed earlier could also aid in electron and hole tunneling across the depletion region in the junction. This trap-assisted tunneling (TAT) leakage current component is much significant under reverse-bias operation (also high built-in electric field) of the diode. One can visualize TAT as electric field-enhanced SRH current. Under this condition, electrons from the valence band of the p-type region hop into a trap in the depletion region and hop out into the conduction band of the n-type region of the diode. Tunneling probability is exponentially dependent on the physical distance of the hop. In this case the two-step trap-assisted hop has a higher probability compared to a single hop from p-type to n-type region (as in

BTBT). In advanced CMOS technologies this component of leakage current under reverse-bias operation is found to be dominant, leading to an increasing leakage current with bias, as opposed to a saturating behavior of the ideal diode current under reverse bias.

All the three discussed leakage phenomena are present in both forward-bias and reverse-bias mode junction currents. Under forward-bias, their impact is seen only for low bias well before the diode turn-on, where model accuracy is not expected. For this reason and to save on model computation time, instead of separate equations a correction to the junction diode current discussed in the previous section is proposed in the form of a nonideality factor. This nonideality factor is captured through parameters NJS used in Eq. (10.3). However, under reverse-bias operation where the leakage currents are significant (in terms of magnitude as well as bias-dependent behavior) compared to the diode reverse saturation current, a semiempirical model is used to capture both the SRH and the TAT leakage currents.

The field-enhanced SRH current or the TAT current is given by [2, 3]

$$I_{jn,TAT} = \int_{\text{depletion region}} K_{SRH} \cdot (1 + \Gamma_{TAT}) \cdot \frac{\left[e^{\frac{V_{jn}}{kT}} - 1 \right]}{\left[e^{\frac{\psi+0.5V_{jn}}{kT}} + \frac{-\psi+0.5V_{jn}}{kT} + 2 \right]} \cdot d\psi \qquad (10.13)$$

where K_{SRH} accounts for the SRH electron/hole trap capture cross-section and trap density in the depletion region, Γ_{TAT} is a field enhancement factor in order to account for the TAT leakage current, and V_{jn} is the bias voltage across the junction. Unfortunately, the integral in Eq. (10.13) does not have any analytic solution and we have to resort to some approximations to obtain one. Although, it has been shown in Ref. [3] that Γ_{TAT} is a field-dependent function, we will approximate as be a constant here. The term in the denominator inside the integral can be replaced with its maximum value that occurs at the metallurgical junction. One can then obtain a closed-form expression for the integral and the SRH + TAT junction current is given by

$$I_{jn,TAT} = K_{SRH} \cdot (1 + \Gamma_{TAT}) \cdot W_{dep} \cdot \left[e^{\frac{V_{jn}}{2kT}} - 1 \right] \qquad (10.14)$$

where W_{dep} is the depletion width. As W_{dep} is a weak function of the bias across the junction compared to the rest of the terms, not much accuracy is lost if one were to assume it to be a constant as well. In the BSIM-CMG model the loss of accuracy due to the assumptions made so far in our derivation is reclaimed by introducing tuning parameters. The reverse junction leakage current for the source-side junction in the BSIM-CMG model is given by

$$\begin{aligned} I_{jn,es,TAT} &= -ASEJ \cdot JTSS \cdot \left[e^{\frac{qV_{es}}{NJTS \cdot kTVTSS - V_{es}}} - 1 \right] \\ &\quad - PSEJ \cdot JTSSWS \cdot \left[e^{\frac{qV_{es}}{NJTSSW \cdot kTVTSSWS - V_{es}}} - 1 \right] \\ &\quad - NF \cdot NFIN \cdot W_{eff0} \cdot JTSSWGS \cdot \left[e^{\frac{qV_{es}}{NJTSSWG \cdot kTVTSSWGS - V_{es}}} - 1 \right] \end{aligned} \qquad (10.15)$$

where JTSS, JTSSWS, and JTSSWGS are the preexponential tuning factors for the bottom area, STI side perimeter, and gate side perimeter junctions, respectively.

NJTS, NJTSW, and NJTSWG are the corresponding nonideality factors whose ideal value as observed in Eq. (10.14) is equal to 2. V_{es} is the junction substrate to source terminal bias voltage. An additional bias-dependent factor of the form $V TX/(V TX - V_{es})$ is multiplied to the term within the exponential to ensure that the effect of the exponential terms rapidly move to zero under forward-bias operation. This is so because the influence of the SRH leakage component has already been taken into account through the nonideality factor NJS in Eq. (10.3). Temperature dependence of the SRH + TAT leakage component has been introduced through the tuning parameters used in Eq. (10.15) and will be discussed later. The drain-side reverse leakage current can be described by an analogous set of equations in a manner similar to earlier.

10.2 Junction diode charge/capacitance model

As described in the introduction of this chapter, the junction capacitance behavior of FinFETs has been observed to be different from that of planar bulk MOSFETs due to the formation of a double junction in the presence of punch-through prevent implant below the fin region. For example, Fig. 10.4 shows that reverse-bias junction capacitance $1/C_{jn}^2$ versus V_{jn} curve for FinFET S/D junctions with PTS deviates from the linear behavior observed without PTS (ideal step junction). The slope of this curve is inversely proportional to the n-type doping at the edge of the depletion region. FinFET S/D junctions tend to show two different slopes and a higher junction

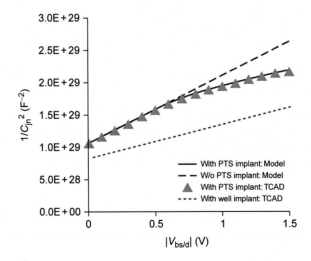

FIG. 10.4

The $p^+|n_{pts}|n_{well}$ junction exhibits two slopes in a $1/C_{jn}^2$ versus V_{jn} plot in comparison to single slope of the ideal step junction. Values of doping used were $p^+ = 3 \times 10^{20} \text{cm}^{-3}$, $n_{pts} = 10^{18} \text{cm}^{-3}$, and $n_{well} = 3 \times 10^{18} \text{cm}^{-3}$ (*Lines*: Model, *Symbols*: TCAD).

capacitance when a PTS implant is used. In what follows we will try to capture this behavior in a new junction diode charge/capacitance model.

10.2.1 Reverse-bias model

In a manner similar to the junction diode current we will restrict our discussion to source-side junction only. The assumptions and derivations for the drain-side junction are similar. First, we will derive a junction charge model for the well-known simple case of a single junction $p|n$ diode capacitance model for reverse bias [2]. From Poisson equation describing the electrostatics of the charge carriers in the p- and n-type regions for a diode, the junction depletion depth is given by

$$W_{\text{dep,jn}} = W_{\text{dep0}} \cdot \left(1 - \frac{V_{\text{es}}}{PBS}\right)^{MJS} \tag{10.16}$$

where V_{es} is the voltage across the source-side junction, W_{dep0} is the zero-bias junction depletion width, parameter PBS represents the source-side junction built-in voltage, and MJS is the junction doping grading coefficient (which is equal to 0.5 for an ideal step-junction). The junction capacitance is then given by

$$C_{\text{jes}} = \frac{\epsilon_0 \epsilon_r}{W_{\text{dep,jn}}} = \frac{C_{\text{jes0}}}{\left(1 - \frac{V_{\text{es}}}{PBS}\right)^{MJS}} \tag{10.17}$$

where ϵ_r is the relative dielectric constant of the junction material and C_{jes0} is the zero-bias junction capacitance. The junction depletion charge density is obtained by integrating the previous equation over voltage as follows:

$$Q_{\text{es}} = \frac{C_{\text{jes0}} PBS}{1 - MJS} \left[1 - \left(1 - \frac{V_{\text{es}}}{PBS}\right)^{1-MJS}\right] \tag{10.18}$$

In order to capture the effect of double junction especially the observed change in slope in the capacitance as shown before, the earlier charge model can be enhanced as follows:

$$Q_{\text{es,rev}} = \begin{cases} C_{\text{j01}} PBS \dfrac{1 - \left(1 - \frac{V_{\text{es}}}{PBS}\right)^{1-MJS}}{1 - MJS} & 0 < V_{\text{es}} < V_{\text{ec}} \\[4mm] C_{\text{j01}} PBS \dfrac{1 - \left(1 - \frac{V_{\text{ec}}}{PBS}\right)^{1-MJS}}{1 - MJS} + C_{\text{j02}} PBS2 \dfrac{1 - \left(1 - \frac{V_{\text{es}} - V_{\text{ec}}}{\phi_{b2}}\right)^{1-MJS2}}{1 - MJS2} & V_{\text{ec}} < V_{\text{es}} \end{cases}$$

$$\tag{10.19}$$

where $C_{\text{j01,2}}$ are the zero-bias capacitance values and PBS and PBS2 are the barrier height of $p'|n_{\text{well}}$ and $p^+|n_{\text{pts}}$ junctions, respectively. MJS and MJS2 represent the gradient of the $p^+|n_{\text{pts}}$ and the $n_{\text{pts}}|n_{\text{well}}$ junctions, respectively. We can observe that the first term in Eq. (10.19) is the same as that for a single junction diode. Eq. (10.19) maintains charge continuity at the cross-over voltage $V_{\text{es}} = V_{\text{ec}}$. The continuity of the first and second derivatives of charge also needs to be ascertained for accuracy in the prediction of higher harmonic power content in the output of a transistor in Analog/

RF circuit simulations. The continuity of the first derivative of charge in Eq. (10.19) (which is also the junction capacitance) at $V_{es} = V_{ec}$ yields

$$C_{j01}\left(1 - \frac{V_{ec}}{PBS}\right)^{-MJS} = C_{j02} \tag{10.20}$$

Ensuring continuity of the second derivative of the charge (first derivative of capacitance) at $V_{ss} = V_{ec}$ gives rise to the following condition:

$$C_{j01}MJS\frac{\left(1 - \frac{V_{ec}}{PBS}\right)^{-1-MJS}}{PBS} = \frac{C_{j02}MJS2}{PBS2} \tag{10.21}$$

These conditions, Eqs. (10.20), (10.21), are factored into the parameter extraction process. In the junction capacitance curve, $1/C_{jn}^2 - V_{es}$ (Fig. 10.4), the first slope region corresponding to depletion edge traversing the PTS implant region is used to extract the values for parameters C_{j01}, PBS, and MJS in a similar way to that for a single junction diode. Among the remaining four parameters (V_{ec}, C_{j02}, PBS2, and MJS2) that correspond to the n_{well} region (the second slope region), Eqs. (10.20), (10.21) provide the flexibility to choose only two of them. We chose parameters C_{j02} and PBS2 that signify the depth of the PTS implant—n_{well} region boundary and n_{well} region doping concentration. The parameters V_{ec} and PBS2 will now be determined by simultaneously solving Eqs. (10.20), (10.21) using the values chosen for C_{j02} and MJS2.

$$V_{ec} = PBS \cdot \left[1 - \left(\frac{C_{j01}}{C_{j02}}\right)^{1/MJS}\right] \tag{10.22}$$

$$PBS2 = \frac{PBS \cdot MJS2 \cdot C_{j02}}{C_{j01} \cdot MJS \cdot \left(1 - \frac{V_{ec}}{PBS}\right)^{-1-MJS}} \tag{10.23}$$

The junction capacitance is then given by the derivative dQ_{es}/dV_{es} as follows:

$$C_{jes,rev} = \begin{cases} C_{j01}\left(1 - \frac{V_{es}}{PBS}\right)^{-MJS} & 0 < V_{es} < V_{ec} \\ C_{j02}\left(1 - \frac{V_{es} - V_{ec}}{PBS2}\right)^{-MJS2} & V_{ec} < V_{es} \end{cases} \tag{10.24}$$

To validate the model, the structure in Fig. 10.2 was simulated using TCAD [4]. A source/drain region p^+ doping was chosen as 3×10^{20} cm^{-3}. The doping for the n_{PTS} and n_{well} regions were chosen to be 10^{18} cm^{-3} and 3×10^{18} cm^{-3}, respectively. The two terminal capacitance across the source/drain and substrate were extracted. From Fig. 10.4 we can observe that the cross-over voltage V_{bc} is around 0.6 V of reverse-bias junction voltage where the slope of the $1/C_{jn}^2$ plot changes reflecting the doping at the edge of the depletion region. The derived model in Eq. (10.24) shows excellent agreement (after parameter tuning) with TCAD simulations for such a junction (Fig. 10.4).

10.2.2 Forward-bias model

The charge/capacitance model in Eqs. (10.19), (10.24) is not valid for voltages V_{es} approaching $-PBS$, that is, when the diode is forward biased. In a realistic diode series resistances of the quasineutral regions dominate and restrict the actual voltage drop across the junction (or rather across the depletion edges on either sides of the metallurgical junction). Eq. (10.24) would numerically result in very large values, making it unwieldy for implementation in a compact modeling framework. For this reason, industry-standard compact models resort to an approximation for the forward-bias region. For example, BSIM4 resorts to using a quadratic equation to describe the junction charge for the forward bias (i.e., junction capacitance is made linear). For a p-FET, the forward-bias junction charge is essentially given by a Taylor series expansion of Eq. (10.24) up to second order around $V_{es} = 0$ V,

$$Q_{es,fwd} = C_{j01} \cdot V_{es} + \frac{C_{j01}MJS}{2 \cdot PBS} V_{es}^2 \quad \text{for } V_{es} < 0 \tag{10.25}$$

Eq. (10.25) taken together with Eq. (10.19) ensures the continuity of junction charge and its first derivative capacitance around $V_{es} = 0$ V. However, there is a discontinuity in the third derivative of the overall junction charge, that is, $d^3 Q_{es}/dV_{es}^3$ at $V_{bs/d} = 0$ V. This discontinuity is not a cause of concern for convergence of a compact model. SPICE simulators require continuity of only up to second derivative of charge for convergence. However, for the RF design community that is interested in accurate prediction of higher-order intermodulation (IM) distortion products to accurately predict out-of-band emission by wireless transmitters, maintaining continuity of charge up to sixth order is desired. For example, in the operation of a passive mixer, the FET operates in a strictly linear region with source-drain voltage, $V_{ds} = 0$ V. In this mixer, if the FET body is tied to ground, the voltage across the source/drain junction is also centered around 0 V during the mixing operation. The discontinuity when Eqs. (10.19), (10.25) are put together will lead to incorrect predictions of higher-order IM products. This can be verified using a simple harmonic-balance simulation setup wherein a single-tone RF stimulation is supplied to the source of an FET with its gate voltage set above the threshold voltage. The power in the harmonic content of the drain current is observed. The slope of the nth-harmonic component power versus input power would be n at low power input. Any discontinuity or issues with model symmetry w.r.t. source and drain would lead to wrong slopes. In Fig. 10.5 we present results from such a test for a model with a symmetric core but using the earlier junction charge/capacitance model. As expected, we see a deviation in the slope for the fourth- and fifth-harmonic output power. In order to rectify the same we propose an alternate model as follows. Instead of a Taylor series expansion around $V_{es} = 0$ V, pushing the transition point further into forward bias helps. We choose a quadratic Taylor series expansion around $V_{es} = k \cdot PBS$ for this as follows:

$$
\begin{aligned}
Q_{es,fwd} &= C_{j01}PBS\frac{(1-k)^{1-MJS}}{1-MJS} + C_{j01}(1-k)^{-MJS} \cdot (V_{es} + k \cdot PBS) \cdots \\
&+ C_{j01}MJS\frac{(1-k)^{-1-MJS}}{2 \cdot PBS} \cdot (V_{es} + k \cdot PBS)^2 \text{ for } V_{es} < -k \cdot PBS
\end{aligned}
\tag{10.26}
$$

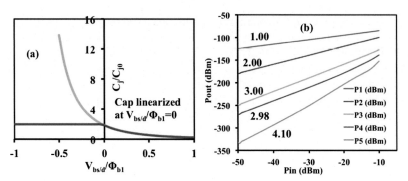

FIG. 10.5

(A) The first derivative of junction capacitance using the linearization in Eq. (10.25).
(B) Results of harmonic balance test for a PMOS passive mixer configuration showing wrong slopes for the fourth- and fifth-harmonic output power.

For implementation purposes in BSIM-CMG, we have chosen $k = 0.9$. This change restores accuracy to higher-order IM products. This modified model was implemented for junction capacitance in BSIM-CMG and results from the harmonic balance simulations for a passive mixer are shown in Fig. 10.6. As expected, we see that this model predicts accurate slopes up to fifth-harmonic content in the MOSFET. The persisting discontinuity at very high forward bias should not be a problem. Given that the diode current behaves exponentially with bias voltage V_{es}, all the voltage drop would be across the parasitic source-drain resistance or the substrate network resistance, restricting the diode junction from experiencing such high voltages.

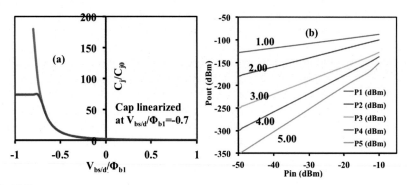

FIG. 10.6

(A) The first derivative of junction capacitance using the linearization in Eq. (10.26) with $k = 0.7$. (B) Results of harmonic balance test for a PMOS passive mixer configuration showing correct slopes for the fourth- and fifth-harmonic output power.

The overall junction charge model is as follows:

$$
Q_{es} = \begin{cases}
C_{j01}PBS\dfrac{(1-k)^{1-MJS}}{1-MJS} + C_{j01}(1-k)^{-MJS}\cdot(V_{es}+k\cdot PBS)\cdots \\
\quad + C_{j01}MJS\dfrac{(1-k)^{-1-MJS}}{2\cdot PBS}\cdot(V_{es}+k\cdot PBS)^2 \hfill V_{es} < -k\cdot PBS \\[6pt]
C_{j01}PBS\dfrac{1-\left(1-\frac{V_{es}}{PBS}\right)^{1-MJS}}{1-MJS} \hfill -k\cdot PBS < V_{es} < V_{ec} \\[6pt]
C_{j01}PBS\dfrac{1-\left(1-\frac{V_{ec}}{PBS}\right)^{1-MJS}}{1-MJS} + C_{j02}PBS2\dfrac{1-\left(1-\frac{V_{es}-V_{ec}}{PBS2}\right)^{1-MJS2}}{1-MJS2} \hfill V_{ec} < V_{es}
\end{cases}
$$

$$(10.27)$$

The junction capacitance given by dQ_{es}/dV_{es} is expressed as

$$
C_{jes} = \begin{cases}
\dfrac{C_{j01}}{(1-k)^{MJS}} + \dfrac{C_{j01}MJS}{PBS(1-k)^{1+MJS}}\cdot(V_{es}+k\cdot PBS) & V_{es} < -k\cdot PBS \\[6pt]
C_{j01}\left(1-\dfrac{V_{es}}{PBS}\right)^{-MJS} & -k\cdot PBS < V_{es} < V_{ec} \\[6pt]
C_{j02}\left(1-\dfrac{V_{es}-V_{ec}}{PBS2}\right)^{-MJS2} & V_{ec} < V_{es}
\end{cases}
\quad (10.28)
$$

Similar to the junction diode current model, junction diode capacitance is split into three components as well: bottom area, STI perimeter, and gate-edge. The derivation so far was a generic description for each of the three components. Separate set of parameters are used to describe the charge for each of components. C_{j01}, in previous equations, is replaced with $ASEJ \cdot CJS$, $PSEJ \cdot CJSWS$, and $NF \cdot NFIN \cdot W_{eff0} \cdot CJSWGS$ for the bottom area, STI perimeter, and gate-edge components, respectively.

Similarly, PBS, PBSWS, and PBSWGS represent the built-in potentials and MJS, MJSWS, and MJSWGS represent the junction doping gradient for the corresponding components. In the BSIM-CMG model C_{j02} is described relative to C_{j01} using a new parameter as $C_{j02} = SJS \cdot C_{j01}$. SJS, SJSWS, and SJSWGS would be the parameters that correspond to the three diode components for this purpose. MJS2, MJSWS2, and MJSWGS2 are the parameters describing the doping gradient of the second junction for all the components. If, in a technology, the effect of the second junction is not that strong, one can conveniently set SJS/SJSWS/SJSWGS to zero. This would convert the model to the default single-step junction diode capacitance.

References

[1] K. Okano, T. Izumida, H. Kawasaki, A. Kaneko, A. Yagishita, T. Kanemura, M. Kondo, S. Ito, N. Aoki, K. Miyano, T. Ono, K. Yahashi, K. Iwade, T. Kubota, T. Matsushitaand,

I. Mizushima, S. Inaba, K. Ishimaru, K. Suguro, K. Eguchi, Y. Tsunashima, H. Ishiuchi, Process integration technology and device characteristics of CMOS FinFET on bulk silicon substrate with sub-10 nm fin width and 20 nm gate length, in: International Electron Devices Meeting Technical Digest (IEDM), 2005, pp. 721–724.

[2] S.M. Sze, K.K. Ng, Physics of Semiconductor Devices, Wiley, New York, 2006.

[3] G.A.M. Hurkx, D.B.M. Klaassen, M.P.G. Knuvers, A new recombination model for device simulation including tunneling, IEEE Trans. Electron Devices 39 (2) (1992) 331–338.

[4] Sentaurus Device, Synopsys, Inc., 2013.

Benchmark tests for compact models

11.1 Introduction

As discussed in Chapter 1, compact models are critically important for circuit simulations. They must meet the requirement of not only the accuracy but also the convergence of the simulations. Circuit simulators, while trying to find the solutions of KCLs and KVLs of a network, continuously use the compact models of the components present in the network. A well-behaved compact model increases the chance of reaching the solution. It is thus very important to test the quality of a compact model before deploying it in a circuit simulator. One simple check on the quality of the compact models is via the principle of asymptotic correctness. This is to check if the model behaves as expected under extreme input conditions. Another more detailed way to evaluate the physical behavior of a compact model is via various benchmark tests developed over the years [1–7]. These tests are especially important for a compact model, which is to be used in RF and analog circuit design. Various tests have been designed to check different aspects of a compact model. We discuss both asymptotic correctness principle and different benchmark tests in this chapter. The BSIM-CMG model is taken as an example and all the benchmark tests are performed on it. It is shown that BSIM-CMG passes the benchmark tests.

11.1.1 Asymptotic correctness

This is a simple principle to check for potential problems in a compact model. As device geometry, temperature, or bias reaches extreme values, the model equations should behave in a physical manner and should not have any numerical problems. The following simple examples illustrate this principle.

The parameter I_s in SPICE Gummel-Poon model [2,8] varies with emitter length L. This dependence is very close to linear but not exactly linear. A very simple way to model and fit the data would be to use

$$I_s = A_0 + A_1 L + A_2 L^2 \tag{11.1}$$

where A_0, A_1, and A_2 are fitting parameters. Typically, measurement data is available for a limited range of Ls and the above expression would fit the data very well for the available limited values of L. However, for extreme values of L it will clearly behave in an unphysical way. This model is a clear violation of the principle of asymptotic

FinFET/GAA Modeling for IC Simulation and Design. https://doi.org/10.1016/B978-0-323-95729-8.00002-7

correctness. In fact, empirical polynomial functions although very commonly used in compact models for quick data fitting, often violate the asymptotic correctness principle.

Another example is the modeling of device geometry dependence of thermal resistance (R_{TH}) in silicon-on-insulator MOSFETs. It is found that RTH decreases with an increase in the area and perimeter of the device [9]. A simple way to model this would be to write

$$R_{TH} = \frac{R_{THA}}{WL} + \frac{R_{THP}}{W + L} \tag{11.2}$$

where R_{THA} and R_{THP} are fitting parameters. Eq. (11.2) will clearly behave in an unphysical way as L or W decreases to an extremely small value as it would predict a physically incorrect very large value of thermal resistance. A better approach is to think in terms of thermal conductance and use

$$R_{TH} = \frac{1}{G_{TH}} = \frac{1}{G_{THA}WL + G_{THP}(W + L)} \tag{11.3}$$

where G_{THA} and G_{THP} are fitting parameters. Eq. (11.3) is well-behaved asymptotically and thus is a better model compared to Eq. (11.2).

11.2 Benchmark tests

11.2.1 Tests for checking physical behavior in weak- and strong-inversion regions

The importance of the weak-inversion region of operation is increasing with the decreasing supply voltages in advanced technology nodes. It is especially important for analog and RF circuits. There are various benchmark tests developed for testing the physical behavior of a compact model in a weak-inversion region. These tests check whether the model behaves physically in weak- and strong-inversion regions. We describe in this section tests such as slope-ratio test, output conductance test, and volume inversion test.

11.2.1.1 Slope-ratio test

This test checks whether a compact model takes care of the different drain-to-source voltage V_{ds} dependence of drain-to-source current I_{ds} in weak- and strong-inversion regions. In weak-inversion region, V_{ds} dependence of I_{ds} can be expressed as

$$I_{ds}\alpha\left(1 - e^{-\frac{V_{ds}}{V_{th}}}\right) \tag{11.4}$$

where V_{th} is the thermal voltage. In strong-inversion (linear condition) region, I_{ds} becomes a linear function of V_{ds}. To see whether the compact model captures this important difference, the slope ratio (SR) is calculated from weak- to strong-inversion regions. S_R is defined as

$$S_R = \frac{(I_{ds2} + I_{ds1})(V_{ds2} - V_{ds1})}{(I_{ds2} - I_{ds1})(V_{ds2} + V_{ds1})} \tag{11.5}$$

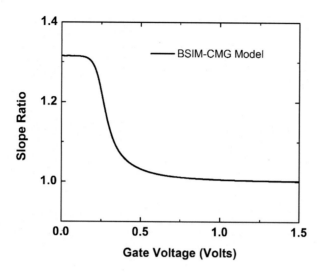

FIG. 11.1

Slope ratio S_R calculated using the BSIM-CMG model at room temperature. $V_{ds1} = 10$ mV and $V_{ds2} = 20$ mV used for the calculation. $T_{fin} = 15$ nm, $L = 1$ μm, and $N_{fin} = 10$.

where $I_{ds1,2}$ are drain-to-source currents at drain-source voltages $V_{ds1,2}$. At room temperature, S_R should smoothly and monotonically decrease from a value close to 1.3–1.0 moving from weak- to strong-inversion regions.

In Fig. 11.1, we show that the BSIM-CMG model passes this test and S_R calculated from the model indeed varies smoothly between the values discussed above. Note that the value of S_R in weak-inversion region is temperature-dependent due to the presence of the thermal voltage V_{th} in Eq. (11.4).

11.2.1.2 Weak-inversion conductance test

This test also checks whether the model captures the different behavior of Ids with V_{ds} in weak- and strong-inversion regions. From Eq. (11.4) it can be seen that in weak-inversion regions,

$$\frac{\partial}{\partial V_{ds}}(g_{ds}\exp(V_{ds}/V_{th})) = 0 \qquad (11.6)$$

where $g_{ds} = \partial I_{ds}/\partial V_{ds}$ is the output conductance. Hence, if $g_{ds} \cdot \exp(V_{ds}/V_{th})$ is plotted versus V_{ds} in a weak-inversion region, it should be absolutely flat. As we move from weak- to strong-inversion region, there is a finite slope seen in the plot of $g_{ds} \cdot \exp(V_{ds}/V_{th})$ versus V_{ds}. In Fig. 11.2, we show that the BSIM-CMG model does behave in this way and passes this test.

11.2.1.3 Volume inversion test

In the subthreshold region of thin-body lightly doped multigate devices, the gate bias moves the energy bands not just at the surface but across the full body thickness of the device [10,11]. This causes inversion to occur in the full body thickness of the

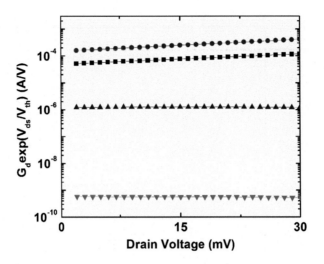

FIG. 11.2

Output conductance in weak- and strong-inversion regions calculated using the BSIM-CMG model. $V_g = 0$–0.6 in steps of 0.2 V. $T_{fin} = 15$ nm, $L = 1\,\mu$m, and $N_{fin} = 10$.

device as opposed to only the surface in the case of bulk MOSFETs. This phenomenon is called volume inversion. It occurs only in the subthreshold region as in strong inversion, the carriers at the surface screen the electric field from reaching deep inside the silicon body. As a result of volume inversion, the subthreshold current in these devices increases with an increase in body thickness. This phenomenon of volume inversion should be captured by the compact model. A compact model can be tested for volume inversion modeling by obtaining the subthreshold current of a long-channel device for various fin thicknesses T_{fin}. All the real device effects should be turned off in the model. The subthreshold current should increase with increasing fin thickness. In Fig. 11.3, we show the simulations from the BSIM-CMG model for devices with fin thicknesses $T_{fin} = 10$, 20, and 30 nm. It is clear from Fig. 11.3 that BSIM-CMG captures the volume inversion phenomenon.

11.2.2 Symmetry tests

Physically, MOSFET is symmetric w.r.t. source and drain terminals. Ideally, this aspect should be reflected by the compact model. However, there can be approximations in the compact model which do not treat drain and source in the same way. For example, the body effect is often modeled as a function of body-to-source voltage, but not body-to-drain voltage. This may cause the violation of source-drain symmetry by the compact model. It has been shown that simulations of circuits like pass gates are not even qualitatively correct if a compact model violates the symmetry requirement [12]. The benchmark tests that can be used to test model symmetry are described in this section, along with results for the BSIM-CMG model for each of these tests.

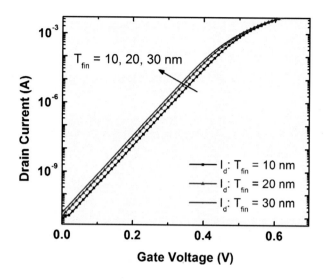

FIG. 11.3

Subthreshold current obtained from the BSIM-CMG model. Current increases with an increase in fin thickness T_{fin} due to volume inversion effect.

11.2.2.1 Gummel symmetry test

The Gummel symmetry test is a standard way of testing the symmetry of the drain-current model. The setup for this test is shown in Fig. 11.4. The voltage source V_x is swept from negative to positive typically for a set of values of V_g. If the model is symmetric, it should have the odd-order derivatives of I_x as even functions and the even-order derivatives as odd functions of V_x. In addition, I_x should be a C-∞ continuous smooth function of V_x. This can be checked by plotting first, second, third, and higher derivatives of I_x w.r.t. V_x obtained from this setup. In Fig. 11.5,

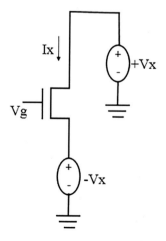

FIG. 11.4

Setup for Gummel symmetry test.

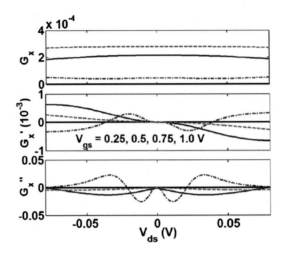

FIG. 11.5

Results of the Gummel symmetry test performed on the BSIM-CMG model. The model shows smooth and continuous drain-current derivatives and passes this symmetry test.

the Gummel symmetry test results for BSIM-CMG are shown. BSIM-CMG clearly passes the Gummel symmetry test.

11.2.2.2 Harmonic balance simulation test

If a compact model has singularities in drain-current derivatives, it will produce unphysical harmonic balance simulation results. Theoretically, the second harmonic is proportional to the square of the input signal, the third harmonic is proportional to the cube of the input signal, and so on [10]. A well-behaved compact model should mimic this behavior when harmonic balance simulations are performed. The setup for harmonic balance simulation is shown in Fig. 11.6. In Fig. 11.7, harmonic balance simulation results from the BSIM-CMG model are shown to be in good agreement with the theory.

11.2.2.3 AC symmetry test

In addition to the drain-current derivatives, the terminal charges obtained from the compact model should also be symmetric. The setup shown in Fig. 11.8 can be used to check the symmetry of the charge model. The AC terminal currents in antiphase (i_g^-, i_d^-, and i_s^-) and phase (i_g^+, i_d^+, and i_s^+) with the corresponding sources are obtained. Then, the quantity δ_{C_g} defined as

$$\delta_{C_g} = \frac{i_g^-}{i_g^+} = \frac{C_{gs} - C_{gd}}{C_{gs} + C_{gd}} \tag{11.7}$$

should be an odd function of V_x. δ_{C_g} checks for the gate-charge model. In a similar way, drain- and source-charge models can be tested by defining $\delta_{C_{sd}}$ given by

$$\delta_{C_{sd}} = \frac{C_{ss} - C_{dd}}{C_{ss} + C_{dd}} \tag{11.8}$$

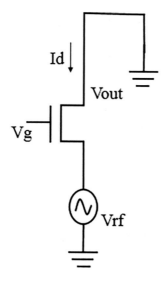

FIG. 11.6

Setup for harmonic balance simulations.

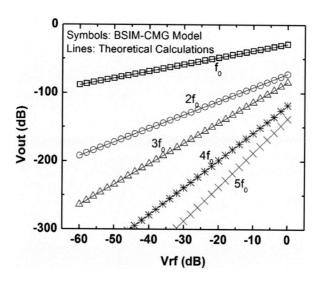

FIG. 11.7

Harmonic balance simulation results showing fundamental f_0, second harmonic $2f_0$, third harmonic $3f_0$, fourth harmonic $4f_0$, and fifth harmonic $5f_0$ from BSIM-CMG model. Model predictions agree very well with the theoretical calculations. $f_0 = 1\,\text{MHz}$. $V_g = 1\,\text{V}$.

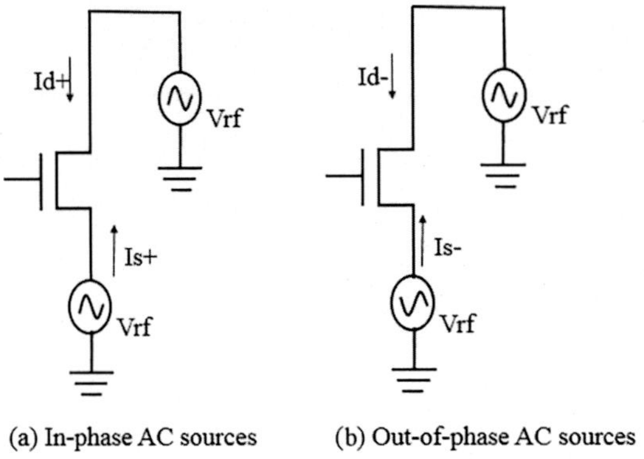

(a) In-phase AC sources   (b) Out-of-phase AC sources

FIG. 11.8

AC symmetry test setup showing AC currents for (A) AC sources are in phase and (B) AC sources are out of phase with each other.

In Fig. 11.9, we show the model results of BSIM-CMG for AC symmetry tests described above. We plot δ_{C_g}, $\delta_{C_{sd}}$, and their first derivatives, which show the symmetry of the charge model as well as its continuous and smooth behavior.

11.2.3 Reciprocity test for capacitances in a compact model

Surface-potential or charge-based compact models model the device's capacitance behavior by developing analytical equations for the terminal charges of the device. The terminal charge Q_i (of the ith terminal) can then be used to evaluate the capacitance between any two terminals i and j of the device by the following expression:

$$C_{ij} = \delta_{ij} \frac{\partial Q_i}{\partial V_j} \quad \text{where } \delta_{ij} = 1 \text{ when } i = j \text{ else} - 1 \tag{11.9}$$

The capacitances obtained from the compact model should behave in a physical way. One such requirement is the reciprocity of capacitances at $V_{ds} = 0$. The model capacitances should follow $C_{ij} = C_{ji}$ at $V_{ds} = 0$. This is shown to be the case for C_{gs}, and C_{sg} obtained from the BSIM-CMG model in Fig. 11.10. This behavior remains the same for C_{gd} and C_{dg} also.

11.2.4 Tests for model of self-heating effect

The self-heating effect is modeled via a thermal network consisting of thermal resistance and thermal capacitance [13]. The voltage across the thermal network is numerically equal to the local channel temperature T_{ch} of the device. To test the implementation of the self-heating model, following simulations should be performed. First, with the self-heating effect turned on, sweep the thermal resistance in the model and obtain the values of drain-current I_d and channel temperature T_{ch} at fixed V_{gs} and V_{ds}. Next, self-heating should be disabled, and for the same V_{gs} and V_{ds} conditions as in

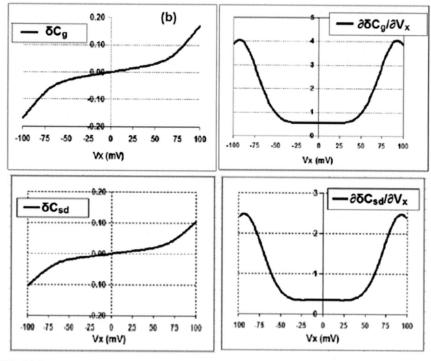

FIG. 11.9

δ_{C_g} and $\delta_{C_{sd}}$ versus V_x calculated from the BSIM-CMG model demonstrating the symmetry of the charge model. Derivatives of δ_{C_g} and $\delta_{C_{sd}}$ show that the charge model in BSIM-CMG is continuous and smooth.

the previous simulation, obtain I_d by sweeping the ambient temperature T_{amb}. For $T_{ch} = T_{amb}$, the drain-current derivatives obtained in the two scenarios should be exactly the same. In Fig. 11.11, we show the results of this test from BSIM-CMG.

11.2.5 Tests for thermal noise model

A complete compact model also includes a model for the thermal noise behavior of the device. A simple test to check the thermal noise model is as follows. Bias the device at $V_{ds} = 0$ and a gate voltage well above the threshold voltage of the device. For such a bias condition, the device is like a resistor of value $R = 1/g_{ds}$. Run the noise simulations at low enough frequencies where capacitances are not important. The current noise spectral density obtained from the model should be equal to $4kTg_{ds}$ A^2/Hz. For $g_{ds} = 10^{-4}$ A/V, the BSIM-CMG model produces 1.65×10^{-24} A^2/Hz as the current noise spectral density at the drain node, which matches the theoretical calculations.

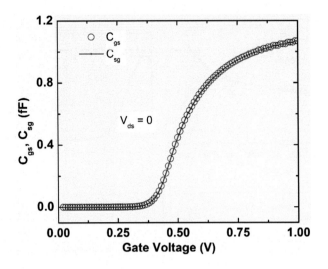

FIG. 11.10

Capacitance (C_{gs}, C_{sg}) versus V_g obtained from the BSIM-CMG model. $C_{gs} = C_{sg}$ at $V_{ds} = 0$ demonstrates that BSIM-CMG has reciprocal capacitances at $V_{ds} = 0$. Other pairs of capacitances also pass the reciprocity test.

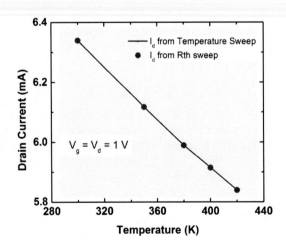

FIG. 11.11

Self-heating test results for the BSIM-CMG model. $T_{fin} = 15$ nm, $L = 30$ nm, and $N_{fin} = 100$. $V_g = V_d = 1$ V. The values of Id obtained in the two different simulation setups match very well. This shows that BSIM-CMG passes the self-heating test.

References

[1] Y. Tsividis, Operation and Modeling of the MOS Transistor, second ed., Oxford University Press, 2003.

[2] C. McAndrew, Practical modeling for circuit simulations, J. Solid State Circ. 33 (3) (1998) 439–448.

[3] Y. Tsividis, K. Suyama, MOSFET modeling for analog circuit CAD: problems and prospects, J. Solid State Circ. 29 (3) (1994) 210–216.

[4] C. McAndrew, H. Gummel, K. Singhal, Benchmarks for compact MOSFET models, in: Proc. SEMATECH Compact Models Workshop, 1995.

[5] X. Li, W. Wu, A. Jha, G. Gildenblat, R. van Langevelde, G.D.J. Smit, A.J. Scholten, D.B. M. Klassen, C.C. McAndrew, J. Watts, M. Olsen, G. Coram, S. Chaudhary, J. Victory, Benchmarking the PSP compact model for MOS transistors, in: Proc. IEEE ICMTS, 2007, pp. 259–264.

[6] C. McAndrew, Validation of MOSFET model source drain symmetry, IEEE Trans. Elect. Dev. 53 (9) (2006) 2202–2206.

[7] G. Gildenblat (Ed.), Compact Modeling Principles, Techniques and Applications, Springer Publications, 2010.

[8] H.K. Gummel, H.C. Poon, An Integral Charge Control Model of Bipolar Transistors, The Bell System Tech. Journal, 1970.

[9] S. Khandelwal, J. Watts, E. Tamilmani, L. Wagner, Scalable thermal resistance model for single- and multi-finger SOI MOSFETs, in: Proc. IEEE ICMTS, 2011, pp. 182–185.

[10] Y. Taur, An analytical solution to a double gate MOSFET with undoped body, IEEE Elect. Dev. Lett. 21 (5) (2005) 245–247.

[11] Y. Taur, X. Liang, W. Wang, H. Lu, A continuous, analytic drain current model for DG MOSFETs, IEEE Elect. Dev. Lett. 25 (2) (2004) 107–109.

[12] K. Joardar, K.K. Gullapalli, C.C. McAndrew, M.E. Burnham, A. Wild, An improved MOSFET model for circuit simulation, IEEE Trans. Elect. Dev. 45 (1) (1998) 104–148.

[13] BSIMSOI4.6 Users' Manual, BSIM Group, 2017. http://www-device.eecs.berkeley.edu/bsim/.

[14] T.H. Lee, The Design of CMOS Radio Frequency Integrated Circuits, second ed., Cambridge University Press, Cambridge, 2004.

BSIM-CMG model parameter extraction

Compact models contain many parameters to model various effects in the device, as described in Chapter 4. Model equations are parameterized for the following reasons: (1) The core model derivation is for an "ideal" device, where, for instance, channel doping profile is known and perfectly uniform, and the mobility is known and its dependence on gate voltage is negligible. As a result, the core model cannot predict an actual transistor's characteristics, even for a long-channel transistor, accurately enough for IC design without some fitting parameters. Parameters are required to match the model to the measured characteristics of the actual device. (2) There are many real-device effects that are practically impossible to model with sufficient accuracy and computational speed without using fitting parameters. In such cases, simple equations based on a good understanding of the device physics can do wonders to capture the complex effects of voltage biases and device geometry on the device behaviors in general, and very accurately with the help of a few fitting parameters. For instance, the output resistance as a function of V_{ds}, V_{gs}, V_{bs}, and L (see Fig. 12.8) could not be accurately modeled by pure "curve fitting" until physics-based model equations were introduced (see Section 5.11). One should not equate the usage of fitting parameters with pure curve fitting. (3) Model parameters are also needed because device geometry, e.g., the shape of the corner or doping profile is not known exactly or has manufacturing variations.

In fact, the judicious introduction of parameters in the model is extremely important for its practical usage. Extracting the values of these parameters is thus a critical step before using the model for circuit design. Indeed, model accuracy and sometimes even its convergence properties heavily depend on the extracted values of the parameters. This chapter discusses the parameter extraction procedure for the BSIM-CMG model. Model results for measured data with channel length varying from 30 nm to 10 μm are shown for the BSIM-CMG model.

12.1 Parameter extraction background

As discussed earlier, a compact model contains many parameters in order to capture various physical effects in the device. The parameter values are found by fitting a variety of measured data like I_d-V_g, I_d-V_d, C_{gg}-V_g, I_d-V_g at multiple ambient temperatures, etc. Extraction of parameter values is thus a multidimensional optimization problem. In fact optimization algorithms like Levenberg-Marquardt Algorithm [1],

Genetic Algorithm [2], Particle Swarm Optimization, etc., are quite regularly used for parameter extraction [3,4]. There are many commercial tools available for parameter extraction like BSIMProPlus [5], ICCAP [6], MBP [7], UTMOST [8], etc. However, in order to effectively use these sophisticated algorithms and tools, an understanding of the model equations along with the parameters and the device characteristics they affect is very important.

Before starting a parameter extraction process, it is important to understand an important classification of parameters. Generally, parameters in a compact model are classified as global and local parameters. This classification can be understood as follows: Consider the case where the parameter extraction is to be performed for a single device. Parameter extraction engineers need only to tune the parameters specific to particular effects like U0, RDSW, etc., without exercising the geometry dependence of these parameters. The parameter set obtained after such extraction is a "local" set as it applies to a specific geometric device. On the other hand, when parameter extraction is performed for a range of channel lengths and widths, the geometry dependence of the "local" parameters needs to be invoked. For instance, mobility parameter U0 is found to decrease with channel length, and this *L*-dependence is modeled via AU0 and BU0. Hence, while U0 being "local" is specific to a device, the parameters AU0 and BU0 can be termed as global parameters. BSIM-CMG Users' Manual [9] describes the various length-scaling equations introduced in the BSIM-CMG model. In a standard CMOS technology, various device geometries for the same flavor of the device, e.g., N-channel high-performance (low threshold voltage) MOSFET, are manufactured to cater to different circuit design needs. A compact model should be able to accurately predict the electrical behavior for varying device geometries. This is possible only when a global parameter set is extracted for the device flavor. In the next section, we discuss the global parameter extraction strategy for the BSIM-CMG model.

12.2 BSIM-CMG parameter extraction strategy

A complete global parameter extraction flow is shown in Fig. 12.1. It starts from parameter initialization to the extraction of noise model parameters. Here, we discuss mainly the drain-current parameter extraction strategy for the BSIM-CMG model at room temperature. It can be divided into the following steps:

1. The first step in the parameter extraction process is to fix the values of parameters that are directly measured or specified by the user. These parameters are device geometry, model selector switches, and those that can be directly taken from the technology information. A list of such parameters in the BSIM-CMG [9] model is given in Table 12.1.

    Here, the parameters with the suffix MOD are model selector switches. These parameters are used in the model either to turn on/off a specific effect like GIDLMOD or to invoke a different set of equations to model the same effect like

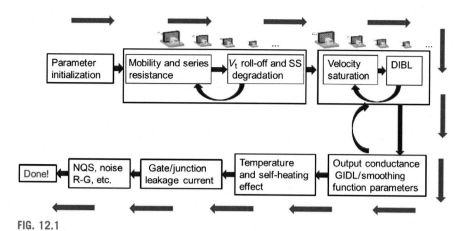

FIG. 12.1

Parameter extraction flow chart for BSIM-CMG model.

Table 12.1 List of parameters that should be set before starting the extraction flow.

Parameter name	Description
EOT	Gate oxide thickness
HFIN	Fin height
TFIN	Fin thickness
L	Fin length drawn
NFIN	Number of fins
NF	Number of fingers in parallel
NBODY	Channel doping concentration
BULKMOD	0: SOI, 1: bulk
GIDLMOD	0: off, 1: on
GEOMOD	0: double gate, 1: triple gate, 2: quadruple gate
RDSMOD	0: internal, 1: external
DEVTYPE	0: PMOS, l: NMOS
NGATE	0: metal gate, >0: poly gate doping

RDSMOD. It is a good practice to turn-off specific effect via these model selector parameters if the effect is not present in the data or is not important for the final application in which the model will be used. This avoids unnecessary model equation evaluations and improves model speed.

2. After setting the above parameters to appropriate values, the global parameters should be initialized. This step is indicated as the parameter initialization step in Fig. 12.1. It consists of three substeps, each setting specific parameters. The following trends in the measured data should be observed for this:

2.1. Plot the resistance $R_d = V_d(\sim 0.05\,\text{V})/I_d(V_g)$ versus L. Make a linear fit to this curve. Extrapolate each straight line and find the intersection point. The lines intersect at $(\Delta L, R_{\text{series}})$. The value of parameters LINT $= \Delta L/2$ and RDSW $= R_{\text{series}}$ are thus obtained. An example of such a curve is given in Fig. 12.2. This step is becoming increasingly important in scaled technology due to their high R_{series} as compared to the intrinsic channel resistance. It is possible and in fact quite common for ΔL to be negative in aggressive production technologies. While not intuitive, this is true because of the light doping in the source and drain just outside the gate edges and the small gate oxide thickness. The gate voltage has a stronger effect on the conductance of this region than the on impurity doping. According to the MOSFET theory, the channel conductance is controlled by the gate voltage while the source-drain conductance is constant. Therefore, this region ought to belong more to the channel more than to the source-drain. A compact model such as BSIM-CMG not only captures this effect but also provides additional parameters to describe the small V_g dependence of the resistance of the source-drain (region outside $L + \Delta L$) as shown in Section 5.6.

2.2. Plot the threshold voltage (V_{th}) difference $\Delta V_{\text{th}} = V_{\text{th}}(\text{short}) - V_{\text{th}}(\text{long})$ versus L for $V_d \sim 0.05\,\text{V}$ and $V_d = V_{\text{dd}}$. V_{th} can be extracted from either a constant current or maximum slope extrapolation algorithm. From this plot, parameters associated with short-channel effect (SCE) and reverse short-channel effect (RSCE) can be extracted. These parameters are: DVT0, DVT1, ETA0, DSUB, K1RSCE, and LPE0. An example plot is shown in Fig. 12.3. SCE is observed as the decrease of V_{th} for short-channel length devices, while RSCE is the opposite. Usually, an increase in V_{th} with

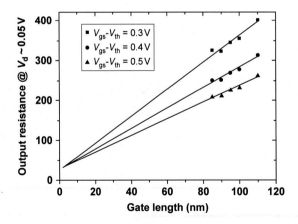

FIG. 12.2

Plot showing V_{ds} $(\sim 0.05)/I_{\text{ds}}$ versus L for different values of V_g. The lines intersect at $(\Delta L, R_{\text{series}})$.

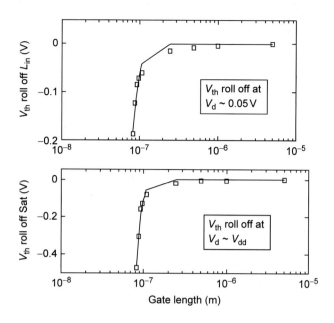

FIG. 12.3

V_{th} roll-off curves for linear and saturation drain-voltage conditions are used to extract SCE and RSCE model parameters.

decreasing channel length (RSCE) is only seen for linear-V_{th} ($V_d \sim 0.05\,\text{V}$), while the saturation-V_{th} ($V_d = V_{dd}$) decreases with channel-length reduction. This different behavior in V_{th} for linear and saturation conditions can be modeled in BSIM-CMG by appropriately tuning RSCE and DIBL parameters

2.3. Plot the subthreshold slope $SS = (d \log_{10} I_d / dV_g)^{-1}$ versus L for both linear and saturation conditions. Subthreshold slope parameters CDSC, CDSCD, and DVT1SS can be extracted from this plot. An example plot is shown in Fig. 12.4. Typically, SS increases as channel length is reduced, indicating a poorer gate control for the short devices. This behavior can be accurately modeled via the SS parameters in the BSIM-CMG model. The adjustment of SS parameters will change the V_{th} versus L fits obtained in the previous step. It is a good practice to keep an eye on the V_{th} fits too while adjusting the SS parameters.

Step 2 is useful not only to initialize the values of global parameters but also to remove outliers from the measured data. While plotting V_{th}, SS and R_{ds} versus L, if a device is seen to be too far off from the generally observed trends, it is possible that the device is an outlier. An outlier device can be because of incorrect measurements or fabrication issues. The model parameters should not be adjusted to fit this device because it is not a representative device for this geometry.

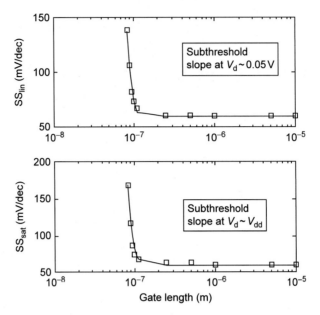

FIG. 12.4

SS versus *L* for linear and saturation conditions are used to extract CDSC, CDSCD, and DVT1SS.

3. Once the scaling parameters are set to reasonable values, the next step focuses on the fitting of linear I_d-V_g curves for long- and short-channel devices. It is divided into the following substeps:

   3.1. First the work-function, interface charge, and mobility parameters are extracted for a long-channel device. Work-function parameter PHIG and interface charge parameter CIT are extracted by fitting the threshold voltage and the subthreshold slope of the I_d-Vg curve, while the mobility parameters are extracted from the I_d-V_g curve at V_g values above V_{th}. The G_m-V_g curve is also very useful in extracting these parameters, especially the mobility degradation parameters like UA, UD, EU, and ETAMOB. Table 12.2 shows the model parameters along with the experimental data from which they should be extracted in this step. Note that U0$_0$, UA$_0$, and UD$_0$ are the global parameters present in length-scaling equations for the local parameters U0, UA, and UD, respectively.

   3.2. Extraction of the above parameters can slightly change the V_{th} versus L and SS versus L behavior of the model for the linear condition. Therefore, V_{th} roll-off and SS scaling parameters set in step 2 should be fine-tuned. Table 12.3 shows the parameters which should be fine-tuned at this stage.

   3.3. Next, the low-field mobility scaling parameters can be extracted from the I_d-V_g characteristics of the long- and medium-channel devices. The parameter names and extraction methodology are indicated in Table 12.4.

Table 12.2 Long-channel work-function, interface charge and mobility parameter names and the experimental data from which they should be extracted.

Extracted parameters	Device and experimental data	Extraction methodology
PHIG, CIT	A long-channel device I_d vs V_g @ $V_d \sim 0.05\,V$	Observe subthreshold region off-set and slope
UO_0, UA_0, UD_0, EU, ETAMOB	A long-channel device I_d vs V_g @ $V_d \sim 0.05\,V$	Observe strong-inversion region Idlin and $G_m$lin

Table 12.3 Model parameters that should be fine-tuned in step 3.2.

Extracted parameters	Device and experimental data	Extraction methodology
DVT0, DVT1, CDSC, DVT2	Both short- and medium-channel devices I_d vs V_g @ $V_d \sim 0.05\,V$	Observe subthreshold region of all devices in the same plot. Optimize DVT0. DVT1, CDSC, DVT2

Table 12.4 Extraction of the low-field mobility scaling parameters for long- and medium-channel devices.

Extracted parameters	Device and experimental data	Extraction methodology
UP, LPA	Long- and medium-channel devices I_d vs V_g @ $V_d \sim 0.05\,V$ $U_0[L] = UO_0 \times (1 - UP \times L_{eff}^{-LPA})$	Observe strong-inversion region Idlin and $G_m$lin, extract U0[L] to get UP, LP, i.e., for each L_i

Up to this stage, subthreshold and strong-inversion current for long- and medium-channel devices for the linear condition are fitted to the data. Low-field mobility is extracted from long-channel devices since short-channel devices suffer from series resistance and enhanced mobility degradation effects.

3.4. After step 3.3, short-channel device parameters for linear condition can be extracted. For this, linear I_d-V_g and G_m-V_g for short- and medium-channel devices should be plotted. Observe the values of UA(L), UD(L), RDSW(L), and $\Delta L(L)$ required to fit the various channel lengths. The L-dependence seen for UA, UD, RDSW, and ΔL can then be fitted using the global parameters AUA, BUA, AUB, BUB, ARDSW, BRDSW, LL, and LLN simultaneously. Typically, L-dependence on ΔL is not required, and it is included in the model for additional flexibility. This step may also require fine-tuning of low-field mobility scaling parameters UA and LPA. An example plot for channel-length dependence of low-field mobility is shown in Fig. 12.5.

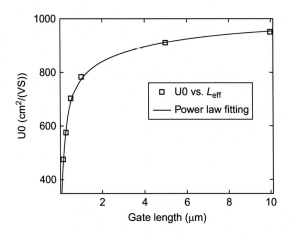

FIG. 12.5

Low-field mobility versus channel length.

This completes step 3, which consisted of fitting linear I_d-V_g from long- to short-channel devices. A sample fitting result obtained after this step is shown in Fig. 12.6.

4. After fitting the linear I_d-V_g characteristics, the saturation I_d-V_g can be fitted. Similar to step 3, this step can also be divided into various substeps:

4.1. First, the DIBL parameters for long- and medium-channel devices should be optimized. Table 12.5 shows the parameters and indicates the data that should be used for their extraction. CDSCD can be tuned to fit the SS for saturation condition without significant impact on the linear SS. ETA0 can be extracted from fitting the saturation V_{th}.

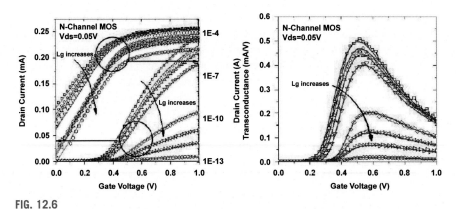

FIG. 12.6

Sample fitting results obtained with the BSIM-CMG model after extraction of linear I_d-V_g parameters for devices with channel lengths varying from 10 μm to 30 nm.

Table 12.5 DIBL effect parameters and the experimental data from which they should be extracted.

Extracted parameters	Device and experimental data	Extraction methodology
ETA0, DSUB, CDSCD	Short- and long-channel devices I_d vs V_g @ $V_d \sim V_{dd}$	Observe subthreshold region of all devices in the same plot. Optimize ETA0, DSUB, CDSCD

Table 12.6 Velocity saturation parameters for long- and medium-channel devices.

Extracted parameters	Device and experimental data	Extraction methodology
$VSAT_0$, $VSAT1_0$, $PTWG_0$, $K\,SATIV_0$, $MEXP_0$	Long- and medium-channel devices I_d vs V_g @ $V_d \sim V_{dd}$	Observe strong-inversion region $I_d$sat, $G_m$sat, $I_d V_d$

4.2. The velocity saturation parameters for long- and medium-channel length devices should be extracted after optimizing DIBL parameters. These parameters are shown in Table 12.6. Here, $PARAM_0$ is the global parameter modeling the geometry dependence of the local parameter PARAM. VSAT1 is used in the output conductance model, while VSAT is used in the calculation of saturation voltage V_{dsat} (see Section 5.7). Typically, VSAT should be equal to VSAT1, they are provided separately in the BSIM-CMG model for additional flexibility in fitting. PTWG is an empirical parameter introduced in the BSIM-CMG model to improve the model fitting of saturation G_m-V_g characteristics.

4.3. After long- and medium-channel devices' fitting, velocity saturation parameters for the short-channel devices should be extracted. The length scaling of the velocity saturation parameters should be used for this purpose. The parameter names along with the extraction methodology are given in Table 12.7.

A sample model result obtained after this step is shown in Fig. 12.7 in linear and semilog scale.

5. Once saturation I_d-V_g characteristics have been fitted, the output characteristics should be looked at. For long-channel device, output characteristics should already fit well to the measured data. A fine-tuning on long-channel output characteristics can be done using the MEXP parameter which models the transition between linear and saturation regions. For short-channel devices, the output conductance parameters should be tuned as shown in Table 12.8. The output conductance parameters are for channel-length modulation, DIBL, and SCEs. Each of these effects impacts specific regions of the I_d-V_d characteristics as

Table 12.7 Velocity saturation parameters for the short-channel devices.

Extracted parameters	Device and experimental data	Extraction methodology
AVSAT, AVSAT1, APTWG, BVSAT, BVSAT1, BPTWG	Short- and medium-channel devices I_d vs V_g @ $V_d \sim V_{dd}$	(a) Observe strong-inversion region of $I_d$sat and $G_m$sat. Find VSAT1[L_i]=X_i, VSAT[L_i]=Y_i, PTWG[L_i]=Z_i to fit data (b) Extract AVSAT1, BVSAT1 from (L_i, X_i); AVSAT, BVSAT from (L_i, Y_i) APTWG, BPTWG from (L_i, Z_i)

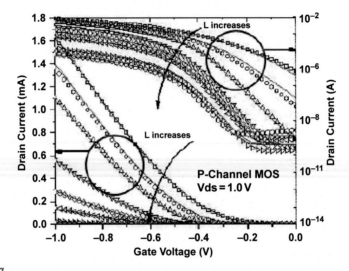

FIG. 12.7

Fitting results obtained for saturation I_d-V_g characteristics with the BSIM-CMG model for PMOS devices with channel lengths varying from 10 µm to 30 nm.

discussed in Section 5.11. Accurate modeling of output conductance is facilitated by the CLM, DIBL, and SCE model parameters. For further improvement in fitting, L-dependence of MEXP can be used. A sample fitting for I_d-V_d and G_d-V_d after this step is shown in Fig. 12.8 for L=90-nm NMOS device.

6. Once the drain-current model parameters for all the devices have been extracted, the parameters for additional effects like GIDL, gate leakage, temperature dependence, etc., can be extracted from the region of device characteristics they

Table 12.8 BSIM-CMG parameters for fitting the output characteristics.

Extracted parameters	Device and experimental data	Extraction methodology
MEXP[L], PCLM, PDIBL1, PDIBL2, DROUT, PVAG	Long- and short-channel devices I_d vs V_d @ different V_g	Observe strong-inversion region I_d vs V_d and G_d vs V_d @ different V_g

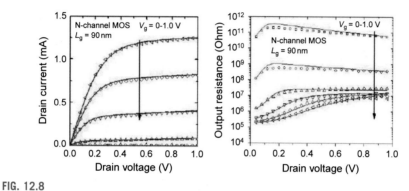

FIG. 12.8

Fitting results of I_d-V_d and G_d-V_d using the BSIM-CMG model.

dominate. Gate-leakage current parameter extraction may slightly change the fitting of large area (wide long-channel) devices, while the GIDL model parameters may slightly affect the off-state current predicted by the model. The temperature dependence of device characteristics is modeled by temperature dependence of key model parameters like mobility, threshold voltage, etc. (see Chapter 12). The temperature parameters can be easily extracted from device characteristics measured at multiple ambient temperatures.

This completes the guideline steps for the extraction of BSIM-CMG drain-current model parameters. It was shown that the BSIM-CMG model parameters can be extracted systematically to fit devices with channel lengths varying from long- to very short-channel lengths [10,11]. More advanced model extraction like extraction of RF model parameters can follow the steps indicated in Ref. [9] for the BSIM6 model.

12.3 Conclusion

In conclusion, parameter extraction steps for the BSIM-CMG model were discussed in this chapter. These are guideline steps and there can be variations on these depending on the availability or nonavailability of certain measurement data. The significance of modeling specific regions of device characteristics also depends on the final application of the model. This may also alter the parameter extraction flow. In any case, after parameter extraction, it is important to do a sanity check on the extracted values. The parameter values should lie in a sensible range and scale with the device geometry and ambient temperature in a reasonable way. It is a good practice to run the benchmark tests discussed in Chapter 10 on the model card obtained after the parameter extraction.

References

[1] K. Levenberg, A method for the solution of certain non-linear problems in least squares, Q. Appl. Math. 2 (1944) 164–168.

[2] D.E. Goldberg, Genetic Algorithm in Search, Optimization and Machine Learning, Addison-Wesley Professional, 1989.

[3] A.M. Chopde, S. Khandelwal, R.A. Thakker, M.B. Patil, K.G. Anil, Parameter extraction for MOS Model 11 using particle swarm optimization, in: Proceedings of the International Workshop on Physics of Semiconductor Devices, 2007, pp. 253–256.

[4] J. Watts, C. Bittner, D. Heaberlin, J. Hoffman, Extraction of compact model parameters for ULSI MOSFETs using a genetic algorithm, in: Proceedings of the International Conference on Modeling and Simulation of Microsystems, 1999, pp. 176–179.

[5] BSIMProPlus, Proplus. http://www.proplussolutions.com/en/pro1/Advanced-SPICE-Modeling-Platform- - -BSIMProPlus.html#.

[6] PathWave Device Modeling (ICCAP) Modeling Handbook, Keysight, 2021, Available from: http://www.home.agilent.com/en/pc-2112961/model-builder-program-silicon-focused-device-modeling-software?nid=-33185.0.00&cc=US&lc=eng.

[7] PathWave Model Builder (MBP), Keysight, 2017, Available from: http://edadownload.software.keysight.com/eedl/mbp/2017/pdf/Using_MBP.pdf.

[8] UTMOST Users' Manual, Silvaco, 2015. Available from: https://www.ucourse.ir/downloads/utmost4_users1.pdf.

[9] BSIM-CMG Users' Manual, BSIM Group, 2024. Available from: http://www-device.eecs.berkeley.edu/bsim/.

[10] S. Yao, T. Morshed, D.D. Lu, S. Venugopalan, W. Xiong, C.R. Cleaelin, A.M. Niknejad, C. Hu, Global parameter extraction for a multi-gate MOSFET compact model, in: Proc. ICMTS, 2010, pp. 9.1–9.4.

[11] S. Venugopalan, K. Dandu, S. Martin, R. Taylor, C. Cirba, X. Zhang, A.M. Niknejad, C. Hu, A non-iterative physical procedure for RF CMOS compact model extraction using BSIM6, in: Proc. CICC, 2012, pp. 1–4.

Temperature dependence

It is well known that any change in operating temperature will affect device characteristics and, hence, change circuit performance. In fact, the major reason for not increasing the operating frequency of microprocessors is the power density, which leads to a huge increase in temperature, thereby affecting performance and reliability. For any model, it is essential to provide accurate device characteristics for the entire temperature range. Ideally, the model should be able to predict this behavior for all temperatures but it is not possible practically. Normally, most compact MOSFET models are valid and tested for −50°C to 150°C, which is also the normal operation range of the majority of the circuits. An accurate temperature model requires accurate modeling of the temperature dependence of the different device parameters, such as threshold voltage, mobility, parasitic source/drain resistances, source/drain junction parameters, etc. Here, we briefly describe the temperature-dependence of BSIM-CMG.

The nominal parameter extraction is assumed to have been performed at TNOM, which is a model parameter. Generally, the value of TNOM is taken to be 300.15 K as nominal extraction is done at room temperature.

13.1 Semiconductor properties

Basic properties of intrinsic and extrinsic semiconductors change with temperature and are discussed in detail in device physics books [1]. Here, we will discuss a few of these, especially those involving model parameters, as they may need to be modified during parameter extraction.

13.1.1 Bandgap temperature dependence

Bandgap of semiconductors decreases with an increase in temperature [2]. The spacing between atoms increases when the amplitude of the atomic vibrations increases due to increased thermal energy. The increased interatomic spacing decreases the potential seen by the electrons in the semiconductor, which in turn reduces the size of the energy bandgap [3]. The temperature dependence of bandgap is modeled empirically in the literature using Varshni's expression [4]:

$$E_g = BG0SUB - \frac{TBGASUB \cdot T^2}{T + TBGBSUB} \tag{13.1}$$

FinFET/GAA Modeling for IC Simulation and Design. https://doi.org/10.1016/B978-0-323-95729-8.00006-4

where T is the temperature in Kelvin. BG0SUB, TBGASUB, and TBGBSUB are the model parameters.

The direct modulation of the interatomic distance, such as by applying high compressive (tensile) stress, also causes an increase (decrease) of the bandgap. This principle is used in all of the modern integrated circuits—strained silicon transistor.

13.1.2 Temperature dependence of N_C, V_{bi}, and Φ_B

The temperature dependence of the effective density of states, source-drain built-in potential, and Φ_B is given by

$$N_c = NC0SUB \cdot \left(\frac{T}{300.15}\right)^{3/2} \tag{13.2}$$

$$V_{bi} = \frac{kT}{q} \cdot \ln\left(\frac{NSD \cdot NBODY}{n_i^2}\right) \tag{13.3}$$

$$\Phi_B = \frac{kT}{q} \cdot \ln\left(\frac{NBODY}{n_i}\right) \tag{13.4}$$

13.1.3 Temperature dependence of intrinsic carrier concentration

The intrinsic carrier concentration is expressed as

$$n_i = \sqrt{N_C N_V} \cdot \exp\left(-\frac{E_g}{2kT}\right) \tag{13.5}$$

where N_C and N_V are the effective densities of states in the conduction and valence bands, respectively, and are a function of temperature ($\propto T^{3/2}$). The main temperature dependence in n_i comes from the exponential part, which causes n_i to increase with increase in temperature. This is why semiconductors become "intrinsic" at very high temperatures as n_i can become larger than the dopant concentration. On the other extreme (at very low temperatures), there may be a freeze-out effect, which causes dopant nonionization. This is why undoped III–V semiconductor materials are preferred over silicon at cryogenic temperatures to avoid freeze-out.

Taking out temperature dependence from Eq. (13.5) and denoting n_i as the parameter NI0SUB at $T = $ TNOM, we can write

$$n_i = NI0SUB \cdot \left(\frac{T}{TNOM}\right)^{3/2} \cdot \exp\left(\frac{BG0SUB}{2k \cdot TNOM} - \frac{E_g}{2k \cdot T}\right) \tag{13.6}$$

where BG0SUB is a parameter, which refers to the bandgap at $T = $ TNOM.

13.2 Temperature dependence of threshold voltage

The threshold voltage (V_{th}) decreases with an increase in temperature due to a shift in the fermi energy level and decrease in the bandgap. Although the temperature dependence of semiconductor properties is captured through their parameters as described in the earlier section, we do not use it for threshold voltage. The main reason for this approach is to decouple different effects and thereby parameters to provide flexibility in model fitting.

This dependence of threshold voltage is almost linear for long-channel devices [5] for wide range of temperature and modeled as [6–8]

$$V_{\text{th}}(T) = V_{\text{th}}(TNOM, L, V_{\text{ds}}) + \left(KT1 + \frac{KT1L}{L}\right)\left(\frac{T}{TNOM} - 1\right) \tag{13.7}$$

The term $\frac{KT1L}{L}$ is introduced to aid in better fitting. The temperature dependence of the bandgap E_{g}, the surface potential ϕ_{s} at threshold, and the intrinsic carrier concentration n_i are all accounted for by KT1, which uses a simple linear equation. This is done to simplify parameter extraction and make the model easier to fit. The temperature dependence of E_{g} and n_i is maintained while evaluating the saturation current of the source/drain junction.

13.2.1 Temperature dependence of DIBL

Normally, the drain voltage dependence of the threshold voltage, that is, DIBL in MOSFET is independent of temperature [6]. However, to provide flexibility for better fitting across drain bias especially for short-channel devices, the TETA0 parameter is provided.

$$ETA0(T) = ETA0 \cdot (1 - TETA0 \cdot (T - TNOM)) \tag{13.8}$$

$$ETA0R(T) = ETA0R \cdot (1 - TETA0R \cdot (T - TNOM)) \tag{13.9}$$

The parameter TETA0R can be used when $ASYMMOD = 1$.

13.2.2 Temperature dependence of body effect

The substrate bias effect may also have temperature dependence, which is captured by

$$K0(T) = K0 + K01 \cdot (T - TNOM) \tag{13.10}$$

$$K1(T) = K1 + K11 \cdot (T - TNOM) \tag{13.11}$$

$$K0SI(T) = K0SI + K0SI1 \cdot (T - TNOM) \tag{13.12}$$

$$K1SI(T) = K1SI + K1SI1 \cdot (T - TNOM) \tag{13.13}$$

$$K1SAT(T) = K1SAT + K1SAT1 \cdot (T - TNOM) \tag{13.14}$$

13.2.3 Subthreshold swing

The subthreshold swing is very sensitive to temperature as the diffusion current dominates in this region of operation. The off-current increases exponentially with decrease in temperature. The subthreshold slope varies almost linearly with temperature and is modeled as

$$\Theta_{SS} = 1 + TSS \cdot (T - TNOM) \tag{13.15}$$

The temperature-dependent subthreshold factor Θ_{SS} is multiplied by the nV_t term in (5.31), which is then used in surface potential calculation.

13.3 Temperature dependence of mobility

There has been lot of research on the accurate modeling of the temperature dependence on mobility [9, 10]. The characteristics of carrier mobility in the inversion layer are governed by three major scattering mechanisms: phonon scattering, surface scattering, and Coulomb scattering (including ionized impurity scattering and interface charge scattering). Each of these scattering mechanisms dominate over others in different temperature ranges [11], for example, phonon scattering becomes the dominant mechanism at temperatures above 250 K, while Coulomb scattering becomes important for very low temperatures. The temperature dependence of low field mobility is modeled as

$$\mu_0(T) = U0 \cdot \left(\frac{T}{TNOM}\right)^{UTE} + UTL \cdot (T - TNOM) \tag{13.16}$$

where T is the temperature in Kelvin and TNOM is the temperature at which the nominal model parameters are extracted. In the mobility model, ETAMOB, UA, UC, UD, and UCS also have temperature dependence and are modeled by the following equations:

$$ETAMOB(T) = ETAMOB \cdot [1 + EMOBT \cdot (T - TNOM)] \tag{13.17}$$

$$UA(T) = UA + UA1 \cdot (T - TNOM) \tag{13.18}$$

$$UC(T) = UC \cdot [1 + UC1 \cdot (T - TNOM)] \tag{13.19}$$

$$UD(T) = UD \cdot \left(\frac{T}{TNOM}\right)^{UD1} \tag{13.20}$$

$$UCS(T) = UCS \cdot \left(\frac{T}{TNOM}\right)^{UCSTE} \tag{13.21}$$

The parameters U0, ETAMOB, UA, UC, UD, and UCS are extracted at a nominal temperature.

13.4 Temperature dependence of velocity saturation

The carrier velocity saturates at high field, as discussed and modeled in Chapter 6. The electron velocity in inversion layer decreases with increase in temperature due to increased scattering caused by lattice vibrations. The temperature dependence of saturated velocity (v_{SAT}) in the inversion layer is linear as demonstrated experimentally in the literature [6–8], which is modeled as follows in BSIM-CMG:

$$VSAT(T) = VSAT \cdot (1 - AT \cdot (T - TNOM)) \tag{13.22}$$

$$VSAT1(T) = VSAT1 \cdot (1 - AT \cdot (T - TNOM)) \tag{13.23}$$

$$VSAT1R(T) = VSAT1R \cdot (1 - AT \cdot (T - TNOM)) \tag{13.24}$$

$$VSATCV(T) = VSATCV \cdot (1 - AT \cdot (T - TNOM)) \tag{13.25}$$

$$PTWG(T) = PTWG \cdot (1 - PTWGT \cdot (T - TNOM)) \tag{13.26}$$

$$PTWGR(T) = PTWGR \cdot (1 - PTWGT \cdot (T - TNOM)) \tag{13.27}$$

$$MEXP(T) = MEXP \cdot (1 + TMEXP \cdot (T - TNOM)) \tag{13.28}$$

The saturated velocity is a weak function of temperature as opposed to other quantities in MOSFET. This is why MOSFET will not provide significant gain when used at low temperatures as saturation current does not increase significantly.

13.4.1 Temperature dependence of the nonsaturation effect

The temperature dependence of the nonsaturation effect is also modeled linearly as follows:

$$A1(T) = A1 + A11 \cdot (T - TNOM) \tag{13.29}$$

$$A2(T) = A2 + A21 \cdot (T - TNOM) \tag{13.30}$$

13.5 Temperature dependence of leakage currents

The temperature dependence of leakage currents is empirically modeled as follows.

13.5.1 Gate current

$$AIGBINV(T) = AIGBINV + AIGBINV1 \cdot (T - TNOM) \tag{13.31}$$

$$AIGBACC(T) = AIGBACC + AIGBACC1 \cdot (T - TNOM) \tag{13.32}$$

$$AIGC(T) = AIGC + AIGC1 \cdot (T - TNOM) \tag{13.33}$$

$$AIGS(T) = AIGS + AIGS1 \cdot (T - TNOM) \tag{13.34}$$

$$AIGD(T) = AIGD + AIGD1 \cdot (T - TNOM) \tag{13.35}$$

$$igtemp = \left(\frac{T}{TNOM}\right)^{IGT} \tag{13.36}$$

This *igtemp* is multiplied with the gate current equation.

13.5.2 Gate-induced drain/source leakage

$$BGIDL(T) = BGIDL \cdot (1 + TGIDL \cdot (T - TNOM)) \tag{13.37}$$

$$BGISL(T) = BGISL \cdot (1 + TGIDL \cdot (T - TNOM)) \tag{13.38}$$

13.5.3 Impact ionization

$$ALPHA0(T) = ALPHA0 + ALPHA01 \cdot (T - TNOM) \tag{13.39}$$

$$ALPHA1(T) = ALPHA1 + ALPHA11 \cdot (T - TNOM) \tag{13.40}$$

$$ALPHAII0(T) = ALPHAII0 + ALPHAII01 \cdot (T - TNOM) \tag{13.41}$$

$$ALPHAII1(T) = ALPHAII1 + ALPHAII11 \cdot (T - TNOM) \tag{13.42}$$

$$BETA0(T) = BETA0 \cdot \left(\frac{T}{TNOM}\right)^{IIT} \tag{13.43}$$

$$SII0(T) = SII0\left(1 + TII\left(\frac{T}{TNOM} - 1\right)\right) \tag{13.44}$$

13.6 Temperature dependence of parasitic source/drain resistances

For FinFET, GAAFET or, in general, any short-channel device, parasitic source/drain resistances significantly decrease the drive current. The accurate modeling of these parasitic resistances has already been discussed in details in Chapter 7. The contact resistance, diffusion resistance, and spreading resistance at the edge of the inversion layer may have different temperature coefficients due to different materials and/or doping combinations. All of these resistances are taken into account in the R_{ds} model, which increases linearly with increasing temperature [6, 8] and is modeled as

For RDSMOD = 1 and ASYMOD = 0,

$$RSWMIN(T) = RSWMIN \cdot (1 + PRT \cdot (T - TNOM)) \tag{13.45}$$
$$RDWMIN(T) = RDWMIN \cdot (1 + PRT \cdot (T - TNOM)) \tag{13.46}$$
$$RSW(T) = RSW \cdot (1 + PRT \cdot (T - TNOM)) \tag{13.47}$$
$$RDW(T) = RDW \cdot (1 + PRT \cdot (T - TNOM)) \tag{13.48}$$

For RDSMOD = 0 and 2,

$$RDSWMIN(T) = RDSWMIN \cdot (1 + PRT \cdot (T - TNOM)) \tag{13.49}$$

$$RDSW(T) = RDSW \cdot (1 + PRT \cdot (T - TNOM)) \tag{13.50}$$

$$RSWMIN(T) = RSWMIN \cdot (1 + PRT \cdot (T - TNOM)) \tag{13.51}$$

$$RDWMIN(T) = RDWMIN \cdot (1 + PRT \cdot (T - TNOM)) \tag{13.52}$$

For RDSMOD = 2,

$$R_{s,geo}(T) = R_{s,geo} \cdot (1 + PRT \cdot (T - TNOM)) \tag{13.53}$$
$$R_{d,geo}(T) = R_{d,geo} \cdot (1 + PRT \cdot (T - TNOM)) \tag{13.54}$$

For RDSMOD = 1 and ASYMOD = 1, the following parameters may be used to bring asymmetry in the temperature dependence of drain and source side resistances.

$$RSDR(T) = RSDR \cdot (1 + TRSDR \cdot (T - TNOM)) \tag{13.55}$$
$$RSDRR(T) = RSDRR \cdot (1 + TRSDR \cdot (T - TNOM)) \tag{13.56}$$
$$RDDR(T) = RDDR \cdot (1 + TRDDR \cdot (T - TNOM)) \tag{13.57}$$
$$RDDRR(T) = RDDRR \cdot (1 + TRDDR \cdot (T - TNOM)) \tag{13.58}$$

13.7 Temperature dependence of source/drain diode characteristics

The *p-n* junction diode has temperature dependence in both of its components, that is, current and capacitance. Both of these need to be modeled under zero-bias conditions. The temperature dependence in the DC current comes from saturation current, which in turn gets it from the temperature dependence of the intrinsic carrier density (n_i) and the energy bandgap (E_g) [5, 12, 13]. The temperature dependence of the junction capacitance depends on the temperature dependence of the dielectric constant of silicon and the junction's built-in potential [5, 13].

13.7.1 DC model

The saturation current is comprised of the generation current caused by the thermal generation of electron-hole pairs inside the depletion region and the diffusion current due to minority carriers diffusing across the depletion region [1]. Both of these currents are strong functions of temperature as follows:

$$I_{generation} \propto n_i \propto T^{3/2} \exp\left(-\frac{E_g}{2kT}\right) \tag{13.59}$$

$$I_{diffusion} \propto n_i^2 \propto T^3 \exp\left(-\frac{E_g}{kT}\right) \tag{13.60}$$

As explained in Chapter 10, both of these are combined in the single current equation as shown here (parameter names used from source side junctions).

$$I_{\text{saturation}} \propto T^{XTIS} \exp\left(-\frac{E_g}{NJS \cdot kT}\right) \tag{13.61}$$

or

$$I_{\text{saturation}} \propto \exp(\ln(T^{XTIS})) \exp\left(-\frac{E_g}{NJS \cdot kT}\right) \tag{13.62}$$

or

$$I_{\text{saturation}} \propto \exp\left(\frac{-\dfrac{E_g}{kT} + XTIS \cdot \ln(T)}{NJS}\right) \tag{13.63}$$

where the parameter XTIS is the junction current temperature exponent and NJS is the junction emission coefficient.

The source and drain side junction saturation currents are modeled as follows.

Source side

The source side junction saturation current is given by

$$I_{\text{sbs}} = J_{\text{ss}}(T) \cdot ASEJ + J_{\text{sws}}(T) \cdot PSEJ + J_{\text{swgs}}(T) \cdot W_{\text{eff0}} \cdot NFINtotal \tag{13.64}$$

where $ASEJ$ is the area of the bottom of the source and $PSEJ$ is the perimeter. J_{ss} is the saturation current density (per unit area) at the bottom of the junction, J_{sws} is the saturation sidewall current density (per unit length), and J_{swgs} is the gate sidewall saturation current density (per unit length). Using Eq. (13.63), the temperature dependence of the source side current densities can be written as follows.

$$J_{\text{ss}}(T) = JSS \cdot \exp\left[\frac{\dfrac{E_{g,\text{TNOM}}}{kTNOM} - \dfrac{E_g}{kT} + XTIS \cdot \ln\left(\dfrac{T}{TNOM}\right)}{NJS}\right] \tag{13.65}$$

$$J_{\text{sws}}(T) = JSWS \cdot \exp\left[\frac{\dfrac{E_{g,\text{TNOM}}}{kTNOM} - \dfrac{E_g}{kT} + XTIS \cdot \ln\left(\dfrac{T}{TNOM}\right)}{NJS}\right] \tag{13.66}$$

$$J_{\text{swgs}}(T) = JSWGS \cdot \exp\left[\frac{\dfrac{E_{g,\text{TNOM}}}{kTNOM} - \dfrac{E_g}{kT} + XTIS \cdot \ln\left(\dfrac{T}{TNOM}\right)}{NJS}\right] \tag{13.67}$$

Drain side

Similarly, for the drain side, we have

$$I_{\text{sbd}} = J_{\text{sd}}(T) \cdot ADEJ + J_{\text{swd}}(T) \cdot PDEJ + J_{\text{swgd}}(T) \cdot W_{\text{eff0}} \cdot NFINtotal \tag{13.68}$$

where *ADEJ* is the area of the bottom of the drain and *PDEJ* is the perimeter. J_{sd} is the saturation current density (per unit area) at the bottom of the junction, J_{swd} is the saturation sidewall current density (per unit length), and J_{swgd} is the gate sidewall saturation current density (per unit length). Using Eq. (13.63), the temperature dependence of the drain side current densities can be written as follows:

$$J_{sd}(T) = JSD \cdot \exp\left[\frac{\frac{E_{g,TNOM}}{kTNOM} - \frac{E_g}{kT} + XTID \cdot \ln\left(\frac{T}{TNOM}\right)}{NJD}\right] \tag{13.69}$$

$$J_{swd}(T) = JSWD \cdot \exp\left[\frac{\frac{E_{g,TNOM}}{kTNOM} - \frac{E_g}{kT} + XTID \cdot \ln\left(\frac{T}{TNOM}\right)}{NJD}\right] \tag{13.70}$$

$$J_{swgd}(T) = JSWGD \cdot \exp\left[\frac{\frac{E_{g,TNOM}}{kTNOM} - \frac{E_g}{kT} + XTID \cdot \ln\left(\frac{T}{TNOM}\right)}{NJD}\right] \tag{13.71}$$

13.7.2 Capacitance

The temperature dependence of the junction capacitances is modeled through temperature-dependent zero-bias capacitances per unit area or per unit length and temperature-dependent junction built-in potentials.

Source side

On the source side, the temperature dependence of the junction capacitances and potentials is defined as

$$CJS(T) = CJS[1 + TCJ(T - TNOM)] \tag{13.72}$$

$$CJSWS(T) = CJSWS[1 + TCJSW(T - TNOM)] \tag{13.73}$$

$$CJSWGS(T) = CJSWGS[1 + TCJSWG(T - TNOM)] \tag{13.74}$$

$$PBS(T) = PBS(TNOM) - TPB(T - TNOM) \tag{13.75}$$

$$PBSWS(T) = PBSWS(TNOM) - TPBSW(T - TNOM) \tag{13.76}$$

$$PBSWGS(T) = PBSWGS(TNOM) - TPBSWG(T - TNOM) \tag{13.77}$$

where *CJS* is the junction capacitance at the bottom, *CJSWS* is the junction capacitance at the source sidewall, *CJSWGS* is the junction capacitance at the gate sidewall, *PBS* is the junction potential at the bottom, *PBSWS* is the junction capacitance at the source sidewall, and *PBSWGS* is the junction capacitance at the gate sidewall.

Drain side

Similarly, on the drain side, we have

$$CJD(T) = CJD[1 + TCJ(T - TNOM)] \tag{13.78}$$

$$CJSWD(T) = CJSWD[1 + TCJSW(T - TNOM)] \tag{13.79}$$

$$CJSWGD(T) = CJSWGD[1 + TCJSWG(T - TNOM)] \tag{13.80}$$

$$PBD(T) = PBD(TNOM) - TPB(T - TNOM) \tag{13.81}$$

$$PBSWD(T) = PBSWD(TNOM) - TPBSW(T - TNOM) \tag{13.82}$$

$$PBSWGD(T) = PBSWGD(TNOM) - TPBSWG(T - TNOM) \tag{13.83}$$

where CJD is the junction capacitance at the bottom, $CJSWD$ is the junction capacitance at the source sidewall, $CJSWGD$ is the junction capacitance at the gate sidewall, PBD is the junction potential at the bottom, $PBSWD$ is the junction capacitance at the source sidewall, and $PBSWGD$ is the junction capacitance at the gate sidewall.

13.7.3 Trap-assisted tunneling current

The temperature dependence for the trap-assisted tunneling saturation current has been modeled as follows.

Source side

On the source side, we have

$$J_{\text{tss}}(T) = JTSS \cdot \exp\left(\frac{E_{g,\text{TNOM}} \cdot XTSS \cdot \left(\frac{T}{TNOM} - 1 \right)}{kT} \right) \tag{13.84}$$

$$J_{\text{tssws}}(T) = JTSSWS \cdot \exp\left(\frac{E_{g,\text{TNOM}} \cdot XTSSWS \cdot \left(\frac{T}{TNOM} - 1 \right)}{kT} \right) \tag{13.85}$$

$$
\begin{aligned}
J_{\text{tsswgs}}(T) &= JTSSWGS \times \left(\sqrt{\frac{JTWEFF}{W_{\text{eff0}}}} + 1 \right) \\
&\times \exp\left(\frac{E_{g,\text{TNOM}} \cdot XTSSWGS \cdot \left(\frac{T}{TNOM} - 1 \right)}{kT} \right)
\end{aligned} \tag{13.86}
$$

where J_{tss} is the current density for the junction bottom, J_{tssws} is for the junction sidewall, and J_{tsswgs} is for the gate sidewall.

Drain side

Similarly, on the drain side, we have

$$J_{\text{tsd}}(T) = JTSD \cdot \exp\left(\frac{E_{\text{g,TNOM}} \cdot XTSD \cdot \left(\frac{T}{TNOM} - 1\right)}{kT}\right) \tag{13.87}$$

$$J_{\text{tsswd}}(T) = JTSSWD \cdot \exp\left(\frac{E_{\text{g,TNOM}} \cdot XTSSWD \cdot \left(\frac{T}{TNOM} - 1\right)}{kT}\right) \tag{13.88}$$

$$J_{\text{tsswgd}}(T) = JTSSWGD \times \left(\sqrt{\frac{JTWEFF}{W_{\text{eff0}}}} + 1\right)$$
$$\times \exp\left(\frac{E_{\text{g,TNOM}} \cdot XTSSWGD \cdot \left(\frac{T}{TNOM} - 1\right)}{kT}\right) \tag{13.89}$$

where J_{tsd} is the current density for the junction bottom, J_{tsswd} is for the junction sidewall, and J_{tsswgd} is for the gate sidewall.

The body diode nonideality factors are also dependent on the temperature and modeled as follows.

Source side

For the source side, we have

$$NJTS(T) = NJTS \times \left(1 + TNJTS \cdot \left(\frac{T}{TNOM} - 1\right)\right) \tag{13.90}$$

$$NJTSSW(T) = NJTSSW \times \left(1 + TNJTSSW \cdot \left(\frac{T}{TNOM} - 1\right)\right) \tag{13.91}$$

$$NJTSSWG(T) = NJTSSWG \times \left(1 + TNJTSSWG \cdot \left(\frac{T}{TNOM} - 1\right)\right) \tag{13.92}$$

where $NJTS$ corresponds to the junction bottom, $NJTSSW$ corresponds to the sidewall, and $NJTSSWG$ corresponds to the gate sidewall.

Drain side

Similarly, on the drain side, we have

$$NJTSD(T) = NJTSD \times \left(1 + TNJTSD \cdot \left(\frac{T}{TNOM} - 1\right)\right) \tag{13.93}$$

$$NJTSSWD(T) = NJTSSWD \times \left(1 + TNJTSSWD \cdot \left(\frac{T}{TNOM} - 1\right)\right) \qquad (13.94)$$

$$NJTSSWGD(T) = NJTSSWGD \times \left(1 + TNJTSSWGD \cdot \left(\frac{T}{TNOM} - 1\right)\right) \qquad (13.95)$$

where *NJTSD* corresponds to the junction bottom, *NJTSSWD* corresponds to the sidewall, and *NJTSSWGD* corresponds to the gate sidewall.

13.8 Self-heating effect

Power density has been increasing with new technology nodes. The power is dissipated in the channel in the form of heat. As discussed and modeled earlier, the increase in temperature affects device parameters such as mobility threshold voltage, etc., which results in a decrease in transistor current. The self-heating effect depends on the power dissipated and the thermal conductivity of the materials used in the device. This is the reason this effect is more prominent in high-voltage/power devices, that is, LDMOS, IGBT, etc., as well as in SOI MOSFETs due to the low-thermal conductivity of the SOI substrate compared to bulk silicon. FinFETs and GAAFETs also show a self-heating effect as the drive current is large and there is not much space to release this heat quickly away from the device.

The self-heating effect is modeled using a thermal network [14] as shown in Fig. 13.1 by invoking SHMOD = 1. The power dissipated in the transistor is fed into this thermal network and the voltage drop across the network increases in temperature due to the self-heating effect. This increase in temperature is added to the device temperature. SPICE optimizes the temperature increase and current decrease due to the self-heating effect till convergence is achieved. The increased iterations by SPICE do cause speed penalty.

The values of R_{th} and C_{th} scale with width and are obtained using the following equations:

$$R_{th} = \frac{RTH0}{WTH0 + F_{pitch} \cdot NFIN} \qquad (13.96)$$

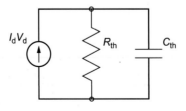

FIG. 13.1

Thermal network for modeling of the self-heating effect.

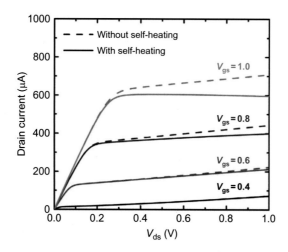

FIG. 13.2

I_D-V_D simulation with and without simulation. The drain current decreases due to the self-heating effect and may cause negative g_{ds}.

$$C_{th} = CTH0 \cdot (WTH0 + F_{pitch} \cdot NFIN) \tag{13.97}$$

where RTH0 and CTH0 are model parameters corresponding to normalized thermal resistance and thermal capacitance, respectively. The parameter WTH0 refers to the minimum width for thermal resistance calculation.

Fig. 13.2 shows the simulation of drain current with drain voltage with and without a self-heating effect. When self-heating is turned on using the SHMOD = 1, the drain current gets reduced compared to the SHMOD = 0 case. If the value of thermal resistance (RTH0) is sufficiently large, it may even cause a decrease in current with increasing drain voltage, thereby causing negative g_{ds}. The benchmark test to check the correct implementation of the self-heating effect has been discussed in Chapter 11.

13.9 Validation range

The model is valid for temperatures ranging from $-50°C$ to $200°C$. For temperatures much lower than $-50°C$, the device physics changes considerably, and modeling for such low temperatures is a separate challenging issue.

13.10 Model validation on measured data

The temperature effects discussed in previous sections are validated on measured data of SOI FinFETs with $L = 60nm$, TFIN = 17nm, HFIN = 60nm, and NFIN = 20. The temperature is swept from $-50°C$ to $200°C$ in steps of $50°C$.

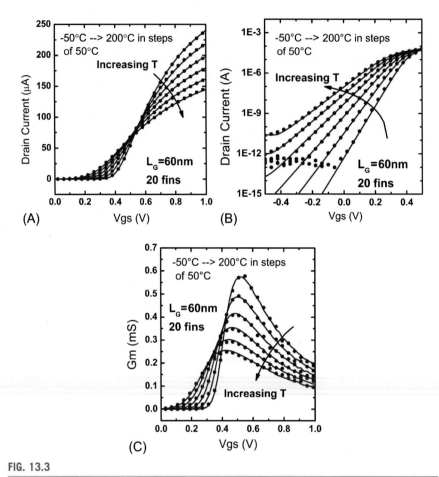

FIG. 13.3

(A) Drain current on a linear axis, (B) drain current on a log axis, and (C) transconductance versus gate bias in the linear region of operation ($V_{ds} = 50$mV) for $T = -50°$C to 200°C in steps 50°C. The experimental data are from the SOI FinFET with $L = 60$nm, TFIN $= 17$nm, and HFIN $= 60$nm, NFIN $= 20$.

Fig. 13.3 shows the drain current in linear regions for different temperatures. The threshold voltage for different temperatures matches perfectly. At low gate voltage, mobility is strongly affected by temperature while at high gate voltage, series resistances dominate the linear current characteristics. The model also matches the zero-temperature-coefficient, which is important in designing temperature invariant circuits. Fig. 13.4 shows the drain current in the saturation region for different temperatures. In saturation, the current is dominated by velocity saturation and the model shows an excellent match with experimental data.

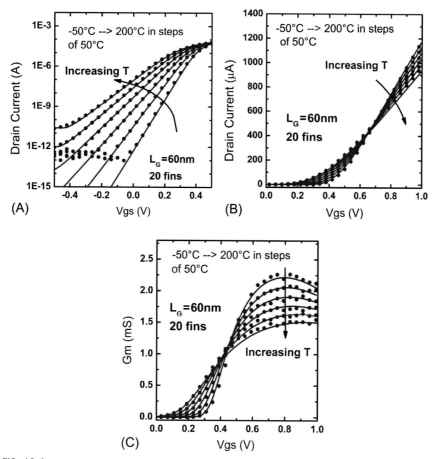

FIG. 13.4

(A) Drain current on a linear axis, (B) drain current on a log axis, and (C) transconductance versus gate bias in the saturation region of operation ($V_{ds} = 1.0V$) for $T = -50°C$ to $200°C$ in steps $50°C$. The experimental data are from the SOI FinFET with $L = 60nm$, TFIN = 17nm, HFIN = 60nm, and NFIN = 20.

References

[1] C. Hu, Modern Semiconductor Devices for Integrated Circuits, Prentice Hall, 2010.

[2] H.Y. Fan, Temperature dependence of the energy gap in semiconductors, Phys. Rev. 82 (6) (1951) 900–905.

[3] B.V. Zeghbroeck, Principles of Semiconductor Devices [Online], Available from: http://ecee.colorado.edu/bart/book/contents.htm.

[4] Y.P. Varshni, Temperature dependence of the energy gap in semiconductors, Physica 34 (1) (1967) 149–154.

[5] S.M. Sze, Physics of Semiconductor Devices, Wiley, New York, 1981.

[6] Y. Cheng, et al., Modeling temperature effects of quarter micrometer MOSFETs in BSIM3v3 for circuit simulation, Semicond. Sci. Technol. 12 (1997) 1349–1354.

[7] J.H. Huang, et al., BSIM3 Manual (Version 2.0), University of California, Berkeley, 1994.

[8] Y. Cheng, et al., BSIM3 Version 3.1 User's Manual, University of California, Berkeley, 1997. Memorandum No. UCB/ERL M97/2.

[9] C.L. Huang, G.S. Gildenblat, Measurements and modeling of the n-channel MQSFET inversion layer mobility and device characteristics in the temperature range 60–300K, IEEE Trans. Electron Devices 37 (1990) 1289–1300.

[10] M.S. Liang, J.Y. Choi, P.K. Ko, C. Hu, Inversion-layer capacitance and mobility of very thin gate-oxide MOSFETs, IEEE Trans. Electron Devices 33 (1986) 409.

[11] Y. Cheng, C. Hu, MOSFET Modeling and BSIM3 User's Guide, Kluwer Academic Publishers, 1999.

[12] G. Massobrio, P. Antognetti, Semiconductor Device Modeling With SPICE, McGraw-Hill, Inc., New York, 1993.

[13] N. Arora, MOSFET Models for VLSI Circuit Simulation, Springer-Verlag, Wien, NY, 1994.

[14] P. Su, S.K.H. Fung, S. Tang, F. Assaderaghi, C. Hu, BSIMPD: a partial-depletion SOI MOSFET model for deep-submicron CMOS designs, in: Proceedings of the IEEE Custom Integrated Circuits Conference, 2000, pp. 197–200.

Cryogenic temperature modeling

14

The exploration of transistor operation under cryogenic temperatures dates back to the early 1970s [1, 2]. Such extreme temperature conditions have found relevance in space and defense electronic systems, with the traditional use of cryogenic CMOS (Cryo-CMOS) primarily confined to these domains. Furthermore, Cryo-CMOS has garnered attention in high-performance computing applications due to its exceptional electrical characteristics, including steep subthreshold switching, high ON-state current, minimal parasitics, and remarkably low thermal noise. Nevertheless, the challenges of cryogenic technology optimization and cooling costs persist. More recently, another pivotal application of Cryo-CMOS has arisen in the context of quantum computing systems. Quantum computing represents a potentially transformative paradigm shift in computing, with the potential for "Quantum Supremacy" over classical computing algorithms, enabling exponentially faster solutions to seemingly impossible problems. At the heart of a quantum computer lies the qubit, akin to a conventional binary computer's bit, highlighting the fundamental unit in quantum computation. While there are several ways a qubit can be constructed, significant attention within the industry has been devoted to superconducting Josephon junctions and silicon quantum dots. These qubits are susceptible to thermal noise; even the slightest perturbation can disrupt their quantum states. Consequently, they operate at cryogenic temperatures, typically in the mK range using dilution refrigerators. However, essential tasks like read-out, control, and continuous error correction necessitate electrical connections to the external world, currently achieved through coaxial cables. These cables span the temperature gradient from 10–20 mK to 300 K and must maintain a considerable length to shield the quantum system from thermal noise. As the number of qubits scales up, managing this increasingly bulky and intricate wiring infrastructure becomes a formidable challenge. To address the challenge of managing a large number of electrical connections required for scaling up the qubit count, researchers have introduced the concept of cryogenic control ICs based on CMOS technology [3, 4]. These chips can be strategically placed in close proximity to the qubits within the cryogenic environment of the dilution refrigerator, at ≈ 4 K. This arrangement offers several advantages, including enhanced thermal isolation for the qubits and reduced signal latency [5, 6]. The ultimate objective is to bridge the temperature differential between the qubits and CMOS control circuitry, aiming for their seamless integration onto a single chip.

To realize the Cryo-CMOS VLSI design, computationally efficient and robust compact models are needed, which can capture the device effects observed at these

FinFET/GAA Modeling for IC Simulation and Design. https://doi.org/10.1016/B978-0-323-95729-8.00010-6

extreme temperatures. These models must effectively capture the nuanced device behaviors that manifest at extreme cryogenic temperatures, introducing new effects that are inconsequential under near-room temperature conditions. In this chapter we discuss the modeling intricacies of these cryogenic-specific effects. Furthermore, it is noteworthy that industry-standard temperature compact models traditionally cater to a relatively narrow temperature range centered at approximately room temperature (typically spanning from approximately −40°C to 125°C). Extending this range to encompass the cryogenic regime necessitates a fundamental reconfiguration of conventional temperature models.

14.1 Core model including band tail states

It has been experimentally observed across several different transistor technologies that the subthreshold swing (SS) shows a deviation from the usual linear temperature dependency (kT/q) as the temperature is lowered down to the cryogenic regime, as shown in Fig. 14.1 [7–9]. It can be seen that the SS starts to saturate at a swing greater than kT/q at low T. Sometimes, a rise in the SS is also observed at very low T [7, 10]. Researchers have attributed this phenomenon to the presence of band tail states at the conduction/valance band edges. These are disorder-induced states that can be present due to residual impurities, defects, etc., which can disturb the periodic order of the lattice. We use a compact modeling approach to take into account the impact of band tail states in the core model of BSIM-CMG as follows. Let us assume a toy model for

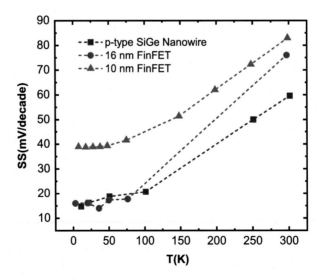

FIG. 14.1

Subthreshold swing as a function temperature for different transistor technologies.

the density of states (DOS) with exponential tail states at the conduction band edge [11, 12] for an n-type MOSFET, as shown in Fig. 14.2A and given as

$$g(E) = \begin{cases} g_{t0} \exp\left(\dfrac{E - E_0}{k \cdot \text{TLOW}}\right) & E < E_0 \\ g_{b0}\sqrt{E - E_c} & E \geq E_0 \end{cases} \qquad (14.1)$$

where $k \cdot$ TLOW is a constant that determines the extent of the tail ($k =$ Boltzmann's constant and TLOW represents a characteristic band tail temperature), and g_{b0}

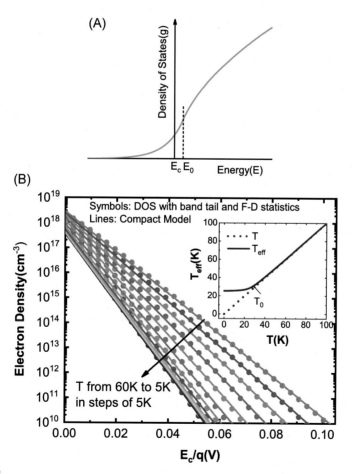

FIG. 14.2

(A) A toy model for density of states including band tail at the conduction band edge. (B) Comparison of the electron density obtained from our model with that calculated numerically from Eq. (14.5). The *inset* of (B) shows effective temperature used in the compact model as a function of temperature. TLOW = 25 K and DTLOW = 15 K are used.

$$g_{b0} = \frac{1}{2\pi^2}\left(\frac{2m^*}{\hbar^2}\right)^{3/2} \tag{14.2}$$

where m^* is the DOS effective mass of the electron and $\hbar$ is reduced Plank's constant. E_0 and g_{t0} can be expressed in terms of E_c, g_{b0}, and $k \cdot \text{TLOW}$ assuming the continuity of g and its first derivative at $E = E_0$ as [11]

$$E_0 = E_c + \frac{k \cdot \text{TLOW}}{2} \tag{14.3}$$

and

$$g_{t0} = g_{b0}\sqrt{\frac{k \cdot \text{TLOW}}{2\exp(1)}} \tag{14.4}$$

The electron density can now be calculated as

$$n(E_c, T) = \int_{E_v}^{\infty} g(E)f(E)dE \tag{14.5}$$

where $f(E)$ is the Fermi-Dirac (F-D) distribution function. For low cryogenic temperatures ($T \ll \text{TLOW}$), the electron concentration (n) is dominated by the electrons in the band tail states.

$$n(E_c, T) \approx \int_{E_v}^{E_c} g_{t0}\exp\left(\frac{E - E_0}{k \cdot \text{TLOW}}\right)f(E)dE \tag{14.6}$$

Within this range of temperatures, $f(E)$ can be simplified as a Heaviside step function, denoted as $\theta(E_f - E)$, where E_f represents the electron Fermi level. This approximation is applicable for computing the electron density up to a moderate inversion level in the channel of the transistor, specifically when E_f is a few kT less than E_c. Subsequently, we get

$$\begin{aligned} n(E_c, T) &\approx \int_{E_v}^{E_f} g_{t0}\exp\left(\frac{E - E_0}{k \cdot \text{TLOW}}\right)dE \\ &= g_{t0}k\text{TLOW}\left[\exp\left(\frac{E_f - E_0}{k \cdot \text{TLOW}}\right) - \exp\left(\frac{E_v - E_0}{k \cdot \text{TLOW}}\right)\right] \end{aligned} \tag{14.7}$$

The second exponential term in Eq. (14.8) can be neglected as compared to the first term during transistor operation, starting from the subthreshold region and beyond. Using Eqs. (14.3), (14.4) in Eq. (14.7), we obtain

$$n(E_c, T) = N_c(\text{TLOW})\exp\left(\frac{E_f - E_c - \Delta E_c}{k \cdot \text{TLOW}}\right) \tag{14.8}$$

where $N_c(\text{TLOW}) = \sqrt{\pi/2}g_{b0}(k \cdot \text{TLOW})^{3/2}$ and $\Delta E_c = \frac{k \cdot \text{TLOW}}{2}\left[\log_e(\pi/2) + 1\right]$. From Eq. (14.8), it is evident that for temperatures significantly smaller than TLOW, n can be equated to the electron density expression obtained using Maxwell-Boltzmann (M-B) statistics at $T = \text{TLOW}$ with effective DOS $= N_c(\text{TLOW})$ and a small constant correction term ΔE_c in E_c. Furthermore, it is established that at temperatures significantly higher than TLOW, the value of n is predominantly governed by the band electrons. In this case, the F-D statistics can be approximated by M-B statistics (up to a moderate inversion level). As a result, we obtain the conventional expression of n as

$$n(E_c, T) = N_c(T) \exp\left(\frac{E_f - E_c}{kT}\right) \tag{14.9}$$

The BSIM-CMG core model assumes the M-B statistics for charge carriers (Chapter 3). Considering the previous analysis, the following compact modeling strategy can be employed to emulate the influence of F-D statistics and the band tail effect at cryogenic temperatures while making minimal adjustments to the core model calculations.

1. The calculation of charge density is performed using the established formulation of BSIM-CMG in Chapter 3, which assumes the conventional expression for electron density based on M-B statistics, as given in Eq. (14.9).
2. Instead of using T directly, an effective temperature T_{eff} is employed. T_{eff} replaces T in the N_c expression, and the core model calculations of charge density in BSIM-CMG (thermal voltage is replaced by kT_{low0}/q).

To calculate T_{eff}, we first define a temperature variable, T_{low0} as

$$T_{low0} = \frac{T + \text{TLOW} + \sqrt{(T - \text{TLOW})^2 + 0.25 \cdot \text{DTLOW}^2}}{2} \tag{14.10}$$

For $T < \text{TLOW}$, T_{low0} is gradually clamped to TLOW. The smoothing of the transition can be adjusted by the smoothing parameter DTLOW. This term will result in the saturation of SS w.r.t T below TLOW.

As mentioned earlier, sometimes a rise in the SS with temperature reduction is also observed in experimental data at very low temperatures. To capture this effect, a linear temperature function, T_{low1} is used when T goes below another smaller temperature TLOW1 as

$$T_{low1}(T) = \frac{\text{KLOW1} \cdot (\text{TLOW1} - T)}{2} \\ + \frac{\sqrt{[\text{KLOW1} \cdot (\text{TLOW1} - T)]^2 + 0.25 \cdot \text{DTLOW1}^2}}{2} \tag{14.11}$$

Finally, T_{eff} is calculated as

$$T1 = T_{low0}(T) + T_{low1}(T) - T_{low0}(T_{nom}) - T_{low1}(T_{nom}) + T_{nom} \tag{14.12}$$

$$T_{eff} = \frac{T + T1 + \sqrt{(T - T1)^2 + 0.25 \cdot 0.04}}{2} \tag{14.13}$$

The terms in $T1$ other than T_{low0} and T_{low1} ensure that T_{eff} becomes exactly equal to T_{nom} at $T = T_{nom}$. This, in turn, ensures that the data fitting done for $T = T_{nom}$ does not get disturbed while tuning the T_{eff} model parameters for fitting the data at other temperatures.

3. We introduce an additional term for threshold voltage correction, which is added to the gate voltage.

$$\Delta V_{\text{th,temp}} = \frac{KT11}{1 + \exp\left(\dfrac{T - TVTH}{KT12}\right)} - \frac{KT11}{1 + \exp\left(\dfrac{T_{\text{nom}} - TVTH}{KT12}\right)} \tag{14.14}$$

where KT11, KT12, and TVTH are temperature-independent parameters. Eq. (14.14) enables the correction of the surface potential at low temperatures ($T <$ TVTH). This equation provides a means to account for various factors including (i) band tail effect (represented by ΔE_c term in Eq. 14.8), (ii) discrepancies between electron densities derived from M-B and F-D statistics in the subthreshold region, and (iii) presence of interface traps[a] . Because $\Delta V_{\text{th, temp}}$ may arise from multiple factors, the temperature threshold TVTH, in general, can differ from TLOW. It is important to note that, in the traditional BSIM temperature formulation, $\Delta V_{\text{th, temp}}$ is modeled as a linear function of temperature, which fails to accurately capture the aforementioned effects. A simulation example illustrating $\Delta V_{\text{th, temp}}$ is shown in Fig. 14.3.

Fig. 14.2B presents a comparison between the calculated values of n using the above compact modeling approach (utilizing T_{eff} instead of T and incorporating the correction for E_c from Eq. 14.14 in the M-B based n expression given by Eq. 14.9) and the numerically computed values from Eq. (14.5), as a function of E_c. T_{eff} used in the compact model is also displayed in the inset. Fig. 14.2B demon-

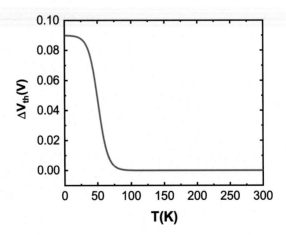

FIG. 14.3

Threshold voltage offset ($\Delta V_{\text{th, temp0}}$) as a function of temperature.

---

[a]At low T, where E_f is closer to E_c, there is an increased number of fixed charged trap states below E_f (assuming acceptor-type states), leading to an additional shift in V_{th}.

strates a high level of agreement between our model and the numerical simulations across a wide temperature range employed.

14.2 Revisiting temperature power laws of mobility

The mobility degradation model due to a vertical electric field in the channel (at zero body bias) in BSIM-CMG is given by [13]

$$\mu_{\text{eff}} = \frac{\mu_0}{1 + \text{UA}E_v^{\text{EU}} + \dfrac{\text{UD}}{\left[\frac{1}{2}\left(1 + \frac{U_{\text{DS}}q_{\text{is}} + U_{\text{DD}}q_{\text{id}}}{q_0}\right)\right]^{\text{UCS}}}} \tag{14.15}$$

Here, q_{is} denotes the normalized inversion charge density on the source side, while q_{id} represents the normalized inversion charge density on the drain side. q_0 is a temperature-independent constant. E_v is the effective vertical electric field, which can be expressed in terms of the inversion charge densities q_{is} and q_{id} [13]. The second and third terms in the denominator of Eq. (14.15) account for the mobility degradation due to phonon/surface roughness scattering and Coulomb scattering (CS), respectively. The coefficients μ_0, UA, UD, and UCS are temperature-dependent mobility coefficients and in BSIM models they assume conventional temperature power laws as T_r^γ, where $T_r = T/T_{\text{nom}}$ and γ represents a temperature-independent constant (Section 13.3) [13–15]. Additionally, the parameter EU is assumed to be independent of temperature. To accurately capture the drain bias dependence of CS, two additional temperature-dependent mobility coefficients, U_{DS} and U_{DD}, are introduced in Eq. (14.15). For $U_{\text{DS}} = U_{\text{DD}} = 0.5$, Eq. (14.15) simplifies to the BSIM-CMG expression of μ_{eff} [13].

Fig. 14.4A and B illustrates a comparison between the conventional BSIM mobility models (termed as conventional models) and experimental data from Refs. [16, 17], respectively. When the temperature is reduced, lattice vibrations, and consequently, phonon scattering, decrease significantly. As a result, the peak mobility increases. However, the current temperature power law scheme with constant γ fails to accurately capture this temperature variation in the experimental data, especially for the mid-temperature range.

Furthermore, as the temperature decreases, the surface roughness component increases due to the reduction in the thermal velocity of electrons [18]. This leads to an increased slope in the μ_{eff} curves as temperature decreases in the high electric field/strong inversion region. The slope of mobility in this high E_v region is determined by the parameters μ_0, UA, and EU. As observed in Fig. 14.4A and B, the conventional mobility model struggles to fit the slopes accurately in the high E_v region. The influence of EU can be distinguished from μ_0 and UA by employing a log-log plot such as shown in Fig. 14.4A. For a large E_v in Eq. (14.15), the CS term and 1 in the denominator can be neglected, allowing μ_{eff} to be approximated as

$$\mu_{\text{eff}} \approx \frac{\mu_0}{\text{UA}E_v^{\text{EU}}} \tag{14.16}$$

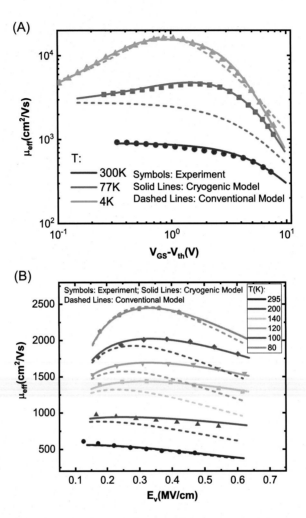

FIG. 14.4

Modeling the impact of temperature on mobility degradation due to a vertical field. (A) Mobility as a function of gate drive voltage. (B) Mobility as a function of the effective vertical field.

(A) Experimental data is from K.V. Seendripu, A 4 K CMOS Self-Calibration Circuit for Josephson Flash-Type Analog-to-Digital Converters (Ph.D. thesis), EECS Department, University of California, Berkeley, 1996; (B) experimental data is from C.-L. Huang, G.S. Gildenblat, Measurements and modeling of the n-channel MOSFET inversion layer mobility and device characteristics in the temperature range 60-300 K, IEEE Trans. Electron Devices 37 (5) (1990) 1289–1300.

As demonstrated in Eq. (14.17), the derivative of the log-log plot of μ_{eff} is now solely dependent on EU as

$$\frac{d\log\mu_{\text{eff}}}{d\log E_{\text{v}}} = -\text{EU} \tag{14.17}$$

The experimental data presented in Fig. 14.4A provides clear evidence that in the high $V_{\text{gs}} - V_{\text{th}}$ or E_{v} (which depends linearly on the gate voltage V_{gs} in strong inversion) region, the slope EU must be a function of T. This contradicts the assumption made in the conventional mobility models used in BSIM, where EU is treated as a temperature-independent parameter.

At low electric field levels, the CS component becomes noticeable as charge carriers are affected by the Coulomb potential of the unscreened fixed ions in the bulk and interface charges. This effect becomes more prominent at low cryogenic temperatures, where the thermal velocity of carriers is small and interaction between the electrons and CS centers increases significantly [19]. Fig. 14.4A and B demonstrates that the temperature dependence of the CS component, both in terms of magnitude and slope, cannot be adequately explained by simple temperature power laws.

To accurately represent the mobility behavior across a broad temperature range, including the cryogenic regime, it is crucial to revise the temperature dependence of the mobility coefficients μ_0, UA, U_{C}, UD, and EU. The temperature power laws governing these coefficients are modified to incorporate a linear temperature dependence in the power rather than a constant power, resulting in the following expressions:

$$\mu_0(T) = \mu_0(T_{\text{nom}}) \cdot T_{\text{r}}^{\,\text{U01} + \text{U02} \cdot T_{\text{r}}} \tag{14.18}$$

$$\text{UA}(T) = \text{UA}(T_{\text{nom}}) \cdot T_{\text{r}}^{\,\text{UA1} + \text{UA2} \cdot T_{\text{r}}} \tag{14.19}$$

$$\text{UD}(T) = \text{UD}(T_{\text{nom}}) \cdot T_{\text{r}}^{\,\text{UD1} + \text{UD2} \cdot T_{\text{r}}} \tag{14.20}$$

$$\text{UCS}(T) = \text{UCS}(T_{\text{nom}}) \cdot T_{\text{r}}^{\,\text{UCS1} + \text{UCS2} \cdot T_{\text{r}}} \tag{14.21}$$

Here, U01, U02, UA1, UA2, UD1, UD2, UCS1, and UCS2 are temperature-independent parameters. Moreover, to align with the experimental slope of μ_{eff} versus E_{v} in the high-field regime, EU is modified to be a linear function of temperature, expressed as

$$\text{EU}(T) = \text{EU}(T_{\text{nom}}) + \text{EU1} \cdot (T_{\text{r}} - 1) \tag{14.22}$$

where EU1 is a temperature-independent model parameter. In addition, through extensive fitting of published data, it has been determined that both U_{DS} and U_{DD} exhibit an exponential dependence on temperature, given by

$$U_{\text{DS}}(T) = \text{UDS} \cdot [\exp(\text{UDS1} \cdot (T_{\text{r}} - 1)) - 1] \tag{14.23}$$

$$U_{\text{DD}}(T) = \text{UDD} \cdot [\exp(\text{UDD1} \cdot (T_{\text{r}} - 1)) - 1] \tag{14.24}$$

As depicted in Fig. 14.4A and B, the proposed temperature mobility model provides significantly improved accuracy in describing the temperature characteristics when compared to the conventional temperature models. Our modified formulation also allows for greater flexibility in fitting a wide range of experimental temperature data.

14.3 Bias dependence of subthreshold swing

Experiments have demonstrated that as temperature decreases, the SS exhibits an increased dependence on V_{gs} or I_D [20] (see SS vs I_D curves in Fig. 14.8A). This behavior has been attributed to the presence of localized trap energy states near the band edges, which coexist with delocalized tail states participating in conduction [20, 21]. The bias dependency in SS can also be partially attributed to the V_{gs}-dependent CS mobility, where the scattering effect is particularly strong at cryogenic temperatures. The V_{gs} dependency arises from the screening of fixed charge centers by mobile charge carriers [16, 17, 19, 22, 23]. As discussed in Section 14.2, the CS dominates in the low temperature and low V_{gs} regime where inversion charge density is low (Fig. 14.4). The considerable dependency of mobility on V_{gs}, as observed in Fig. 14.4A and B, can cause distortions in the subthreshold characteristics of $I_D - V_{gs}$, consequently affecting the SS.

Fig. 14.5 depicts the influence of CS on SS using the mobility model described in Eq. (14.15). In this specific illustration, the CS is modulated by adjusting the parameter UD in Eq. (14.15). As shown in Fig. 14.5, an increase in CS leads to a greater dependency of SS on I_D, indicating that the SS becomes increasingly influenced by changes in I_D as the CS is intensified.

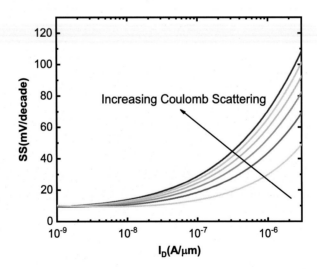

FIG. 14.5

Impact of CS on the I_D dependence of SS. CS parameter UD in Eq. (14.4) is varied from 0 (no CS) to 0.125 to obtain the *curves* shown.

14.4 Temperature dependency of current saturation

The saturation of current with respect to V_{ds} is determined by two main factors: drain side velocity saturation and channel pinch-off effects. These effects also play a crucial role in determining the drain saturation voltage (V_{dsat}). In this section, we will examine the temperature dependencies of these effects.

The carrier drift velocity v_d can be described as a function of the longitudinal electric field in the channel E_l as

$$v_d = \frac{\mu_{eff} E_l'}{1 + \dfrac{E_l'}{ESAT}} \tag{14.25}$$

where

$$E_l' = \frac{E_l}{\left[1 + \left(\frac{E_l}{ESAT}\right)^{MEXP}\right]^{1/MEXP}} \tag{14.26}$$

Here, $ESAT = 2VSAT/\mu_{eff}$, where VSAT denotes the saturation velocity parameter. The parameter MEXP in Eq. (14.26) facilitates a smooth transition from the linear region to the saturation region. In Section 13.3, both VSAT and MEXP were assumed to vary linearly with temperature. Fig. 14.6A illustrates experimental v_d-E_l curves at different temperatures for a bulk silicon sample [24], where $\mu_{eff} = \mu_0$. While the values of v_d in a MOSFET with an inversion layer may differ from those in bulk silicon, the temperature trends of the $v_d - E_l$ characteristics are not expected to exhibit significant differences. To determine the true temperature dependence of the velocity saturation model parameters, they are first extracted by fitting the model described by Eqs. (14.25), (14.26) to the experimental data for each specific temperature. The parameter μ_0 for each temperature is obtained by analyzing the slope of the corresponding $v_l - E_l$ curve at low E_l. Parameters MEXP and VSAT are extracted by fitting the linear to saturation transition region and the saturation region, respectively, of the $v_l - E_l$ curves. The extracted parameters, indicated by symbols, are presented in Fig. 14.6B–D. It is evident from Fig. 14.6C and D that neither VSAT nor MEXP exhibits a linear dependency on temperature. The temperature trends of all the model parameters can be directly correlated with the experimental $v_d - E_l$ curves displayed in Fig. 14.6A. In the low E_l region, as the temperature decreases, lattice scattering diminishes, which explains the increase in μ_0 shown in Fig. 14.6B. This behavior is well captured by the modified temperature power law in Eq. (14.15). In the low-field regime, the increase in μ_0 also leads to a higher v_d for the same applied E_l. This accounts for the decrease in ESAT with decreasing temperature, as illustrated in Fig. 14.6C. As E_l is further increased, v_d begins to saturate. The relationship between V_{dsat} and

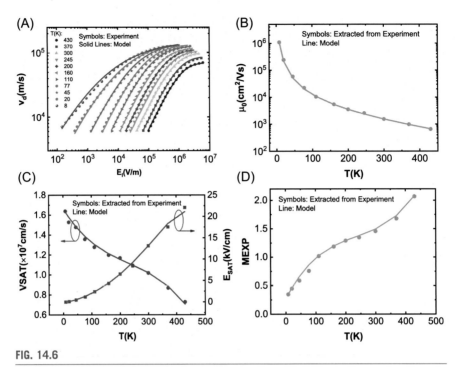

FIG. 14.6

Modeling temperature effects in drift velocity saturation. (A) Drift velocity as a function of electric field for different temperatures. (B) Mobility. (C) Saturation velocity parameter VSAT and saturation voltage parameter ESAT. (D) Linear to saturation region smoothing parameter MEXP as functions of temperature.

(A) Experimental data is extracted from C. Jacoboni, C. Canali, G. Ottaviani, A. Alberigi Quaranta, A review of some charge transport properties of silicon, Solid-State Electron. 20 (2) (1977) 77–89, ISSN 0038-1101.

temperature is shown in Fig. 14.6C. To accurately capture the nonlinear behavior of the experimentally extracted V_{dsat} across a wide range of temperature values, a cubic polynomial temperature dependency is used.

$$\text{VSAT}(T) = \text{VSAT}(T_{nom}) + \text{AT} \cdot \Delta T + \text{AT1} \cdot \Delta T^2 + \text{AT2} \cdot \Delta T^3 \quad (14.27)$$

where $\Delta T = T - T_{nom}$. AT, AT1, and AT2 are temperature-independent model parameters. Fig. 14.6C demonstrates that our model aligns well with the nonlinear temperature dependency observed in VSAT.

Furthermore, with temperature reduction, the distinction between the linear region and saturation region in the $v_d - E_1$ characteristics gradually diminishes, resulting in a smoother transition process. At low temperatures, even a small E_1 can energize the electrons to high speeds. The enhanced interaction of these warm electrons with the lattice (or exchange of acoustic phonons) results in an early onset of the nonlinear E_1 dependence regime [25]. Consequently, it is necessary to increase the smoothing parameter MEXP as the temperature decreases, as illustrated in

Fig. 14.6D. To accurately model the temperature dependency of MEXP, a cubic polynomial function in T is employed.

$$\text{MEXP}(T) = \text{MEXP}(T_{\text{nom}}) + \text{TMEXP} \cdot \Delta T + \text{TMEXP2} \cdot \Delta T^2 + \text{TMEXP3} \cdot \Delta T^3 \quad (14.28)$$

where TMEXP, TMEXP2, and TMEXP3 denote temperature-independent model parameters. This model accurately fits the experimentally extracted MEXP values across the entire temperature range, as demonstrated in Fig. 14.6D.

The complete model fitting of the experimental $v_d - E_1$ data using Eq. (14.25) with temperature-dependent parameters obtained from Eqs. (14.18), (14.27), (14.28) is depicted in Fig. 14.6A (represented by lines). As shown, the model shows excellent agreement with the experimental data across the entire temperature range, spanning from a high temperature of 430 K down to cryogenic temperatures of 8 K.

Let us now formulate temperature dependency of the pinch-off effect. The channel pinch-off effect in BSIM is taken into account when calculating V_{dsat} using the long channel pinch-off voltage expression: $\text{KSATIV}(V_{\text{gs}} - V_{\text{th}})$. Here, V_{th} represents the threshold voltage and KSATIV is a user-defined parameter [13]. In the BSIM formulation, KSATIV is assumed to be a constant. However, through extensive data fitting of I_D at high V_{ds} across various technology devices, ranging from room temperature to cryogenic temperatures, it is found that KSATIV exhibits a similar temperature dependency as VSAT and MEXP, that is,

$$\text{KSATIV}(T) = \text{KSATIV} \ (T_{\text{nom}}) + \text{KSATIVT} \cdot \Delta T + \text{KSATIVT2} \cdot \Delta T^2 + \text{KSATIVT3} \cdot \Delta T^3 \quad (14.29)$$

where KSATIVT, KSATIVT2, and KSATIVT3 are temperature-independent model parameters.

14.5 Temperature dependency of source/drain resistance

In Section 13.6, we assumed a linear relation between the resistivity and temperature. However, as shown in Fig. 14.7, at low temperatures, silicon resistivity has a different (lower) slope. To capture this effect, a dual-slope resistivity model is used. For high temperatures, the series resistance temperature dependence is given by a linear relation with a slope PRT:

$$r_{\text{ds,temp0}} = 1 + \text{PRT} \cdot (T - T_{\text{nom}}) \quad (14.30)$$

For low temperatures, the series resistance temperature dependence is again given by a linear relation with a slope PRT1:

$$r_{\text{ds,temp1}} = 1 + \text{PRT1} \cdot (T - T_1) \quad (14.31)$$

where

$$T_1 = \text{TR0} - \frac{\text{PRT}}{\text{PRT1}} \cdot (\text{TR0} - T_{\text{nom}}) \quad (14.32)$$

Here, TR0 denotes the parameter for the corner temperature, that is, the intersection point of the two starlight lines in Fig. 14.7.

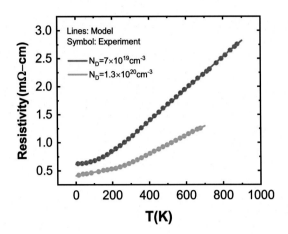

FIG. 14.7

Silicon resistivity as a function of temperature for different doping concentrations [26].

Finally, temperature dependence of series resistances is obtained by smoothly joining the high temperature and low temperature straight lines using a smoothing parameter SPRT as

$$r_{\text{ds,temp}} = \frac{r_{\text{ds,temp0}} + r_{\text{ds,temp1}} + \sqrt{\left(r_{\text{ds,temp0}} - r_{\text{ds,temp1}}\right)^2 + 0.25 \cdot \text{SPRT}^2}}{2}$$
$$- \frac{PRT1 \cdot (T_{\text{nom}} - T_1) + \sqrt{\left[PRT1 \cdot (T_{\text{nom}} - T_1)\right]^2 + 0.25 \cdot \text{SPRT}^2}}{2} \quad (14.33)$$

$$R_{\text{D/S}}(T) = R_{\text{D/S}}(T_{\text{nom}}) \cdot r_{\text{ds,temp}} \quad (14.34)$$

where $R_{\text{D/S}}$ represents any source/drain resistance given in Section 13.6.

14.6 Parameter extraction down to cryogenic temperatures

For a multiple temperature (T) fitting, device characteristics at the nominal temperature (TNOM) are fitted first using the procedure described in Chapter 12. Once TNOM characteristics are fitted well, the temperature models in the cryogenic mode can be used to fit the device characteristics for multiple temperatures down to the cryogenic temperature range.

14.6.1 Drain current fitting in linear region (I_{ds} vs V_{gs} at low V_{ds})

Step 1. Subthreshold region fitting: In this step, the V_{gs} dependence of the drain current I_{ds} in subthreshold region is extracted.

- **TLOW, DTLOW, TLOW1, KLOW1, DTLOW1**: Observe the SS at different temperatures. TLOW is used to extract the saturation of SS w.r.t temperature in the cryogenic temperature range. DTLOW is used to smoothly transition to the SS saturation region at low T from the usual SS behavior governed by the Boltzmann law ($\ln(10)kT/q$) at high T. At very low temperatures, the SS may again start to rise with T and TLOW1 is the temperature below which this happens. DTLOW1 is used to capture the smoothness of this SS transition and KLOW1 is used to capture the rate at which the SS rises w.r.t T (assuming linear dependency on T).

- **TVTH, KT11, KT12, KT1**: Observe threshold voltage offset at different temperatures. TVTH is approximately the temperature below which the V_{th} offset model ($\Delta V_{th,\,temp0}$) becomes nonzero. KT11 is used to capture the extent of V_{th} offset required at low T. KT12 is used for smoothing out the transition between $\Delta V_{th,\,temp0}$ at low T and high T. KT1 can be used to introduce a linear temperature dependency term in $\Delta V_{th,\,temp0}$, if needed. An example model simulation of $\Delta V_{th,\,temp0}$ is shown in Fig. 14.3.

Step 2. Near-threshold region fitting: In this step, the V_{gs} dependence of the drain current I_{ds} in a low-to-intermediate V_{gs} region is extracted, especially, the increased V_{GS} dependence in SS at cryogenic temperatures observed, which we attribute to the CS.

- **UD1, UD2, UCSTE, UCSTE1, UDS, UDS1, UDD, UDD1**: Observe near-threshold region of I_{dlin} and g_{mlin} in log and linear scales at different temperatures to extract these parameters. These parameters can also be used to capture the increased SS dependency of V_{gs}, if needed. The experimental cryogenic data also suggests that the impact of CS is smaller at high V_{DS}, and therefore source and drain charge densities in the Coulomb mobility model require different weight factors, UDS_{eff} and UDD_{eff}, respectively (Eq. 14.15). UDS and UDS1 can be used to capture the temperature dependency of the weight factor UDS_{eff} assigned to the source charge density in the CS model. Similarly, UDD and UDD1 capture the temperature dependency of the weight factor UDD_{eff} assigned to the drain charge density. I_{ds} versus V_{gs} at $V_{ds} = 50\,\text{mV}$ can be used first to extract an approximate value of $UDS_{eff} + UDD_{eff}$ since $q_{is} \approx q_{id}$. I_{ds} versus V_{gs} at $V_{ds} = V_{DD}$ can later be used to differentiate between values of UDS_{eff} and UDD_{eff} and the related parameters.

Step 3. Strong inversion region fitting: In this step, the V_{gs} dependence of the drain current I_{ds} in intermediate-to-high V_{gs} region is extracted.

- **UTE, UTE1, UA1, UA2, EU1, EMOBT, PRT, PRT1, TR0, SPRT**: Observe the strong inversion region of I_{dlin} and g_{mlin} in a linear scale at different temperatures to extract these parameters.

14.6.2 Drain current fitting in saturation region

Step 4. Temperature dependency of DIBL.

- **TETA0**: Observe the threshold voltage offset of I_{ds} versus V_{gs} at $V_{ds} = V_{DD}$ at different T.

Step 5. Drain bias dependency of the CS at low temperatures.

- **UDS, UDS1**: At high V_{ds}, q_{id} becomes negligible. Observe near-threshold region of $I_{ds} - V_{gs}$ and $g_m - V_{gs}$ at $V_{ds} = V_{DD}$ in log and linear scales at different temperatures to extract UDS and UDS1. Adjust UDD and UDD1 now such that the I_{dlin} fitting performed earlier remains unaffected.

Step 6. Temperature dependency of the velocity saturation model and CLM.

- **AT, AT2, KSATIV1, KSATIV2, PTWGT**: Observe the strong inversion region of $I_{ds} - V_{gs}$ and $g_m - V_{gs}$ at $V_{ds} = V_{DD}$, and at different T as well as the saturation region of $I_{ds} - V_{ds}$ at different V_{gs} for multiple T.
- **PCLMT**: Observe slope of the saturation transition region of $I_{ds} - V_{ds}$ at different V_{gs} for multiple T.
- **TMEXP, TMEXP2**: Observe the linear to saturation transition region of $I_{ds} - V_{ds}$ at different V_{gs} for multiple T.

The examples of experimental data fitting with the cryogenic model in a wide range of temperature down to the cryogenic range are shown in Figs. 14.8–14.10. Figs. 14.8 and 14.9 show the model validation for a p-type FinFET device. Both the deviation of the SS from the Boltzmann limit as well as its increased bias dependency with temperature reduction can be noticed in Fig. 14.8. A satisfactory fitting of the subthreshold region in Fig. 14.8A validates the effectiveness of the band tail temperature model (Section 14.1) as well as the CS models (Sections 14.2 and 14.3). Fig. 14.8A shows g_m versus V_{gs} validation for proving the efficacy of the mobility temperature models (Section 14.2). Similarly, Fig. 14.9 shows the model validation for a short channel 10 nm p-FinFET experimental data. Fig. 14.10 further shows the model validation for an n-type FinFET. The model agrees well with both $I_d - V_{gs}$ as well as $I_d - V_{ds}$ characteristics.

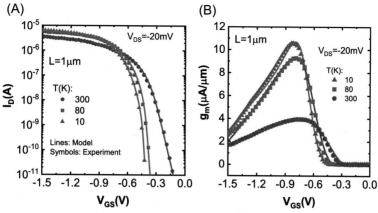

FIG. 14.8

Model validation with p-type FinFET data from room 300 K to 10 K. (A) $I_D - V_{GS}$ characteristics and (B) $g_m - V_{GS}$ characteristics. A single global set of parameters is used to fit the experimental data [27] for the entire temperature range. For the experimental device, $L = 1\mu m$, EOT = 1.5 nm, fin thickness = 25 nm, and fin height = 65 nm.

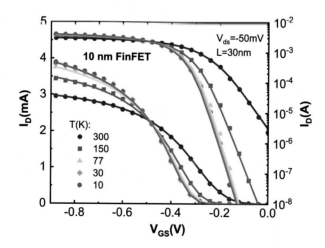

FIG. 14.9

Model validation with 10 nm technology p-type FinFET data from 300 K to 10 K. A single global set of parameters is used to fit the experimental data [9] for the entire temperature range.

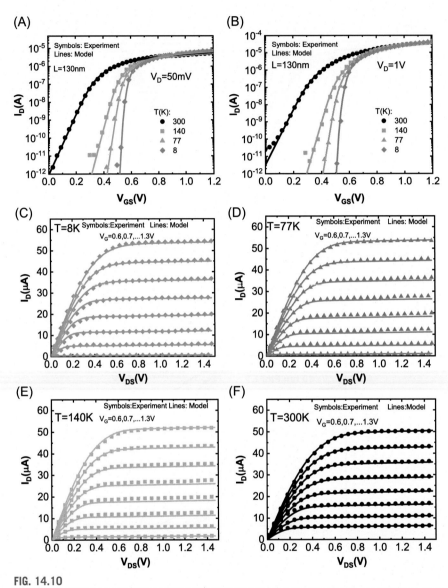

FIG. 14.10

Model validation with n-type FinFET data for different biases and temperatures. $I_D - V_{DS}$ characteristics: (A) $V_{DS} = 50$ mV and (B) $V_{DS} = 1$ V. $I_D - V_{DS}$ characteristics for different V_{GS} for (C) $T = 8$ K, (D) $T = 77$ K, (E) $T = 140$ K, and (F) $T = 300$ K. A single global set of parameters is used to fit the experimental data [28].

14.7 Summary

We have developed accurate and reliable temperature models for key variables in the BSIM-CMG model, including channel charge density, threshold voltage shift, low-field mobility, and current saturation. These temperature models have undergone extensive validation using published experimental data from various sources, focusing on multigate technologies and covering a wide temperature range from room temperature down to cryogenic temperatures as low as 4 K. Additionally, we have proposed a mechanism involving decreasing mobility due to CS to explain the observed current dependency of the SS at temperatures below 100 K. These temperature models significantly improve the accuracy and applicability of the BSIM-CMG model for circuit design in cryogenic temperature conditions, making it suitable for various applications such as quantum computing and high-performance computing.

References

[1] H.C. Nathanson, C. Jund, J. Grosvalet, Temperature dependence of apparent threshold voltage of silicon MOS transistors at cryogenic temperatures, IEEE Trans. Electron Devices 15 (6) (1968) 362–368.

[2] S.S. Sesnic, G.R. Craig, Thermal effects in JFET and MOSFET devices at cryogenic temperatures, IEEE Trans. Electron Devices 19 (8) (1972) 933–942.

[3] B. Patra, R.M. Incandela, J.P.G. van Dijk, H.A.R. Homulle, L. Song, M. Shahmohammadi, R.B. Staszewski, A. Vladimirescu, M. Babaie, F. Sebastiano, E. Charbon, Cryo-CMOS circuits and systems for quantum computing applications, IEEE J. Solid-State Circuits 53 (1) (2018) 309–321.

[4] B. Patra, J.P.G. van Dijk, S. Subramanian, A. Corna, X. Xue, C. Jeon, F. Sheikh, E. Juarez-Hernandez, B.P. Esparza, H. Rampurawala, B. Carlton, N. Samkharadze, S. Ravikumar, C. Nieva, S. Kim, H. Lee, A. Sammak, G. Scappucci, M. Veldhorst, L. M.K. Vandersypen, M. Babaie, F. Sebastiano, E. Charbon, S. Pellerano, A scalable Cryo-CMOS 2-to-20 GHz digitally intensive controller for 4×32 frequency multiplexed spin qubits/transmons in 22 nm FinFET technology for quantum computers, in: IEEE International Solid-State Circuits Conference (ISSCC), 2020, pp. 304–306.

[5] F. Jazaeri, A. Beckers, A. Tajalli, J.-M. Sallese, A review on quantum computing: from qubits to front-end electronics and cryogenic MOSFET physics, in: International Conference "Mixed Design of Integrated Circuits and Systems", IEEE, 2019, pp. 15–25.

[6] E. Charbon, Cryo-CMOS electronics for quantum computing applications, in: European Solid-State Device Research Conference (ESSDERC), 2019, pp. 1–6.

[7] B.C. Paz, M. Cassé, S. Barraud, G. Reimbold, M. Vinet, O. Faynot, M.A. Pavanello, Low temperature influence on performance and transport of Ω-gate p-type SiGe-on-insulator nanowire MOSFETs, Solid-State Electron. 0038-1101, 159 (2019) 83–89. SI: EUROSOI 2018.

[8] H.-C. Han, F. Jazaeri, A. D'Amico, A. Baschirotto, E. Charbon, C. Enz, Cryogenic characterization of 16 nm FinFET technology for quantum computing, in: ESSCIRC 2021—IEEE 47th European Solid State Circuits Conference (ESSCIRC), 2021, pp. 71–74.

[9] S.K. Singh, S. Gupta, R.A. Vega, A. Dixit, Accurate modeling of cryogenic temperature effects in 10-nm bulk CMOS FinFETs using the BSIM-CMG model, IEEE Electron Device Lett. 43 (5) (2022) 689–692.

[10] Y. Liu, L. Lang, Y. Chang, Y. Shan, X. Chen, Y. Dong, Cryogenic characteristics of multinanoscales field-effect transistors, IEEE Trans. Electron Devices 68 (2) (2021) 456–463.

[11] J.F. Wager, Real- and reciprocal-space attributes of band tail states, AIP Adv. 7 (12) (2017) 125321.

[12] A. Beckers, D. Beckers, F. Jazaeri, B. Parvais, C. Enz, Generalized Boltzmann relations in semiconductors including band tails, J. Appl. Phys. 129 (4) (2021) 045701.

[13] BSIM-CMG Technical Manual, 2021, Available from: http://bsim.berkeley.edu/models/bsimcmg/.

[14] C. Lombardi, S. Manzini, A. Saporito, M. Vanzi, A physically based mobility model for numerical simulation of nonplanar devices, IEEE Trans. Comput. Aided Des. Integr. Circuits Syst. 7 (11) (1988) 1164–1171.

[15] BSIM-IMG Technical Manual, 2020, Available from: http://bsim.berkeley.edu/models/bsimimg/.

[16] K.V. Seendripu, A 4K CMOS Self-Calibration Circuit for Josephson Flash-Type Analog-to-Digital Converters (Ph.D. thesis), EECS Department, University of California, Berkeley, 1996.

[17] C.-L. Huang, G.S. Gildenblat, Measurements and modeling of the n-channel MOSFET inversion layer mobility and device characteristics in the temperature range 60-300 K, IEEE Trans. Electron Devices 37 (5) (1990) 1289–1300.

[18] K. Chain, J.H. Huang, J. Duster, P.K. Ko, C. Hu, A MOSFET electron mobility model of wide temperature range (77-400 K) for IC simulation, Semicond. Sci. Technol. 12 (4) (1997) 355–358.

[19] D.S. Jeon, D.E. Burk, MOSFET electron inversion layer mobilities—a physically based semi-empirical model for a wide temperature range, IEEE Trans. Electron Devices 36 (8) (1989) 1456–1463.

[20] H. Bohuslavskyi, A.G.M. Jansen, S. Barraud, V. Barral, M. Cassé, L.L. Guevel, X. Jehl, L. Hutin, B. Bertrand, G. Billiot, G. Pillonnet, F. Arnaud, P. Galy, S. De Franceschi, M. Vinet, M. Sanquer, Cryogenic subthreshold swing saturation in FD-SOI MOSFETs described with band broadening, IEEE Electron Device Lett. 40 (5) (2019) 784–787.

[21] A. Beckers, F. Jazaeri, C. Enz, Inflection phenomenon in cryogenic MOSFET behavior, IEEE Trans. Electron Devices 67 (3) (2020) 1357–1360.

[22] A. Hairapetian, D. Gitlin, C.R. Viswanathan, Low-temperature mobility measurements on CMOS devices, IEEE Trans. Electron Devices 36 (8) (1989) 1448–1455.

[23] J.S. Duster, Z.H. Liu, P.K. Ko, C. Hu, Temperature effects of the inversion layer electron and hole mobility of MOSFETs from 85K to 500K, in: International Conference on Solid State Devices and Materials (SSDM), Makuhari, 1993, pp. 835–837.

[24] C. Jacoboni, C. Canali, G. Ottaviani, A. Alberigi Quaranta, A review of some charge transport properties of silicon, Solid-State Electron. 0038-1101, 20 (2) (1977) 77–89.

[25] W. Shockley, Hot electrons in germanium and Ohm's law, Bell Syst. Tech. J. 30 (4) (1951) 990–1034.

[26] P.W. Chapman, O.N. Tufte, J.D. Zook, D. Long, Electrical properties of heavily doped silicon, J. Appl. Phys. 34 (11) (1963) 3291–3295.

[27] H. Achour, R. Talmat, B. Cretu, J.-M. Routoure, A. Benfdila, R. Carin, N. Collaert, E. Simoen, A. Mercha, C. Claey, DC and low frequency noise performances of SOI p-FinFETs at very low temperature, Solid-State Electron. 0038-1101, 90 (2013) 160–165.

[28] S. O'uchi, K. Endo, M. Maezawa, T. Nakagawa, H. Ota, Y.X. Liu, T. Matsukawa, Y. Ishikawa, J. Tsukada, H. Yamauchi, W. Mizubayashi, S. Migita, Y. Morita, T. Sekigawa, H. Koike, K. Sakamoto, M. Masahara, Cryogenic operation of double-gate FinFET and demonstration of analog circuit at 4.2 K, in: IEEE International SOI Conference (SOI), 2012, pp. 1–2.

Index

Note: Page numbers followed by *f* indicate figures and *t* indicate tables.